For Geoffrey

People I was
to know all
Them, + knowing whom has
brought the greatest pleasure
to our lives.

George

(42)

£12-99

C6

[£33-NEW]

Dying to Know

Dying to Know

Scientific Epistemology and Narrative
in Victorian England

GEORGE LEVINE

The University of Chicago Press
Chicago and London

George Levine is the Kenneth Burke Professor of English at Rutgers University, where he directs the Center for the Critical Analysis of Contemporary Culture. An editor of the journal *Victorian Studies* from 1959 to 1968, Levine is the author of *Boundaries of Fiction, The Realistic Imagination, Darwin and the Novelists,* and *Lifebirds.* He is the editor of *Realism and Representation* and *Aesthetics and Ideology,* among others.

The University of Chicago Press, Chicago 60637
The University of Chicago Press, Ltd., London
© 2002 by The University of Chicago
All rights reserved. Published 2002
Printed in the United States of America

11 10 09 08 07 06 05 04 03 02   1 2 3 4 5
ISBN: 0-226-47536-0 (cloth)

Library of Congress Cataloging-in-Publication Data

Levine, George Lewis.
   Dying to know : scientific epistemology and narrative in Victorian England / George Levine.
     p. cm.
   Includes bibliographical references and index.
   ISBN 0-226-47536-0 (cloth : alk. paper)
   1. English prose literature—19th century—History and criticism.
2. Literature and science—Great Britain—History—19th century.
3. Descartes, Rene, 1596–1650—Influence.   4. Knowledge, Theory of, in literature.   5. Science in literature.   6. Science—Philosophy.
7. Narration (Rhetoric).   I. Title.
   PR788.S33 L48 2002
   828'.80809356—dc21

                                        2001006417

∞ The paper used in this publication meets the minimum requirements of the American National Standard for Information Sciences—Permanence of Paper for Printed Library Materials, ANSI Z39.48-1992.

Alla mia Vita Nuova
agli amici ed alla famiglia che l'hanno resa possibile

CONTENTS

Acknowledgments / ix

Introduction / Dying to Know / 1

1 / The Narrative of Scientific Epistemology / 17

2 / Dying to Know Descartes / 44

3 / Carlyle, Descartes, and Objectivity: Lessen Thy Denominator / 66

4 / Autobiography As Epistemology: The Effacement of Self / 85

5 / My Life As a Machine: Francis Galton, with Some Reflections on A. R. Wallace / 104

6 / Self-Effacement Revisited: Women and Scientific Autobiography / 126

7 / The Test of Truth: *Our Mutual Friend* / 148

8 / *Daniel Deronda:* A New Epistemology / 171

9 / The Cartesian Hardy: I Think, Therefore I'm Doomed / 200

10 / Daring to Know: Karl Pearson and the Romance of Science / 220

11 / The Epistemology of Science and Art: Pearson and Pater / 244

Epilogue / Objectivity and Altruism / 268

Notes / 285

Index / 317

ACKNOWLEDGMENTS

This is the first book I've written whose title preceded it. The title suggests a pursuit that I felt myself living out in the writing—pursuit of completed knowledge that kept asymptotically verging away. I have pursued the various clues to utter comprehensiveness through what ultimately came to seem zany urges rather than reasoned ambition. Certainly, the pursuit had a chimerical aspect to it, as though I were obliged to explore everything, and it has taken so long in development that it really did seem as though it would leave me, literally, dying to know. But here it is: not compendious and universal but fragmentary and perhaps arbitrary. In its incompleteness, nevertheless, it owes a great deal to many innocent of its failures who have done what they could to help me forestall the literal implication of the title.

I must begin by lamenting that it was only in the last weeks of the writing of this book that I had the privilege of reading the manuscripts of two extraordinary works that would have influenced my thought much more had I been lucky enough to see them sooner. Amanda Anderson's *The Powers of Distance* and Christopher Herbert's *Victorian Relativism* brilliantly reread the Victorians, tracing two distinct yet closely related intellectual traditions whose lineaments are oddly similar to those of the tradition I attempt to trace in this book.

I have been lucky enough, in a term as Avalon Professor at Northwestern University, to get to know Christopher Herbert well, and to know him is to learn from him and, yes, to love him. What I owe Carolyn Williams, my colleague at Rutgers and a constant adviser on my obsessions with objectivity, can only be hinted here. I have always been dying to know what she thinks about my ideas, and she has given them and me life. Suzy Anger was a generous sympathizer in the project and an acute critic of it, while also a constant source of new ideas and information. Theodore Porter has helped me enormously, with a truly unusual scholarly generosity, in my work on Karl Pearson. Joe Vining, whose passion for knowledge and justice and meaning marks all his work, has taught me much and inspired me yet more.

To Carol Clover I owe all I know about "Loki," the "author" of Karl Pearson's novel. Lorraine Daston was generous enough to let me see several of her essays on objectivity in manuscript form. Among the many good people who have helped me along the way with editorial advice or sage counsel about objectivity, I would like to thank Bernard Lightman, Patricia O'Hara, Howard Horwitz, Jim Phelan, Oscar Kenshur, and Bruce Robbins. Stephen Kellert put me on to a fresh way to think about the logical positivists. Paul Meyer, a wonderful friend and mathematician, will be surprised to find himself mentioned here among those who have mattered most to this book, but our extraordinary conversations between birds on long and wonderful birding trips contributed both to my mental health and to my recognition of the importance of precision in thinking where my mind ran all too quickly to easy answers.

I owe a great deal, as well, to Rutgers University, which for more than thirty years has supported my work and given me the time and assistance for travel and thinking. Barry Qualls, colleague, friend, and now also dean, has been generous beyond thanks. Richard Foley—alas, now moved to another university—has been a superb and supportive dean with the extraordinary undeanlike quality of a deep commitment to the best intellectual work. The Center for the Critical Analysis of Contemporary Culture has been my home, thanks to the foresight of the late Ed Bloustein, and of Nat Pallone; and there Link Larsen has been my indispensable friend and aide, while former associates such as Susan Gal, Andy Abbott, and Neil Smith have been constant sources of intellectual nourishment. Many years ago, John Woodcock of Indiana University suggested to me that scientific autobiographies often found ways to avoid making the scientist himself the overt subject. Among the very good people who have helped me along the way I must thank Alan Rauch, Steve Amarnick, and Tanya Agathocleous. The English departments at the University of Utah and Northwestern University were my generous hosts during productive quarters in which I tried out some of these ideas.

As ever, I owe much to my family, who, during the many years of the writing of this book, have enriched my life well beyond epistemology: to Marge, who opened the space for these ruminations and has unaccountably stuck with me through a career's worth of academic obsessions; to Rachel and Dale, who tolerate my computer failures and my paternal eccentricities; to their children, the splendid and life-giving Aaron and Benjamin, who emerged during the years of the writing of this book; and to David and Becca, who married during that time and brightened my life as a consequence.

This work would never have come close to completion without support from the Rockefeller Foundation Center at Bellagio, where the chapters on Karl Pearson took initial shape; and I am in even greater debt to the Bogliasco Foundation for residence at the Liguria Center in Bogliasco, where a wonderful staff and a wonderful place made a thorough rewriting of the whole manuscript possible. When in Italy, I found that dying to know is the fullest experience of living.

INTRODUCTION

Dying to Know

> "[A]ll who rightly touch philosophy,
> study nothing else than to die, and to be dead."
> PLATO AS QUOTED BY PATER

> Art is more powerful than knowledge,
> because it desires life, whereas knowledge attains
> as its final goal only—annihilation.
> NIETZSCHE

The apparent extravagance of the Platonic formulation may disguise its accuracy as a description of the great traditions of Western philosophy and of the dominant commonsense understanding of what is entailed in knowing. The implication that one must prepare oneself in life by achieving a purity that will approach the final state of death lingers in the way we think now. So, too, does the implication that one only achieves purity in death. Although I noticed the quotation from Plato long after I had begun to worry the question of the relation of epistemology to the way people have lived and thought and constructed narratives, it leaped off the page at me when I came upon it again. For it merged immediately with what I knew would be this book's presiding metaphor, a commonplace colloquialism.

"I'm dying to know" is what virtually all of us say when we are particularly curious about something. Its gossipy quality and its hyperbole do not belie, however, a persistent Platonic passion of Western culture. That is, to put it at its crudest, there is something in our culture—and the "our" here riskily embraces all of Western culture and most of Western history—that drives it to find things out, even at the risk of life. While obviously we would much prefer not to die, knowledge has been taken to be worth the price. Aspiration to it has shaped some of the dominant myths of our culture—from Adam's fall, in which we sinned all, to Socrates' principled death, to the Faustus myth and its Frankensteinian variations, to the unfortunately insufficiently mythical creation of the atomic bomb. These stories are so integral to our way of being that it is perhaps easy not to notice that they are reflected in that innocent idiom, "dying to know." With it, I want to foreground the narrative basis of our culture's large and abstract commitment to knowledge and truth, to shift the grounds of discussions of disinterest and objectivity from philosophical haggling to human and ethical concerns. From that point of view, the ideal is both impossible and necessary.

The metaphor of dying is dangerous, in more ways than one. But what I try to mean by it throughout is, first, a passion for knowing so intense that one would risk one's life to achieve it; and second, a willingness to repress the aspiring, desiring, emotion-ridden self and everything merely personal, contingent, historical, material that might get in the way of acquiring knowledge. Part of the irony of the conditions I will be exploring is that distrust of the material and the personal is an aspect not only of the long Platonic traditions of Western thought but of the empiricism and the naturalism that sought to displace them. It is a view built into the idea that the senses that are the gateway to all knowledge must be disciplined and checked in order to provide that knowledge; and all knowledge is comprised within the realm of the natural. Empiricism, from Bacon forward, chastises the senses as much as it depends on them. Descartes's move toward certain knowledge seems to depend on the assertion of a living cogito. But that self is defined by the purgation from it of everything that is contingent, temporal, social, inherited, human. The epistemological ventures of modernity are thick with paradox—materiality entails the incorporeal, the self gains its power by annihilating itself.

Part of the paradox of dying to know is, of course, that one cannot know anything when one is dead. The phrase implies, then, a kind of liminal position, at the edge of nonbeing, and it implies a persistent tragedy: only in death can one understand what it has meant to be alive. The continuing aspiration to get it straight, to understand what it means, to transcend the limits imposed by the limiting self, depends on the elimination of the self. The world out there that it chooses to know is only knowable when it is too late. This book derives in part from my sense that this pervasive paradox is startlingly attractive as well as obviously disturbing. The epistemological paradox clearly corresponds to deep and recurrent elements of the human psyche and—I would also suggest—the constitution of the world. The "desire for death" is a central aspect of modern civilization, but it is also and particularly an aspect of modern epistemology. Or perhaps it might be said that modern scientific epistemology is a perfect medium through which the desire for death might get expressed.[1]

Although I take the problem to transcend the limits of any defined historical period, this book focuses on the nineteenth century. As John Kucich has put it, "it was perfectly clear even to nineteenth-century observers—as it is to us—that some momentous inflation of the cultural cachet of honesty took place in Victorian England."[2] Kucich's concern is not, for the most part, with science, and only peripherally with epistemology, but the condition he describes here is confirmed in the direction of science by Peter Galison. Galison argues that "in the nineteenth century—or, more specifically, after about 1830— . . . the desired character of the natural philosopher inverted to one of self-abnegation," becoming "more saint-like in self-denial than powerful in genial interpretation."[3] Galison points out that for nineteenth-century scientists, the responsibility to be "objective," to gain access to objects of knowledge and thus to allow the facts to speak for themselves, took priority over fullness of understanding. Even interpretation, that is to say, violated for the nineteenth-century scientist the demands of moralized objectivity. Self-denial, a surrender of the self to the thing studied, became a priority of that time. While, as Christopher Herbert demonstrates impressively, a recognition of the dependence of knowledge on the position and condition of the knower developed alongside the scientific commitment to "objectivity," the "relativism" of writers like William Hamilton and Dean Mansel and then of writers and thinkers up to the end of the century, like Walter Pater and the statistician Karl Pearson, was also highly moralized and required a similar kind of self-abnegation.[4] Even a brilliantly original thinker like William Whewell, who claimed in a stunning sentence in the midst of some very serious explication, "there is a mask of theory over the whole face of nature,"[5] sustained a commitment to the ideals of self-abnegation. In fact, with the recognition of the inescapable presence of the interpreting self, it was incumbent on all scientists and thinkers the more rigorously to repress their own biases. Honesty in representation, not intellectual power and ingenuity, is the greatest virtue.

In the nineteenth century, as Galison and Lorraine Daston have shown, a traditional activity of attempting to attain to "truth to nature" was assimilated to an ideal of "self-abnegation" in order to "let nature speak for itself."[6] The emphasis on renunciation that one finds in Carlyle—"Selbst-Tödung," the idea of which he claims to have got from Goethe, but which clearly derives at least equally from the rigid Calvinism of his childhood—is not so different after all from Keats's "negative capability," or from the celebration of Baconian method that marks most justifications of science in the early nineteenth century.[7] Nor is it surprising that John Stuart Mill would have turned to Carlyle for his saving "anti-self-consciousness principle," which he claims allowed him to imagine the possibility of happiness by not consciously seeking it or thinking about his own happiness. Mill's epistemology, Carlyle or not, is built not only on hostility to ideology, but on the effort to put aside interest entirely. "We cannot believe a proposition only by wishing, or only by dreading, to believe it," he asserts.[8] He thought it was important to make the feelings "powerless."[9] One must seek the truth even when it threatens to produce unpleasant results. The nineteenth-century intellectual enterprise of pushing aside the church and religion as authority in knowledge (displacing it with the newly invented institution of science itself) largely depends on this (proclaimed) willingness to suffer the consequences of finding out that the world is not only not made for us, but that it may well be without intention, meaning, or direction. Facing the amorality of the world entailed, so the implication went, a higher morality than that of traditional religion.

The voice of science was the voice of this morality, and perhaps a single example here will suggest something of the tone and narrative implications of the scientific argument. In an 1854 essay about the process of scientific knowing, John Tyndall, one of the most famous and forceful of the "scientific naturalists," wrote the following:

> The first condition of success is patient industry, and honest receptivity, and a willingness to abandon all preconceived notions, however cherished, if they be found to contradict the truth. Believe me, a self-renunciation that has something lofty in it, and of which the world never hears, is often enacted in the private experience of the true votary of science. And if a man be not capable of this self-renunciation—this loyal surrender of himself to Nature and to fact, he lacks, in my opinion, the first mark of a true philosopher. Thus the earnest prosecutor of science, who does not work with the idea of producing a sensation in the world, who loves the truth better than the transitory blaze of to-day's fame, who comes to his task with a single eye, finds in that task an indirect means of the highest moral culture. And although the virtue of the act depends upon its privacy, the sacrifice of self, this upright determination to accept the truth, no matter how it may present itself—even at the hands of a scientific foe, if necessary—carries with it its own reward. When prejudice is put under foot and the stains of personal bias have been washed away—when a man consents to lay aside his vanity and to become Nature's organ—his elevation is the instant consequence of his humility.[10]

If Tyndall does not talk about death, he does talk about an ideal of self-abnegation, a more than Keatsian negative capability, without any self-interest, with no desire but for the truth. In Tyndall's implicit narrative, self-denial produces "elevation," and with that, the full story of dying-to-know's paradoxical nature is told.

In discursive form, the notion of self-annihilation producing knowledge has a rational force that it loses in narrative because the position it recommends as ideal for learning is all but equivalent to death; in narratives, outside the sphere of the miracles that magic or religion can perform, dead people are usually too dead to know anything at all. Several elements of this passage are relevant to the arguments of this book. First, the scientific stance is moralized, that is, the power to observe accurately becomes a moral as well as an epistemological virtue—it requires "patient industry," "honest receptivity," and "the sacrifice of self." Second, epistemology here has many of the characteristics of narrative—it is understood as a story of a protagonist who by way of "patient industry" and "humility" denies personal bias, achieves real knowledge, and instantly finds that unsought for "elevation." Aware of the shape of his story and the characterization of its protagonist, Tyndall later anticipates his audience sniggering at the idea "that I regard the man of science as a heroic, if not indeed angelic character."[11] Third, success in science requires the banishment of all preconceived notions. There is no room in science for the personal, and even the senses are not to be entirely trusted. Fourth, to conduct one's science successfully, one has to renounce personal desire, abase oneself before the facts. In Tyndall's wonderful phrase, the scientist must become "Nature's organ." And fifth, there is the paradoxical consequence of the refusal of fame and of one's own desire and personal interests. The religious forms of resurrection are repeated in epistemological success deriving from the denial of the self who seeks the knowledge in the first place. Except it die, it cannot know.

As against this nineteenth-century passion for selfless dedication to knowledge, our contemporary social science and cultural theory has been producing a long series of books exploding the notion of objectivity in various scientific or quasi-scientific intellectual disciplines.[12] While the very possibility of "objectivity," no less "truth," has been under attack for lots of good and maybe lots of not so good reasons, from poststructuralist theories, the new pragmatism, various forms of gender and ethnic studies, postcolonialism, new historicism, and even, perhaps especially, science studies, I have found myself drawn more sympathetically to these nineteenth-century ideals, and fascinated by the paradox apparently underlying the whole project of Western epistemology.

The persistence and popularity (and even the shudder of horror that accompanies this view in science fiction film versions of it) suggest that the self-abnegation that is part of the history of scientific epistemology must disguise some kind of interest. While my argument throughout is that this suicidal narrative of knowledge is unsatisfactory and often dangerous and needs reimagining, I want also to insist that nineteenth-century aspirations toward knowledge were not merely, as they have often been "exposed" as being, disguises of egoistic aggressions, reflections of surreptitious ideologies, disreputable programs of power, intimations of personal and culturally deep prejudices. The connection between knowledge and death, which I have turned into the metaphor "dying to know," emerged precisely out of reasonable efforts to avoid being those things. Part of the narrative of scientific epistemology is a preliminary clearing away of the ground in order to make increased knowledge possible—Bacon's "idols" are the most famous and obvious instance. If disinterest is ultimately impossible, even the crudest common sense will recognize that Mill wasn't entirely wrong, and that a deep interest in a particular outcome is likely to obstruct a clear reading of the case. How can one trust the paid spokespeople for products advertised on television, for example? Even if pure disinterest is impossible, interest is not trustworthy; and if one can be interested in one reading of the world but concede the greater validity of an opposed one, that is a sign that something approaching disinterest is common in everyone's experience.

This is not to discount the critiques. It seems merely blinkered any longer to think, for example, that Darwinian theory is unconnected to Malthusian political conservatism, to laissez-faire economics and to Darwin's own Whiggish family and community biases,[13] nor can I think about *David Copperfield* again without a certain embarrassment at what seems now almost blatant celebration of self-discipline and self-sacrifice in the interest of bourgeois identity and bourgeois ideology.[14] Moreover, the long Western tradition of aspiration toward an ideal extracorporeal condition obviously must be read in relation to questions of class: only the wealthy can afford to imagine the superiority of the unmoved mover to the worker. Scientific epistemology is bound up in a traditional religious and aristocratic enterprise.

But there remains the question of what is left after the ideology and the disguised egoism is subtracted, and the answer is—as I hope this book will show—"plenty." The moves toward self-abnegation, the determination to push the truth even to the death (and John Ruskin in effect makes precisely this argument in *Unto this Last*),[15] are difficult to avoid. Yes, the last thing women in Victorian Britain needed was to be lectured on the virtues of self-abnegation and self-sacrifice, and feminist critics who have admired George Eliot's independence as writer and thinker have found such lecturing particularly disturbing. Yet those virtues were the conditions, as George Eliot understood it, for her own intellectual successes. Although her relation to system and abstraction was complicated by her insistence that everything depended upon particulars, she was representative of a dominant intellectual movement when she wrote:

> The great conception of universal regular sequence, without partiality and without caprice—the conception which is the most potent force at work in the modification of our faith, and of the practical form given to our sentiments—could only grow out of that patient watching of external fact, and that silencing of preconceived notions, which are urged upon the mind by the problems of physical science.[16]

The turn to science, the language of impartiality, the stance of "patient watching," and particularly the "silencing of preconceived notions" iterates once again the ideals of scientific epistemology, its ethical force, and its implicit story. The virtues of patience, the determination to thwart desire and preconceptions might, of course, be transferred to the gender questions that have reasonably disturbed feminist critics of George Eliot. But the epistemological problems remain after the ideology is accounted for; if science can be invoked to keep women in their place, the silencing of preconceived notions remains a condition of adequate knowledge—and for finding a way for women to get out of that place.

And so the narrator's famous intervention in the seventeenth chapter of *Adam Bede* seems to me reasonably representative of what this epistemological tradition entails:

> I aspire to give no more than a faithful account of men and things as they have mirrored themselves in my mind. The mirror is doubtless defective; the outlines will sometimes be disturbed; the reflection faint or confused; but I feel as much bound to tell you, as precisely as I can, what that reflection is, as if I were in the witness-box narrating my experience on oath.[17]

The project, then, depends on the narrator's willingness to suspend her self-interest, recognizing all the time that there are limits both to the authority of what she (he, in the novel) can know and to the powers to get it right in relating it. But once she is aware that "the mirror is doubtless defective" she can give herself to the complications of truth, always being aware of limits.

There is not only nothing wrong with this; it does seem to me a condition of good knowing and of good narration. It is true that, as Richard Rorty might see it, George Eliot is trapped in the language of representation, the notion of an inside and an outside, of a reality out there to be discovered, of traditional Platonic and Cartesian dualism. But her primary concern is the exploration of the conditions of a world that is larger than any of its citizens and that will not bend to human will. It is obviously true, as George Eliot herself surely understood, that nobody is exempt from limitation: everyone, located historically, socially, personally must perforce have a temporally and spatially limited perspective, will be driven by unrecognized passions (or even recognized ones), and will make false claims for integrity and self-sacrifice—this is virtually the dominant theme of *Middlemarch*. In other words, the critique of epistemology that insists on the inevitability of distortions and limits is at least partially right, from the perspective of narrative. Only Utopia is nowhere.

But it is wrong in denigrating the utopian epistemological pursuit. All epistemologists know that the pursuit is difficult, probably impossible. Self-delusion about it is common and has been, historically, very dangerous. Deliberate falsification in the name of the impersonal, the ideal, the universal has perpetrated the most serious crimes. But in the course of worrying my metaphor of death and the questions of epistemology and epistemology's relation to the way we lived then and live now, I have been increasingly impelled toward the idea that the ideal and its particular incarnations need to be separated. Narrative helps toward that separation because it requires attention to embodied particulars and can make no universal claims.

When I claim that my arguments are not philosophical but literary, I mean that my focus will be on the oddly necessary interaction of narrative and epistemology, that is, on how philosophy behaves when it is embodied and its ideas take on the life of metaphor and mean something in the lived experience of people located in particular times and places. Ideas live in culture not disembodied, but as actions, attitudes, assumptions, moral imperatives. Each epistemological enterprise is written into a context that requires special justifications for the activity; the requirements, in a way, can be understood as partly producing the activity (so, for example, the fate of Galileo had much to do with Descartes's turn from his doing science to his theorizing about it).

In chapter 4 I will be looking closely at Darwin's *Autobiography*, in which he establishes what Linda Peterson sees as the basic pattern of scientific autobiography.[18] But here I would offer another example: the separation of mind and body implicit in Darwin emerges rather differently in Einstein's *Autobiographical Notes*. For one thing, Einstein writes these notes as an "obituary," and focuses only on his "thought." At one point, having settled into an analysis of the limitations of Newtonian mechanics, he interrupts: "'Is this supposed to be an obituary?' the astonished reader will likely ask. I would like to reply: essentially, yes. For the essential in the being of a man of my type lies precisely in *what* he thinks and *how* he thinks, not in what he does or suffers."[19] While this may be taken as a refusal to make public the complications of his personal life, it is also an assertion that he aspired to separate completely the man who suffers from the scientist who creates. Discussing the self and one's ideas is like an "obituary."

Within this tradition of self-repression, as scientists imagine themselves surrendering to the force of a world larger than they, personality is a burden, just as T. S. Eliot, in his most influential modernist moments, celebrated the artist's release from personality. Einstein's description of his movement from the "deep religiousness" of his early youth to a commitment to science interestingly recapitulates the major motifs of late nineteenth-century objectivist epistemology. He found himself under the impression that "youth is intentionally being deceived by the state through lies" and feeling mistrust "of every kind of authority." Descartes's Great Doubt takes another shape, as does T. H. Huxley's view that the mark of the scientist is the refusal to accept anything on authority. So the initial move for all of these writers—in what I will be describing as the narrative of scientific epistemology—is rebelliousness, but as in Einstein's case, also a quest for a finally stabilizing certainty:

> It is quite clear to me that the religious paradise of youth, which was thus lost, was a first attempt to free myself from the chains of the "merely personal," from an existence dominated by wishes, hopes, and primitive feelings. Out yonder there was this huge world, which exists independently of us human beings and which stands before us like a great, eternal riddle, at least partially accessible to our inspection and thinking. The contemplation of this world beckoned as a liberation, and I soon noticed that many a man whom I had learned to esteem and to admire had found inner freedom and security in its pursuit. The mental grasp of this extra-personal world with the frame of our capabilities presented itself to my mind, half consciously, half unconsciously, as a supreme goal. Similarly motivated men of the present and of the past, as well as the insights they had achieved, were the friends who could not be lost. The road to this paradise was not as comfortable and alluring as the road to the religious paradise; but it has shown itself reliable, and I have never regretted having chosen it.[20]

But before settling into a focus on the particular narratives that will make the substance of this book, it will be useful to discuss more abstractly here the problems the narrative engages. The narrative of "dying to know" has many manifestations, all of them implying some separation between living self and knower. To put this problem in a philosophically more respectable way, the question is like the one that Thomas Nagel raises in his *View from Nowhere:* "how to combine the perspective of a particular person inside the world with an objective view of that same world, the person and the viewpoint included."[21] Nagel strongly affirms that the real existence of rationality and objectivity sets up the problem, as I want to consider it. That is, in putting the question in terms of space, Nagel opens up the paradox: the "perspective" of the knower requires being inside and outside at the same time. It is much easier to think one's way through such a problem abstractly (not, of course, that it can ever be easy) than when there are bodies involved. Even the self-abnegating Darwin, who always tended to exaggerate his inadequacies, has a place from which he stands and a quite serious stake in making himself look good.

In reaction to the critiques of objectivity I have already mentioned, Nagel claims in a later book that the philosophical effect of such critique is the dominant view that "understanding and justification come to an end," not with objective principles whose validity is "independent of our point of view," but "within our point of view—individual or shared—so that ultimately, even the apparently most objective and universal principles derive their validity and authority from the perspective and practice of those who follow them."[22]

Where narrative forces an emphasis on the lived incoherences and high costs of the strict philosophical argument, there remain other living questions: what happens after having satisfyingly (maybe complacently) rejected objectivity as a myth? The distrust of "interest" remains in place—that distrust is a source of many of the critiques of objectivity: that is, the claim, meant to be a rejection, that disinterest is not really disinterested, that objectivity is only a disguise for interest. Yet there have to be ways to decide that some things are true and some things are not—even if we decide that *true* is not the operative word, but, say, *valid*, or *the case*. The necessity persists not on the grounds of needing to produce a coherent theory, or of wanting to demonstrate the coherence of things, but because action requires knowledge—or luck—if it is to work at all. That is, what we do depends on what we believe to be true, and it's better to be right. "Truth," "objectivity," "epistemological validity" is always a pragmatic, not merely a theoretical issue. It is fundamentally impractical to act as though "the case" is not a condition that exists regardless of my perception of it. Objectivity may well have been exposed as a fraud. Much of the work it did, however, needs to be done anyway.

We are, the argument reasonably goes, utterly limited creatures, located in place and time and thus confined by perspectives, desires, prejudices that may be disguised or changed but not dismissed. Late twentieth-century theory has often radically challenged even the notion that there are stable selves with which to have those varying perspectives or to be prejudiced. Feminist theory, some of which has certainly refused to go the postmodernist way, has been largely dedicated to tearing off the disguises of objectivity with which the situated has been represented as universal.[23] The ideals of Enlightenment rationality have, since the 1960s, had a peculiarly difficult time.

Nagel's work and that of Susan Haack in philosophy, among much other, is part of an attempt at recuperation, but it is not only philosophical realists who worry the question of "objectivity." Recent debates have much complicated the relativist-objectivist dualism, and there have been interesting critiques of the sometimes self-contradictory and more frequently self-subversive aspects of antiobjectivist stances, even in the fields of cultural studies, and certainly from many aspects of literary theory.[24] To take one strong example (from philosophy once more), John Dupré, in his *Disunity of Science,* finds no conflict between the determination to shift emphasis from the formal work of science to a focus on the goals and motives of scientists *and* a belief that science can describe "the way things objectively are."[25] Committed himself to the work of cultural critics of science, he yet finds that

> [b]y asserting that all scientific belief should be explained in terms of the goals, interests, and prejudices of the scientist, and denying any role whatever for the recalcitrance of nature, it leaves no space for the criticism of specific scientific beliefs on the grounds that they do reflect such prejudices rather than being plausibly grounded in fact. (p. 12)

In other words, the effect of such an emphasis tends, ironically, to be conservative. For such a program of critique to be effective, Dupré claims, it "must be understood in interaction with some account of justifiable belief" (p. 13).[26]

Novels and narratives are often about questions of justifiable belief: has the protagonist got it right? Why is she being deceived? The relation between cleverness and self-interest, selflessness and correct knowledge, can be recognized as motifs that dominate nineteenth-century fiction, that mark the progress of most *Bildung* narratives. Autobiographies are most often narratives of discovery. It is through narratives, then, that I propose to consider the question of "justifiable belief," and through them that I will be considering the usefulness of even thinking about questions of "truth," discovery, construction, disinterest, objectivity.

The attack on disinterest, the exposure of hidden wickedness in the efforts to universalize, needs to be seen in the perspective of the stories that lie behind the quest for truth (and, indeed, the quest for truth is itself a story).

Within those stories, it is clear, protagonists believe in the validity of their own interests. They are not singular, I am certain, in believing that their own interests are not among the bad ones. Narratives dramatize the likelihood that if everyone else's interest is in bad faith, so, too, is the protagonist's likely to be. And here is one of the problems that the critique of disinterest inevitably raises. While we can go on congratulating ourselves in the usual exposure of situatedness, bad faith, or other people's self-deception, there is something to be learned from watching the ideal of objectivity at work, not through the strong arguments of philosophical "realists" (who come in an almost infinite number of shapes and forms), but by feeling the pressure of "truth" embodied, as novels in particular embody it, by people. The narrative of scientific epistemology folds neatly into the narratives that dominated English literature and culture in the nineteenth century, evidence that the ideal of epistemological disinterest or self-sacrifice or self-annihilation had managed to permeate the culture's consciousness; narrative is its carrier and embodiment.

Locating the epistemological ideal in narrative tends immediately to confirm what contemporary critics have been arguing, the dangers and confusions latent in it; but the ideal often does work worth doing. More important, it is striking how difficult it has been to imagine satisfactory alternatives. Nineteenth-century narratives, at least, suggest that however important it is to affirm the value of personal desire, to recognize the way thought and feeling are entangled, how rationality rarely determines action or even establishes authoritatively what the case is, it feels as though the culture's imagination of what is possible simply requires *something like* objectivity.[27] Nagel's representative articulation of the circularity of the position that denies the possibility of objectivity is once again helpful. He insists, and on this I am entirely in agreement, that "we must distinguish between general philosophical challenges to the objectivity of reason and ordinary challenges to particular examples of reasoning that do not call reason into question."[28] When criticism "exposes" a claim to the real nature of things as a "construction," it is not always clear which of these two challenges is being made. The second seems always legitimate because it can be falsified and it is likely itself to be impelled by some immediate human urgency. The first is itself a ground that makes its own absolute claim to reality, that is, that there are no statements that are not *determined* by nonrational grounds.

Again, narrative, tending to work through particulars, consistently challenges the particular thrusts at "truth" of its protagonists and villains. And narrative is rarely preoccupied with strictly rational grounds; again and again it dramatizes the way decisions and understandings are determined by desire and by the irrational. But in so doing, of course, it creates drama out of the tensions that come from the disparity between what is sensible and what is not: Jane Austen's Emma Woodhouse, for example, shapes her understanding of Mr. Elton and others around her against the understanding that her readers, not blinded by their vanity, can derive from the narrative evidence.

On Nagel's view (surely relativists are tired of hearing this argument), there are no rational grounds on which to accept the idea that there are no rational grounds. "Any claim," Nagel says, "that what is a reason for me is not reason for someone else to draw the same conclusion must be backed up by further reasons, to show that this apparent deviation from generality can be accounted for in terms that are themselves general."[29] Even this very abstract formulation implies an agon—there are, at least, intellectual combatants here, each of whom requires satisfaction. Neither will be satisfied if the other doesn't provide a reason. This, it seems to me, is the way it is. The agon recapitulates the structures of narrative itself because here, too, there is an assumption that the particular must be generalized, as the reader, watching a story unfold, is required to understand more than the characters within the story. The two combatants may not change their minds, may not be persuaded, but the structure of their arguments depends on an assumption that the grounds for belief are shareable. The story of our own excursions into knowledge is determined by an immediate willingness to decide when one has been convinced, when someone else has produced the kinds of reasons that will stick.[30] Epistemology slides into narrative. The structures of narrative both put to the question the idea of some essential truth, and require it.

The animus of this book has changed as it has lived through at least a decade of reflection. If it began as a critique, by way of the metaphor, of the impossible scientific ideal of disinterested knowledge, it has ended in discontent with seemingly complacent contemporary refusals not only of the possibility of objectivity but of the good faith of quests for it. It has thus turned into something of a defense of those impossible strivings toward disinterest and an implicit attack on the view that all attempts at objectivity are disingenuous and politically suspect. I have been struck by how "objectivity," in much academic discourse, has become a curse word, like "positivism" or "essentialism" or "universalism," even as the refusal of community, of shared discourse, of nonviolent grounds for belief has produced the most dreadful consequences. I have been convinced, moreover, that the idea that certain abstract theoretical scientific and philosophical positions have necessary and intrinsic connections with particular ideological positions is dangerously mistaken.

This conviction is one of the leitmotifs of the varied studies that constitute this book. "Ideological essentialism," Oscar Kenshur calls it. My friend Chris Herbert insists on the "ideological indeterminacy of ideas." Without denying that abstract ideas are often invoked for ideologically secretive purposes, Kenshur wants to argue that those purposes are historically rather than intrinsically connected with those ideas.[31] The current good guys, constructivism, relativism, antiessentialism, have been exploitable for political positions diametrically opposed to those that many theorists assume the ideas imply, and—as the history of logical positivism makes clear—objectivism and rationalism have served democratic, antifascist, even socialist purposes. I have found myself wanting to argue by the end of this long journey that the hermeneutics of suspicion and insistence on the primary values of localism and particularism have done what good work they can, and are now—often destructively—playing into the obsessive individualism of contemporary social and economic structures.

Isn't there, after all (and risking accusations of naiveté), something pretty decent and creative in a struggle to get beyond the self? If we have found how often the injunction to self-sacrifice has served the cause of repression and of truncated lives, isn't there still a kind of greatness in the willingness to sacrifice even to death in order to find things out? Is it true that all disinterest is interested? Is it true that altruism is not only counter-Darwinian but impossible? Isn't it true that the provincialism, sectarianism, and distrust that marks so much of the politics of post–cold war global society are showing themselves to be at least as dangerous as the imperious and self-righteous imposition of the "truth" on vast stretches of the globe by a disingenuous Western virtue?

The question is how not to mystify and thus obscure the implications of contingent power that lie in the story of selflessness. Behind the critiques in this book lies my commitment to the idea that while utter selflessness, utter disinterestedness, complete "objectivity," are simply impossible, the possibilities of a world that gets beyond mere contentions for power—that moves from nationalism to cosmopolitanism, from fanatical individualism to a sense of the primacy of the community—reside in the human capacity to work toward these ideals, to act, sometimes, somehow, with a sense of the needs and interests of others. Real detachment, full objectivity is, as Amanda Anderson claims, an "aspiration" rather than a possible achievement. Her position is consistent with Satya Mohanty's view that "human inquiry is by definition historical and meliorative."[32] Anderson's new book, *The Powers of*

Distance, makes a brilliant case for the importance of detachment and for the "cosmopolitanism" that she sees as in effect conditional on detachment.

A last prefatory caveat. Although the extravagance of the epigraph by Plato and even more of the metaphor of "dying" will hover everywhere over the pages of this book, obviously not every act of self-effacement is an equivalent, metaphorical or literal, of death. Obviously, too, there is no single narrative of scientific epistemology. But the extreme formulation is a means to recognize the way in which the insertion of epistemology into narrative changes it, and gives it real and human consequences. At its extremes the ideal of self-effacement becomes, indeed, the reality of dying: if not dying literally in the body, dying in the sense that all those contingent qualities that constitute life for everyone are displaced.

Dying is one consequence of the Faustian pact for knowledge: death both for the aspiring knower, and for the world in which things get known. As I try to suggest at several points in this book, the ultimate ideal impersonality of the knower does historically lead to the decentering of the human. The movement from Galileo, to Darwin, to Freud marks one clear line in this path, but contemporary reductions of the human to hard-wired forces of competing genes is another. The "death" of the inquiring self helps move human knowledge toward the "death"—sometimes quite literal—of other human ideals and wishes.

The story then is complicated. Half of what this book attempts to do is put to the test of narrative the epistemological ideal of self-effacement. This will mean locating in epistemological discourse narrative elements, locating in autobiographies and novels epistemological elements. The other half is a tentative but very serious effort to preserve some of that ideal, to see in it a genuine moral and intellectual triumph that remains crucial to human experience. The narratives test and implicitly criticize the ideal, but there are few narratives that dispense with it. Getting beyond the personal is not merely an illusion; it is a condition of intellectual exchange and of social life; and it is a perilously saving condition toward which the intellectual, moral, and aesthetic ventures of Western society have occasionally inched.

— I —
The Narrative of Scientific Epistemology

In this chapter I want to transgress the boundaries of the nineteenth century in order to consider the consequence of various manifestations of the mutual dependence of epistemological ideas and narrative that is the primary subject of this book. To do this, I will take a rather arbitrary tour through the history of commentary on scientific knowledge, from Descartes to Donna Haraway. While there is obviously no inevitability in the examples, it is important for my argument that however disparate the positions, however friendly or hostile to science, there is a persistent tradition of scientific epistemology that is profoundly affected by narrative and that makes evident the moral urgencies underlying the epistemological enterprises. If there is a certain bravado in juxtaposing Donna Haraway and Bacon, recognition that even they share certain crucial assumptions about what science is will constitute an important part of my case: the combined argument that dying (or at least self-humiliation) is always implicit as a foundation of scientific truth claims, and that there is no way entirely to escape this condition—and perhaps no reason we should try.

Alasdair MacIntyre has argued that " an epistemological crisis is always a crisis in human relationships," and he complains that "the history of epistemology, like the history of ethics itself, is usually written as though it were not a moral narrative, that is, in fact as though it were not a narrative. For narrative requires an evaluative framework in which good or bad character helps to produce unfortunate or happy outcomes."[1] While Descartes will figure importantly in the arguments of future chapters, it is useful to begin with MacIntyre's point that the problem with Descartes's narrative of repudiation of all tradition is that such repudiation is impossible. The very language Descartes thinks in to undertake his move of Great Doubt is part of his intellectual inheritance. Whereas change is possible, complete instantaneous change, MacIntyre argues, is impossible. A total rejection of tradition would have made it impossible for Descartes even to make sense of his rejection.[2]

But Descartes's intellectual extravagance transforms the seeker after truth into a saint. Giving up, as it were, all worldly intellectual goods, Descartes helps inaugurate a scientific epistemology disciplined, austere, self-forgetful, pure, followed through the embodiment (or disembodiment) that narrative gives it. This kind of "heroic" epistemology, as it has been called in particular by feminist scholars,[3] is precisely of the sort that Thomas Nagel has set out to defend. Nagel's argument about the circularity of antiobjectivist positions, when applied to scholarship, works again. Myra Jehlen, for example, points out that in current historiography "it has become routine to stipulate that . . . far from telling the truth, historians are inevitably duplicitous. At the same time . . . recent historians frequently present their accounts as corrections, and by implication as correct."[4] The question, once again, is what authorizes the correction? The exposure of objectivity, if it is to matter at all on the very terms of that exposure, must have some of those qualities that have traditionally been attributed to objectivity. The problem with the rejection of the possibilities of objectivity when considered beyond epistemology, in the realm of human engagement, is that it cuts off also the possibility of learning about the other, since the limits of the knowable are always the limits of the situated perspective, and not only of learning but in the end of caring about that "other" that looms so importantly in contemporary criticism.

Invoking the distinction Nagel makes between a critique that demonstrates the inadequacy of an argument, and a critique that invalidates the rationality that presumably undergirds it, I would suggest that the critique of an objectivity that turns out in fact to be a disguise of interest undercuts its own objectives if after the exposure of particular interests, it goes on to argue—or imply—that the very act of seeking objectivity is invalid. "Objectivity"—the position that allows for the recognition of otherness by someone whose perspective would seem to be limited by personal, family, or class limits—turns out to be a condition of overcoming the repression that many critiques of objectivity seek to expose.

Feminist scholars in particular have been increasingly involved in efforts to rethink the critique of objectivity. Evelyn Fox Keller, for example, has always resisted the full-scale retreat from objectivity. While affirming the importance of critiques of science from the perspective of the personal, she has insisted that although we can often show that scientific impersonality is driven by "prior emotional commitments, expectations, and desires," "any effective critique of science needs to take account of the undeniable successes of science." And she argues that scientists' "shared commitment to the possibility of reliable knowledge of nature, and to its dependence on experimental replicability and logical coherence, is an indispensable prerequisite for the effectiveness of any scientific venture."[5]

It may not seem much to say that scientists begin with a "prior emotional commitment" to the "possibility of reliable knowledge," but faith in such a possibility has been a widespread human need. It is a kind of ethical first commandment of knowledge—the world must make sense to humans. But the need has no place in formal epistemological argument. "Is" and "ought" remain forever separated according to the rules. And yet the ethical is present in the hidden narratives of epistemology—in its inception (driven as it is by the need to satisfy the human desire to know that one *can* know and can use knowledge to make things better), in its object (knowledge—the truth that religious and political tradition say will make us free), and in the process of knowing (that very refusal to require immediate satisfaction, the willingness to die for the truth).

Descartes's epistemological work—the classic Western locus for the foundation of modern epistemology—is first articulated in a narrative, and a narrative that, like many urgent discourses that followed it, attempts to make narrative unnecessary. It seeks to establish knowledge on a firm and universally functional foundation. The continuation of that antinarrative tradition is the object of my interest here. We can see it, for example, in John Herschel's influential formulation in 1830 of the principles of true science. Herschel urged as the one preliminary step to scientific enquiry, "the absolute dismissal and clearing the mind of all prejudice, from whatever source arising, and the determination to stand and fall by the result of a direct appeal to facts in the first instance and strict logical deduction from them afterwards."[6] Partly through the efforts of its own practitioners, partly through a long philosophical tradition that, indeed, predates the Renaissance, science acquired that cold and aloof reputation that places it at the remotest distance from the affairs of the human heart; scientists, the macabre story goes, peep and botanize on their mother's graves. Simple and unexcited as Herschel's prescription is, it yet anticipates the shape of the whole tradition of "dying," or of "heroic epistemology," with which this book is concerned.

The mind must be emptied of all prejudice, a move MacIntyre has already suggested is impossible, and one does not need MacIntyre to recognize that impossibility. The investigator must, in addition, have the Hardyesque willingness to take a good look at the worst and, if it turns out to be the worst, to "stand and fall" by that knowledge. There is always, in other words, an enormous personal risk in the effort to find things out. The "scientific" investigator will not shrink from the effort and will not back off from the results if they turn out not to satisfy emotional needs. This would seem to require not only intellectual but moral strength.

There is, then, a persistent irony that runs through the discourse of distance and detachment. At both ends of the epistemological enterprise there is a powerful moral valence. It is *good* to know, and therefore the enterprise of knowing—even against the initiating myth of the fall into knowledge—is a good enterprise. (It does not take much imagination to notice how such a strikingly moral stance would have been helpful in establishing the authority of science against entrenched religious authorities.) Beyond beginnings and endings, in the very middle of the activity of knowing, there is the injunction of self-sacrifice: to sacrifice anything and everything, particularly one's own desires, in order to know. Only in this position of moral transcendence can knowledge extend from the particular to the universal, valid everywhere and at all times. Epistemology, one might say, is something like a Victorian novel.

The Western epistemological method that makes possible a valid account of the entire world depended on the development of a story whose presence would, ever after, be denied. For the story, aiming always at the greater good, is always also the story of why it is necessary not to tell stories. It is like Carlyle's story of Diogenes Teufelsdröckh (which I will also discuss in chapter 3): it moves tortuously toward a climax in an Everlasting Yea that denies the self affirmed with so much bravado in the Everlasting No, and requires precisely the sort of submission to authority—here the visionary duty—that Teufelsdröckh had resisted throughout his narrative. Stories, after all, move through time and carry with them the burden of the contingent, while truth is universal and timeless. The adventure in the narratives of scientific epistemology lies in the quest to get away from the contingent.

Although there are obviously earlier precedents, the modern version of the story begins more or less officially with Bacon, who was a pervasive presence in the consciousness of theorists of science, and his way of seeing the world was a more or less "official" and certainly a pervasive one in Victorian England.[7] A close look at what he said on the issue suggests that with Bacon, too, the ideals of truth that epistemology under virtually any circumstance achieves are constructed secondarily from logic, but initially and determiningly from narrative. Here is how Bacon sets the stage for these positions in *The Great Instauration:*

> For our own part, from an earnest desire of truth, we have committed ourselves to doubtful, difficult, and solitary ways; and relying on the Divine assistance, have supported our minds against the vehemence of opinions, our own internal doubts and scruples, and the darkness and fantastic images of the mind; that at length we might make more sure and certain discoveries for the benefit of posterity. And if we shall have effected anything to the purpose, what led us to it was a true and genuine humiliation of mind.[8]

It is rather difficult to believe in Bacon genuinely humbling himself, yet this is surely a move of fundamental professional self-legitimation. The grandiose and overtly heroic effort requires humility and courage. Of course, this is not "disinterested" prose. It offers itself as a distinctly personal statement at the start of a work of almost indescribably ambitious scope, and as a piece of narration. It describes a pilgrim's progress (a progress that will reappear in many guises throughout the narratives this book will investigate)—a heroic and pious rejection of the inherited things of this world, a journey necessarily austere, away from temptation and the devious pulls of desire and imagination, into the truth and the future.

Bacon's famous "Idols" are anticipated here. Through the Idols, Bacon launches his investigations, sweeping away the obstacles to knowledge—personal, social, commercial, human. If this is not quite so sweeping as Descartes's Great Doubt was to be, it approximates the later Cartesian move in its effort to remove everything that might divert the knower from a direct engagement with the facts.

The language reverberates with moral implications. The pilgrimage begins with an "earnest desire" and leads the investigator through a series of obstacles, both external and internal, described in such a way that he, the investigator, can only be regarded as brave, fighting against odds, against multitudes. Science makes professional and innovative claims that it is work for "the benefit of posterity." So Bacon begins by *not* assuming that the activity of science is disinterested, for its own sake. Emphasizing the seriousness of the risk and the implicit courage required to take it, the prose implies a moral crusade. Full achievement of the goal is shown to be dependent on a moral quality—"a true and genuine humiliation of mind." Ego and desire must surrender in any confrontation with nature. While Lorraine Daston has argued that Bacon's "humiliation" here and elsewhere is only conventional piety, politically responsive to the culture's expectations, it is not insignificant that Bacon is impelled to make the move of "humiliation" in order even to begin the work of science. The humiliation Bacon requires, as he requires a shift of focus onto the material, natural object, sounds rather like the modern requirement of disinterest in scientific work. To know the truth one must open oneself to it.

As Daston and Peter Galison put it in their discussion of the moralization of knowledge in the nineteenth century, "at issue was not only accuracy but morality as well: the all-too-human scientists must, as a matter of duty, restrain themselves from imposing their hopes, expectations, aesthetics, even ordinary language on the image of nature."[9] Yet by the end of the eighteenth century there was already a strong tradition of celebration of the moral superiority of the scientist. The eulogies delivered in praise of scientists at the Paris Academy of Sciences all through the eighteenth century seemed to convey the message, as Charles B. Paul argues, "that science is not so much a natural as a moral philosophy."[10] Within that tradition, the supreme virtue of science, beyond "mental acuity," is "selfless-devotion."

Bacon's language is itself full of moralizing. Rules of demonstration are, for example, described as "depraved."[11] He speaks of how "man always believes more readily that which he prefers. He, therefore, rejects difficulties for want of patience in investigation; sobriety, because it limits his hope; the depths of nature, from superstition; the light of experiment, from arrogance and pride . . . in short, his feelings imbue and corrupt his understanding in innumerable and sometimes imperceptible ways" (p. 394).

Bacon is here imagining a way of knowing that depends on the capacity to surrender the all-imposing energy of will, feeling, and imagination before the revelatory reality of nature. In the preface to the *The Great Instauration*, he argues that nature " is not to be conquered but by submission" (p. 20). If the prose in praise of submission seems rather unsubmissive itself, that is precisely to the point of what I take the paradox of scientific epistemology to be. Its authority derives from the personal. Bacon's well-known aggression, as described particularly by Evelyn Fox Keller, but well noted by many feminist critics, expresses itself in the determination to "torture" nature into talking. It is "we," that is, Francis Bacon, who work from an "earnest desire for truth." And it is "we" who take great risks and dare to resist "the vehemence of opinions" and the internal weaknesses of the mind in order to "make sure and certain discoveries for the benefit of posterity."

If the Baconian presence did not, in the long run, take hold, the Baconian ethical-epistemological and narrative doctrine did. Bacon established the fundamental pattern of scientific justification and argument, dramatizing himself as a sacrificial and almost chivalric hero, battling against the historically overwhelming forces of the irrational and dangerous populace and of the undisciplined energies of human irrationality. Despite this tacit self-aggrandizement, overt claims for the self are precisely the last things that Bacon wants to make. One can only be a true Baconian hero by way of a true humiliation of mind. As he argues in the *Novum Organum*, this humiliation before God must be accompanied by humiliation before nature. The language of scientific discovery is almost ever after cast in something like the passive—nature impinges on consciousness, which surrenders to its reality. One conquers by way of submission.

More than one commentator has noticed that the conquering is more important than the submission, which is only a means, and has suggested that strategies of self-effacement often disguise egoistic aggressions. This is part of the history of the "dying" metaphor with which I am concerned. "Conquering" nature is the moral mission of science and Bacon's justification for his enterprise: "the point of view is not only the contemplative happiness, but the whole fortunes, and affairs, and powers, and works of men."[12] The driving energy toward the reform of methods of knowing is an ethical goal: improvement of the human condition. So scientific epistemology is driven by and aimed at an ethical ideal: it is entirely in the service of the human. If this objective has often served to forward (and disguise) efforts toward political, racial, imperial dominance, the problem is not with the objective but with the question of who decides how to do it and with what balance of costs. Again, a separation needs to be made between questioning the way in which the work has been done and questioning whether to do it at all. When one asks whether benefiting humankind is *really* the objective of particular works of science, there lies behind the question the assumption that there is a way to distinguish between *real* benefits, adequate methods to achieve those benefits, and mere disguises of exploitation and reaches for power. However disingenuous it may seem, Bacon tries to imagine a world in which human life is democratically easier and more comfortable, where things work better, where intelligence and rationality have a significant place.

The implicit narrative of Bacon's self-construction and quasi-manifesto carries through scientific apologetics at least until the end of the nineteenth century and remains latent in the continuing justifications of modern science—whatever the actual practice may be like—from which narrative is formally banished. Modern science does not, in its professional papers, want to hear the story of the scientists' work on the way to the recorded experiments. The self is banished, along with history and affect; and this banishment allows what appears in professional papers to aspire to and claim universal application. The heavy ethical freight carried in the histories that construct and resolve the various epistemological crises of which MacIntyre talks is clearly visible in the Baconian project.

The parallels between Bacon's narrative and Descartes's remarkably confirm the importance to Western science of narrative justification. Each affirms the presence of contingency even while each argues the possibility of the unconditional. Each rigorously rejects the past as a necessary preliminary to positive scientific activity. However different the "methods" of the two historically influential theorists, their work moves not through a pure discourse of epistemology but through narrative constrained by ethical imperatives. The model for scientific investigation is heroic self-humiliation; the seeker of natural knowledge puts aside worldly things, the idols of theater, cave, and marketplace, and prepares to submit to the blows of reality for the sake of a pilgrimage to the promised land of pure knowledge, human enrichment, and material progress—the purity and the materiality are explicitly linked. Such stories are commonplaces of Western cultural inheritance, from the medieval romances, to the Pilgrim's Progress, to Victor Frankenstein's attempt to create a better race of humanity, to Keats's imaginative absorption of the world he perceived, to John Herschel's defense of science, to the moral progress of Victorian heroes and heroines.

What happens, then, when this epistemological romance emerges in narratives like those of the novel? Bacon's rhetoric catches him in a contradiction between his professions of humility and the enactment of hubris; stories not of the self but of others might avoid such contradictions. When Samuel Smiles sought his examples of exemplary behavior, he often used scientists, and his brief excursus on Michael Faraday conveys legitimately, I believe, the official public notion of what constituted good scientific behavior. "There was one feature in Faraday's character which is worthy of notice," claims Smiles, "one closely akin to self-control: it was his self-denial. By devoting himself to analytical chemistry, he might have speedily realized a large fortune; but he nobly resisted the temptation, and preferred to follow the path of pure science."[13] Resistance to the temptations of money is paralleled by submission to the realities of fact.

William Whewell's important *History of the Inductive Sciences*, perhaps the first full-scale history of science in the West, dramatizes through biography the theory it is meant to imply. The book is a narrative in search of a theory, which, in effect, it will find in Whewell's *Philosophy of the Inductive Sciences*. Whewell saw his "history" as providing the data on which his "philosophy" might be built. As history, then, it is an example of Baconian science: out of the accumulation of facts—and Whewell knew quite well and theorized quite elaborately that facts did not simply tumble randomly from nature into the mind—would emerge a theory of knowledge "as the most promising mode of directing our future efforts to add to its extent and completeness."[14] Near the end of the two volumes of his *Philosophy*, Whewell offers a long section of aphorisms built on a Baconian model.

But for the Kantian Whewell, the mind itself is always a part of what is being studied, a creator. His famous "Fundamental Antithesis of Philosophy," which got him into extended debate with the non-Kantian empiricist John Stuart Mill, affirmed the inevitable contrast between such elements as "things and thoughts," or "theories and facts," and yet insisted that while those apparently antithetical things were always in opposition, the actual boundaries between them were thoroughly blurred. The world is not raw fact but always theorized; the thing in itself is never accessible. "The distinction between Facts and Theories is not tenable," Whewell would argue.[15]

Whewell asserted the importance of imagination and hypothesis, of human active intervention as opposed to mere passive absorption, and this Kantian move troubles the narrative of epistemological self-effacement. Indeed, the Cartesian crisis of whether we are only dreaming haunts Western epistemology and the metaphor that I am exploring: Kantian "idealism"— at least as it was interpreted by Whewell and Karl Pearson—continues into modern positivism, threatening to become a kind of solipsism. What we know is not outside but inside of us. The knowable world is not material fact but our sensations (of it?). In later chapters I will explore some of these Whewellian intimations and ways in which they, too, assimilate to the metaphor of death I am exploring here, sometimes by way of a Paterian ascesis.

But what I want to emphasize is that, even more intensely than Bacon, Whewell was convinced that the enterprise of knowing entailed high moral qualities. Even with his self-consciously creative notion of scientific discovery, Whewell is committed to an ideal of disinterest and imagines the possibility of detachment from bodily and personal desire. Self-effacement is epistemologically heroic. The tradition of religious and moral self-sacrifice glows through his images of intellectual success and particularly in his stunning description of Newton's character:

> The habit to which Newton thus, in some sense, owed his discoveries, this constant attention to the rising thought, and development of its results in every direction, necessarily engaged and absorbed his spirit, and made him inattentive and almost insensible to external impressions and common impulses.... Often, lost in meditation, he knew not what he did, and his mind appeared to have quite forgotten its connection with the body. His servant reported that, on rising in a morning, he frequently sat a large portion of the day, half-dressed, on the side of his bed; and that his meals waited on the table for hours before he came to take them. Even with his transcendent power, to do what he did was almost irreconcilable with the common conditions of human life; and required the utmost devotion of thought, energy of effort, and steadiness of will—the strongest character, as well as the highest endowments, which belong to man.[16]

"He knew not what he did." It is a spiritual exercise, this act of really knowing. And it requires that very separation of mind and body that the Cartesian method seems to affirm. Here the epistemology is embodied, with all its contradiction. For, of course, Newton *has* a body.

Newton is shown, however, to be in touch with the highest sources of spiritual being, constantly attending to what Whewell evocatively calls "the rising thought." The image is Paterian, evoking the intensity and focus required to burn with a hard gemlike (intellectual) flame as one scans the movements of one's own consciousness. The figure drawn here seems also to anticipate, narratively speaking, the absent-minded professor or the far less endearing amoral scientist, but it obviously points more directly toward saintly figures in other more familiar genres—like the far-sighted Dorothea Brooke talking to her sister or to Rosamond Vincy—ready to sacrifice all worldly well-being for the sake of truth. Whewell is quick to point out how Newton was widely regarded "as the greatest example of a natural philosopher . . . his moral qualities, as well as his intellect, have been referred to as models of the philosophical character, and those who love to think that great talents are naturally associated with virtue, have always dwelt with pleasure upon the views given of Newton by his contemporaries; for they have uniformly represented him as candid and humble, mild and good" (ibid.).

The insistence on character makes explicit the connection between epistemology and ethics. Newton's candor and humility, his mildness and goodness (moral traits that repress bodily desire and leave the spirit open to the revelation of the world of fact) are what allowed him to do his revolutionary scientific work. The humility that Bacon had invoked recurs here and everywhere, in fictional narratives about the true hero, in autobiographies that require moral authority if they are to seem credible, in the cultural ideal of the gentleman. It is perhaps the most distinguishing character trait of Darwin: humility that keeps him from imposing himself on nature and that ultimately allows him to become the major propagator of perhaps the most humbling doctrine in the history of Western thought. In effect, Newton as imagined by Whewell gives up the self to the mysteries of nature, and in so doing becomes the model of the scientific hero.

But narratives are not to be trusted entirely. The contingencies of the local narrative necessitate attention to the moral implications of action (or inaction) in the embodied world, and they lead to more possibilities than can be contained in theory. Here, for example, Newton's abstracted behavior suggests more than the focus on spiritual purity can allow. Narrative puts pressure on the distinction between mind and body. It will not have escaped cultural critics here that the purity of Newton's behavior and thought depends on material conditions he apparently can afford to ignore. How can a grown man sit on the edge of his bed all day—unless his material needs are cared for by someone else? Whewell's Newton is a hard worker: "if he had done anything, it was due to nothing but industry and patient thought" (1:417)—a sentiment almost precisely like that of Darwin in his autobiography, one might note. There is nothing outstanding about him except his willingness to work.

Yet the image of Newton sitting on the side of his bed, noticed and described by a background figure, makes it clear that his ordinariness is built on resources not available to most people. The servant who brought the food is ignored. Newton's high moral qualities require support that Whewell and nonnarrative discourse could leave transparent. The location in time and space that narrative requires opens alternative possibilities and begins to put to the question the story Whewell tells about Newton and that all self-effacing narrators tell about themselves.

Narrative shifts attention from the science itself to quality of character. While Newton's work is apparently incompatible "with the common conditions of human life," the story of the incompatibility becomes part of what authorizes the theory. This anecdotal Newton becomes another figure in the narrative of scientific self-legitimation, in which the ethical content disbarred from the science proper is recognized in the activities that produce it. Newton's strategy, like Whewell's, disingenuous or not, gives powerful authority to the knowledge they produce because it implies moral rigor in its production and universal accessibility in its conclusions. Being modest in this context is a way of achieving authority.

Newton, Whewell writes, "could not admit that there was any difference between himself and other men, except in his possession of such habits as we have mentioned, perseverance and vigilance" (1:417). The foreshadowing of Darwin's self-description here is remarkable. And so, too, Descartes, in a sentence that may be the most unbelievable in Western literature, talks of himself as a man of ordinary qualities: "I have never supposed," he writes, "that my mind was in any way out of the ordinary."[17] This is the scientist's trope. What constitutes scientific success is not some rarefied quality of genius but sheer moral self-discipline. Science announces itself as a democratic enterprise, and its heights can be scaled by those who work most diligently and self-effacingly. Intellectual superiority is built on moral superiority.

In the face of such heroism, I want now to intrude a piece of overtly fictional narrative. It is a long way from Newton or from history of science or from epistemological discourse. But it will provide a preliminary example for this book's attempt to excavate from the way we tell stories these presiding ideas about how it is possible to know. Here the mixed sources of scientific activity are not accidentally revealed, and transformation of the abstract enterprise of the quest for natural knowledge into a time- and space-bound human activity with distinctly moral overtones is complete. In a passage from Hardy's odd but in some ways characteristic novel, *Two on a Tower,* one can find a narrative counterpart of heroic epistemology. With Hardyesque extravagance, even a gesture at the comic, the personal implications of that austere ideal of objectivity that makes saints of scientists is foregrounded.

In *Two on a Tower,* Hardy seems almost to be parodying his later cosmically pessimistic self. The developing relation between Swithin St. Cleve and Lady Constantine is made possible by Swithin's surveys of the cosmos through his telescope. His tutorial, as it were, shows that the very strategies of education—Swithin showing Lady Constantine how to use the telescope and educating her about what she is seeing—embody the sense that knowledge cannot be abstracted in some pristine acontextual laboratory. Here it impinges immediately on the lives of emotionally complicated people with other things on their minds. As in *Middlemarch,* Lydgate's scientific enterprise is intricately inwound with his private life, so here the science of astronomy gathers its meaning from the incipient and skewed love relation between Swithin and Lady Constantine. The image of Newton in his bedroom admits human complications only inadvertently, but with Hardy, the education at the telescope quickly becomes part of a complicated ritual of courting. The scene offers one of those exemplary Victorian moments in which, as Edward Dowden puts it, "a fact expressed in terms of feeling affects us as other than the same fact expressed in terms of knowing."[18]

Lady Constantine, unhappily married and older than Swithin, visits him with some fear of impropriety, for a promised tour of the skies. The all too Freudian tower, which she allows him to use for his astronomical observations, is on her property. Showing her Saturn and then some distant "fixed stars" (which are in fact moving "with incredible velocity"), Swithin asks her how many stars she thinks are visible through the microscope. She guesses far too low, and he responds:

> "Twenty millions. So that, whatever the stars were made for, they were not made to please our eyes. It is just the same in everything; nothing is made for man."
>
> "Is it that notion which makes you so sad for your age?" she asked, with almost maternal solicitude. "I think astronomy is a bad study for you. It makes you feel human insignificance too plainly."
>
> "Perhaps it does." "However," he added more cheerfully, "though I feel the study to be one almost tragic in its quality, I hope to be the new Coper-

nicus. What he was to the solar system I aim to be to be to the systems beyond."

Then, by means of the instrument at hand, they traveled together from the earth to Uranus, and the mysterious outskirts of the solar system; from the solar system to a star in the Swan, the nearest fixed star in the northern sky; from the star in the Swan to remoter stars; thence to the remotest visible; till the ghastly chasm which they had bridged by a fragile line of sight was realized by Lady Constantine.

"We are now traversing distances beside which the immense line stretching from the earth to the sun is but an invisible point," said the youth. "When, just now, we had reached a planet whose remoteness is a hundred times the remoteness of the sun from the earth, we were only a two thousandth part of the journey to the spot at which we have optically arrived now."

"O, pray don't; it overpowers me!" she replied not without seriousness. "It makes me feel that it is not worth while to live; it quite annihilates me." [19]

The passage plays out the risk that objectivity will discover things hostile to human interests. Truth, then, is the most dangerous and the highest of virtues. Its dangers were a relatively new thing. As E. A. Burtt puts it, when medieval man "observed a distant object, something proceeded from his eye to that object rather than from that object to his eye. And, of course, that which was real about objects was that which could be immediately perceived about them by human senses." [20] With the telescope as supplement (the kind of prosthetic improvement that Bacon argued for), the stars in Hardy impose their vastness and distance on the eye. The good post-Renaissance scientist surrenders to the stars, and risks the sort of annihilation that Lady Constantine, melodramatically, experiences.

If in Descartes's and Bacon's new alignment of knowledge, the first step to knowledge was dismissal of all traditional intellectual authority, the second was intensification of the empiricist will to surrender to the world. T. H. Huxley famously argued that "[t]he improver of natural knowledge absolutely refuses to acknowledge authority, as such. For him, skepticism is the highest of duties; blind faith the one unpardonable sin." [21] The Renaissance rejection of authority became in the end the condition for the displacement of humanity from its center, a displacement progressively effected by Galileo, Darwin, and Freud. To know, in this post-Galilean world, increasingly is to submit oneself to forces incompatible with human desire.

So Hardy's little contretemps between Swithin and Lady Constantine participates in a long tradition, whose beginnings have been described by

Burtt: "For the dominant trend in medieval thought, man occupied a more significant and determinative place in the universe than the realm of physical nature, while for the main current of modern thought nature holds a more independent, more determinant, and more permanent place than man."[22] As Bertrand Russell was to put it, "The kernel of the scientific outlook is the refusal to regard our own desires, tastes, and interests as affording a key to the understanding of the world."[23]

In *Two on a Tower*, Hardy reinforces the point that the displacement of man from the center of meaning, his transformation from creator into perceiver, makes him an investigator with no special purchase on reality. The quasi-lovers act out the epistemology of self-annihilation, annihilating themselves before the stars that oppress them with their vastness and multitudinousness. "Nothing," says Swithin, "is made for man."

But there is another irony. As Swithin pours on the horrors—astronomers, he says, "who exert their imaginative powers to bury themselves in the depths of the universe merely strain their faculties to gain a new horror"[24]—he becomes more attractive. The more Swithin dwells on the horror, the more Lady Constantine goes for him: "his vast and romantic endeavors lent him a personal force and charm which she could not but apprehend" (p. 59). Half comically imagined as this is, it echoes the Baconian enterprise. Submission is a means to power. Modesty and self-abnegation become indirect instruments for self-aggrandizement. At the very moment that Swithin articulates the sense of human insignificance, he implies that being able to know the vastness of universe could give him the power to be a new Copernicus.

By dropping this epistemology into a love story, Hardy complicates it and makes it half absurd, yet he also affirms its validity and its dangers. The focus shifts from a cosmic philosophy to the people who indulge it: what matters are the lovers, not the ultimate infinite meaninglessness and impenetrability of the universe. Cosmic tragedy becomes a comedy, or farce. "But," says Swithin, after recommending a study of the cosmos to make all things seem unimportant, "you say the universe was not really what you came to see me about. What was it, may I ask, Lady Constantine?" (p. 59) There is no need to trace the development of this narrative in the novel, but in its more or less tragic conclusion, the narrator notes Swithin's failure to understand the implications of Lady Constantine's behavior: "He was a scientist, and took words literally. There is something in the inexorably simple logic of such men, which partakes of the cruelty of the natural laws that are their study" (p. 291). Lady Constantine dies.

Hardy's translation of epistemology back into narrative foregrounds the contradictions and even absurdities of the ideal by juxtaposing to it the human reality. It raises questions about the human role in the constitution of truth and about the degree to which the ideal of truth *ought* to inflect human behavior. Taken straight, the narrative dramatizes, with a combination of irony and belief, the consequences of an epistemology that requires death for knowledge.

For the rest of this chapter, I want to consider several alternative approaches to these narratives of deadly neutrality. Against the almost masochistic objectivist model, William James wrote philosophy like a novel, finding the story of self-sacrificial surrender to an inhuman world both hard to take and untrue. James self-consciously placed at the center of his epistemology the choice-making, knowledge-seeking human agent, and he argued persistently for the possibility that human volition could indeed shape the world. James muses over the fact that "those who have caught the scientific fever . . . pass over to the opposite extreme, and write sometimes as if the incorruptibly truthful intellect ought positively to prefer bitterness and unacceptableness to the heart in its cup."[25] This he finds absurd: "What we say about reality," he claims, "depends on the perspective into which we throw it. The *that* of it is its own; the *what* depends on the which; the which depends on *us*. Both the sensational and the relational parts of reality are dumb. They say absolutely nothing about themselves. We it is who have to speak for them" (p. 118).

In reconstructing epistemology, James simply refuses the self-annihilation of traditional objectivity. We are always in our own consciousness: "When we talk of reality 'independent' of human thinking, then, it seems a thing very hard to find" (p. 119). The egocentric predicament is not paralyzing, but an indication of the centrality of the human. In defending the importance of the satisfaction of desire in any epistemological enterprise, James goes on:

> I had supposed it to be matter of common observation that, of two competing views of the universe which in all other respects are equal, but of which the first denies some vital human need while the second satisfies it, the second will be favored by sane men for the simple reason that it makes the world seem more rational. To choose the first view under such circumstances would be an ascetic act of philosophic self-denial of which no normal human being would be guilty. (p. 171)

To many, this view seemed merely irresponsible despite James's insistence that the will of the knowing subject could not on its own change what had been agreed on by the entire community of scientists and thinkers for a long time. Not only does James intrude will and desire into the epistemological procedure, but he is comfortable with the possibility that there can be no single unifying and universally acknowledged truth about the world. These countermoves to the narrative of objectivity begin with the assumption that all knowledge matters only as it connects with and enhances human life. (This is, we should recall, Bacon's justification of the pursuit of scientific knowledge.) The supreme virtue is not some abstract "Truth," whose demands on human feeling can be brutal, but human well-being. That this sense of the moral significance of knowledge can seem a descent into the chaos promised on the other side of Cartesian foundationalism suggests how deeply engrained in our culture, both as intellectual procedure and moral imperative, is the austere and mortal epistemology of objectivity.

To counter the human austerity of the more traditional notion of objectivity, James gives the example of a man trapped on a mountain climb at a crevasse that he must cross to survive. It is obviously urgent for him to find out whether he is able to leap it. If he believes he cannot leap it, his decision not to jump will make his belief true, and he will die where he sits. If he believes he can leap it and tries, successfully, he survives, and that belief is true. He is right in both cases, the truth being determined by the act of the man rather than a simple "the way things are." Of course, this is a difficult argument. In George Eliot's *Daniel Deronda*, which I will be discussing at length in chapter 8, one might say that Mordecai's desire makes it the case that Daniel Deronda was a Jew all the time without knowing it. (If it hadn't been for Mordecai Deronda would never have investigated, would not have been so satisfied when his mother tells him he is a Jew, after all, and would have led his life as a Christian.) The ethic of self-surrender normally governing George Eliot's epistemology was precisely that sort "of philosophical self-denial of which no human being would be guilty" (p. 117). When the ethic is played out in her earlier narratives, like *The Mill on the Floss*, the deadly consequences become manifest, as do the moral stakes. *Daniel Deronda*, in its deep dissatisfactions with the way things are, attempts to rethink objectivity. Of course, George Eliot never becomes a Jamesian pragmatist, but like James, she cannot settle for the absolute closure on human agency and possibility that a deterministic and passive epistemology seems to lock into place.

While this book will focus on narratives like George Eliot's, it is important here to consider some more modern and more overtly philosophical examples of what I am arguing is the inevitable "narrativizing" of epistemology. Rom Harré, in his *Varieties of Realism*, writes at a moment when much of the charm has come off science's heroic narrative, and Whewell's Newton figure has been transformed in science fiction stories into the scientist who, simply for the sake of knowledge, releases human-devouring monsters on the world. Beyond cinema, by the 1980s, science had been seriously tainted with contingency—with federal funding, with weapons developments, with biological experimentation. It has come under attack from a wide variety of contemporary skeptical discourses (and has in most recent years, of course, launched its counteroffensives),[26] from high cultural theory to cultish ecological critiques.

Shrewdly aware of the kinds of critiques to which science has been subjected recently, Harré is equally willing to accept contemporary skepticism about empirical truth and universal certainty. But his defense of science and of more or less objectivist epistemology is built on a shift of perspective from the object of scientific knowledge to the ways in which that science successfully authorizes itself. In effect, Harré extends the traditional model from the individual to the institution, and in so doing almost inverts it. Epistemological authority depends, in this account, on trust.

Unlike positivist predecessors, Harré assumes that there are many ethical and irrational elements in the practice of science. His theory attempts to reaffirm the basic rationality of science and its continuing generalizable value, but without falling back on the discredited tradition of a timeless and universal enterprise constrained only by the rigors of induction and logic. Realizing, as he puts it, that science is a "practice" as well as a discourse, Harré has to emphasize the presence of living human beings in the center of the epistemological enterprise. The determining categories for scientific epistemology turn out to be moral rather than logical. By placing the scientific community at the heart of the epistemology of science, Harré seems to be moving in the direction of what Philip Kitcher has called "naturalistic epistemology," one that places "the knowing subject firmly back into the discussion of epistemological problems." [27]

Harré begins by attempting to rethink epistemology through a rewriting of the Baconian/Cartesian narrative:

> Science is not just an epistemological but also a moral achievement. In defending the scientific community's just claims to knowledge I am also defending the moral superiority of that community relative to any other human association.
>
> I believe that the scientific community exhibits a model or ideal of rational co-operation set within a strict moral order, the whole having no parallel in any other human activity. And this despite the all-too-human characteristics of the actual members of that community seen as just another social order. Notoriously the rewards of place, power and prestige are

often not commensurate with the quality of individual scientific achievements when these are looked at from a historical perspective. Yet that very community enforces standards of honesty, trustworthiness, and good work against which the moral quality of Christian civilization stands condemned.[28]

As with many other of the epistemological arguments I have thus far cited, this might be taken as wildly extravagant. But like quite different kinds of contemporary critique of science, it brings to the surface the entanglements of epistemology with the ethical; and it thoroughly humanizes epistemology, making it not an encounter with the universal but a process of community consensus, necessarily then historical and narrative. And thus this bold defense of science does not meet the high standards of objectivity that Nagel has insisted on. Against the high moralism of the Cartesian tradition that begins with a heroic rejection of all knowledge based on the authority of the past, Harré posits the necessity of accepting authority, of *trusting* authority. Where nineteenth-century scientists wanted nature to speak to them, to overwhelm them, and as it were drive them to their conclusions, Harré imagines scientists building knowledge by way of trust in a community whose standards transcend the limits of individuals. Authority no longer lies in "experience," but in cumulative shared, tested communal experience.

There is no space here, nor would it be much to the point, to discuss other theories of scientific authenticity that insist on the communal.[29] But this sense of science depending not on a general determination of what essentially constitutes the activity of science but on a recognition of the science as a process in time and in a communal space is not at all exclusive to Harré. David Hull, for example, though arguing less extravagantly for the virtues of science, takes a similar position. The object is the same: to account for scientific progress and validity; like Harré, Hull notes the rather nasty activity of some actual scientists. But he sees this nastiness built into the whole activity of science in a way that does not compromise its ultimate validity. It is an evolutionist model, a process of selection through which competitive scientists, necessarily referring to one another's work, produce replicating arguments. Those scientific arguments that are most widely replicated through citation will in the long run triumph and produce scientific progress. This is not the pure Christian morality that Harré describes, but an almost amoral Darwinian alternative. But once again the community produces a morality (or a "truth") superior to any of its members. The ideal here is not built on self-humiliation, but on a secular reconstruction of that humiliation—produced now not by the individual scientist but by the community of scientists working with an almost laissez-faire-like invisible hand, ultimately rejecting what is not true. Here again the individual willingly participates in a community and surrenders to its standards. Once again, whether or not it is the strict "dying" narrative, the epistemology is engaged in questions of narrative and questions of morality.[30]

Certainly, Harré's theory moves to a new moralism, away from the heroic self-sacrifice of the individual before the intransigence of nature and toward a sacrifice to the trusted community. (In Hull, of course, it is not so much sacrifice of self as combat.) As against Huxley's sort of hostility to authority, this new epistemology begins in trust. Huxley rebelled against an inherited tradition of clerical authority, and he responded to it with an almost theatrical extravagance. But the twentieth-century establishment of the institution of science—largely prepared by Huxley himself—makes this sort of rebellion self-destructive, and Harré must, if he is not to be disingenuous, recognize the degree to which members of the scientific community defer to authority regularly in their work, and have to do so. Scientists depend on authority after all, the authority of the institution itself. In effect, then, Harré confesses that narrative lies behind universalizing and timeless science.

But rejecting what he calls the philosophers' "high redefinition," he does not believe that such a confession undercuts the authority of science. Rather, it establishes science on firmer ground in a world inevitably contingent and interested. His argument is that knowledge requires trust and honesty, and the power of science's epistemological authority is not rational, but ethical: the highest "standards of honesty, trustworthiness, and good work." Where Descartes plunged from skepticism into the Great Doubt and ended with the highest conceivable standards of truth—the clear and distinct—Harré's defense depends on separating science from its normally strict implication in rationality and logic. Yet he brilliantly turns the tables: he shows that the historicist, antiuniversalist critique of science is itself built on the model of epistemology it is attempting to confute. One can only find fault with science because it is implicated in history if one holds to a view that the only authentic knowledge is ahistorical—out of nowhere.

The critiques of science that concern Harré argue that science is not really governed by the strict standards of rationality to which it pretends. If it can be demonstrated, for example, that Darwin's theory of natural selection is structured so as to be consonant with laissez-faire economics, or that his theory of sexual selection is expressive of Victorian attitudes toward gender, the assumption is that something very damaging has been said about the theories. Darwin's science turns out not to be "pure," and its standards of objectivity and detachment are all compromised by its implication in matters of gender, race, or class; thus it has been proven unreliable, ideologically complicit, untrustworthy. Historical and sociological studies that move in such directions have revealed some extremely important things about the practice of science and about scientists. But the basic structure of the arguments, however delicately and intelligently elaborated, in effect employs rather Cartesian assumptions. Why should a critique that derives, itself, from the assumption that there can be no "pure" arguments, move to a negative or skeptical judgment when it discovers impure ones? It must be because of the unstated assumption that if all arguments are impure, all arguments are unreliable. Purity is the only guarantee of the validity of argument. We are back, it seems, to "dying."

Harré's way of handling this sort of critique is worth attending to. He begins with the assumption that impurities are indeed inevitable in any human activity, and therefore it is important to concede it at the start:

> Science is not a logically coherent body of knowledge in the strict, unforgiving sense of the philosophers' high redefinition, but a cluster of material and cognitive practices, carried on within a distinctive moral order, whose main characteristic is the trust that obtains among its members and should obtain between that community and the larger lay community with which it is interdependent.[31]

Even if many reading this passage will gasp at Harré's faith in the moral order of the scientific community, the logical force of the argument remains. There is and must be something other than "pure" distinctness and clarity that confirms scientific argument; and the attack by philosophers and theorists who are guilty of what Harré calls "high redefinition" is invalidated. If every argument is marred, then one asserts nothing in the argument that any given one is marred. The source of the authority of the arguments is, according to Harré, not in the purity of the argument but in the community that works hard to be coherent and consistent in its judgment of particular arguments.

Harré's apparently extravagant and scientifically self-interested argument about science as community seems to belong to an increasingly broad range of thinkers, many of whom do not have as their primary aim a "defense" of science. Helen Longino, for example, in a subtle and beautifully argued essay on the possibilities of feminist science, insists that "scientific objectivity has to be reconceived as a function of the communal structure of scientific inquiry rather than as a property of individual scientists."[32] Longino does not celebrate the moral triumph of the scientific community, but like Harré, she seeks to affirm the validity of science as a practice. Both have a large stake in "objectivity," although neither believes in it in its traditional senses. For Longino, traditional objectivity makes us—as Hardy's narrative dramatizes—"victims of the truth." In order to achieve a feminist science that does not essentialize the feminine, Longino needs to argue that the constraints politics impose on reasoning are inescapable and not grounds for dismissing the thought itself.

The argument from trust and community, thrusting objectivity back into a world with location in time and space, attempts to take into account the critiques emanating from sociology and anthropology of science, which in effect have been attempting to put science back inside the stories from which, in its professionalization, it has attempted to extricate itself. Harré has to return to narrative to reinvest science with the explicit ethical forces that had originally come to it through Bacon and Descartes.

The fact that scientists are marked by "all too human characteristics" need not have the negative consequences for science that many critics have implied. It will not have escaped some readers that there is a striking analogy here between Harré's conception of the scientist and the Roman Catholic Church's conception of the priest, who, in his priestly function, is carrier of the church's spiritual authority, but who is otherwise merely human. Behind every scientific argument, then, there lies an ethical imperative imposed by one of humanity's noblest institutions. The Cartesian anxiety derives from a narrative of absolutes: Harré may be more consistent in banishing such absolutes than the critics he is attempting to refute.

But even in diminishing the claims of science, Harré's new story also ends in a kind of heroism, that startling celebration of the community of science as superior to Christianity. It is here that his project of refiguring objectivity as a condition in space and time, and thus fallible, becomes by indirection a new assertion of virtually absolute moral authority—*because* it is more moral than epistemological, protecting science from the strong critiques that compromise both the abstract ideal of truth and the ostensible disinterest of its practitioners. Harré, then, writes another, if less absolutist narrative of high moral and selfless epistemology, and does not see himself as obliged to junk the authority of scientific epistemology because it is a narrative in disguise.

The decision to regard epistemological authority as deriving from a moral quality that exists not universally but within a defined community is consistent with the moral tradition that has always, more or less surreptitiously, accompanied epistemological arguments. Oddly, his defense of science allies Harré with feminist philosophers like Longino and with yet more radical critics. There are ways, for example, in which the move toward recognizing the narrative and ethical force of epistemology connects with

Donna Haraway's work. Her well-known essay, "Situated Knowledge," belongs in part within the tradition she is rejecting and this book is dedicated to exploring.

Unconventional as it is in the rather formal tradition of the rhetoric of epistemological argument, Haraway's rhetoric leans toward some of the central ideas of that tradition. This is a well-kept secret because Haraway is so outspokenly and overtly *not* disinterested. She insists on the temporality and embodiment of narrative and on its value-laden and political ends. In effect, she gives primacy to the political program over the scientific activity, because, like Hardy and James, she believes that the human is far more interesting and important than "knowledge." Science, she argues, must be made compatible with the moral energies that determine the value of human life. As a result, she says,

> I think my problem and "our" problem is how to have simultaneously an account of radical historical contingency for all knowledge claims and knowing subjects, a critical practice for recognizing our own "semiotic technologies" for making meaning, and a no nonsense commitment to a faithful account of a "real" world, one that can be partially shared and friendly to earth-wide projects of finite freedom, adequate material abundance, modest meaning in suffering, and limited happiness. . . . We need the power of modern critical theories of how meanings and bodies get made, not in order to deny meaning and bodies, but in order to live in meanings and bodies that have a chance for a future.[33]

No scientists I know believe that Haraway wants "a faithful account of a 'real' world," for such a desire would make her both a "realist" and an "objectivist." The rationalist argument is that no account can be "faithful" that begins with a moral or political agenda. And after all, the "real" is in scare quotation marks. Moreover, the passage suggests that meanings are made rather than discovered.

But the tradition of scientific objectivity begins precisely with Haraway's kind of political program, the improvement of the human condition, generally, democratically. Distortion is the thing to be guarded against both in the tradition and in Haraway's version of it. Even the scare quotes around the "real" suggest something of Cartesian skepticism. Moreover, Haraway, too, is asking for "trust." Like Harré, if in different ways for different purposes, she finds that some conditions must be established to make knowledge shareable and authoritative at the same time. "Every culture," Steven Shapin argues, "must put in place some solution to the problem of whom to trust and on what bases."[34] If Haraway refuses to write the old narrative that denies future narratives and affirms the universality of its conclusions, she does require, just as Bacon did, that science should be used to improve the human condition.

But without something like objectivity, she seems to be suggesting, it is impossible to make any but negative claims. The strong epistemological effort of Renaissance thinkers to find a new kind of authority for knowledge did indeed become restrictive and misleading, but it was, in its moment, liberating and indispensable; our recent efforts to locate knowledge inside the range of human contingency have now become a crucial enterprise. The developments through which the culture largely committed itself, as it is still committed, to the idea that the best observer is a dead observer has obviously had its dangers. But the more recent skeptical developments are perhaps equally dangerous, as all observational authority is rejected, all stable positioning is seen as spurious, all thought is contingently contaminated, and thus all thought is subject to radical undermining. Haraway herself recognizes the dangers:

> I, and others, started out wanting a strong tool for deconstructing the truth claims of hostile science by showing the radical historical specificity, and so contestability, of every layer of the onion of scientific and technological constructions, and we ended up with a kind of epistemological electro-shock therapy, which far from ushering us into the high stakes tables of the game of contesting public truths, lays us out on the table with self-induced multiple personality disorder. We wanted a way to go beyond showing bias in science (that proved too easy anyhow), and beyond separating the good scientific sheep from the bad goats of bias and misuse.... We unmasked the doctrines of objectivity because they threatened our budding sense of collective historical subjectivity and agency and our "embodied" accounts of the truth, and we ended up with one more excuse for not learning any post-Newtonian physics and one more reason to drop the old feminist self-help practice of repairing our own cars. They're just texts anyway, so let the boys have them back.[35]

On this account, the cost of expelling objectivity from the feminist conceptual armory is the possibility of making any difference in the real world. Haraway recognizes the need to adjust the liberatory excitement of constructivist thinking to the inevitable contradictions built into any perspectivist and relativist thought that will make claims about its own thoroughgoing applicability.

Wherever one looks, the problem of universal applicability arises, not only in the theoretical formulations but in the narrative embodiments. Haraway looks at the deconstructive project of the philosophers' "high redefinition," and she finds herself excluded from the real world of activity and change—from the world of narrative. So narrative in effect comments negatively not only on the paradoxes of the Cartesian tradition, but on the paradoxes of the deconstructive one. The turn to narrative and away from direct epistemology in my own argument in this book has led me to the view that even the most (absolutely) constructivist discourse contains within it implications of essentialism and universalism—the bad guys in the deconstructive project. For Diana Fuss, this discovery leads to the crucial question, not how to argue one's way out of it but how to use it most effectively.[36] Certainly, Haraway is not asking for the reinstitution of traditional objectivity, but for a recognition of the need for an epistemology that would provide the sort of authority in relation to the "real world" that objectivity used to provide, but without retelling what looks now like a suicidal (or merely paradoxical) story of impersonality, detachment, and disinterest. Like Harré before her but with different reasons and with a different rhetoric, Haraway wants both to accept the critiques of science and to sustain the enterprise of science that makes a difference in the real world.

At least some of the difficulties of this position, narrative and philosophical, have, I hope, been suggested in this chapter. In what follows in the succeeding chapters, I will be exploring different ways in which the paradoxical coexistence of the two poles of these views—a self-denying surrender to the reality of the real, and an imaginative imposition of the self on the world (until at times the world disappears)—play themselves out through the annals of modern epistemology and narrative. We can't, I would argue, have one without the other. The "exposure" of the extremes of these positions, as they emerge in narratives that have, in many ways, shaped the way we have been able to imagine ourselves, can help to an understanding of the paradox and perhaps toward a rethinking of an embodied epistemology.

There is a remarkable and surprising passage in William James's *Pragmatism* that embodies many of the implications of the paradox that underlies this book. It is surprising because one would have expected that the pragmatist James, so chilled by the austerity of the tradition of scientific objectivity, would reject its "realism" and its apparent hostility or indifference to the human. Instead, in an extraordinarily human and humane passage, he finds a figure for the science that has so chilled him, and warmly, passionately defends it:

> When one turns to the magnificent edifice of the physical science, and sees how it was reared; what thousands of disinterested moral lives of men lie buried in its mere foundations; what patience and postponement, what choking down of preference, what submission to the icy laws of outer fact are wrought into its very stones and mortar; how absolutely impersonal it stands in its vast augustness—then how besotted and contemptible seems every little sentimentalist who comes blowing his voluntary smoke-wreaths, and pretending to decide things out of his private dream.[37]

While it is necessary to remember that James imaginatively constructed this homage to the epistemology of self-annihilation on the way to a position that requires the insertion of will to get to the truth at all, he would not have been willing to be a "little sentimentalist," and the passage is an impressive reminder that current efforts at reimagining scientific objectivity must not be the "blowing of voluntary smoke-wreaths." The heroic narrative of scientific epistemology emerges again in James's defense, and without irony.

Yet James is unabashedly narrative. The question, as we incorporate the heroic narrative into a larger one, is not how to eliminate narrative in the interests of pure science but how to come to terms with its inevitability, to make it visible, to understand that it is not only not incompatible with science—a mere blowing of smoke rings—but a condition of its work. Can the morality of knowing—which has hitherto emphasized the heroic refusal to admit desire—be made compatible with human need? A very Jamesian question. Haraway, Longino, Keller, Shapin, Harré, all imagine an epistemology that is both objectivist and human, an epistemology that inserts the knowing subject back into the discussion of epistemology. Science is sanctioned not by abstract rules of thought or by the individual heroism of self-annihilating figures, but by human interchange, the community of trust that Harré describes.

Keeping the human in is part of the point of the explorations of narrative that I have begun in this chapter and that I undertake in the rest of this book. Keller defines objectivity "as the pursuit of a maximally authentic, and hence maximally reliable, understanding of the world around oneself." But she seeks something more and calls for a "dynamic objectivity," "a pursuit of knowledge that makes use of subjective experience ... in the interests of a more effective objectivity."[38] The dynamic observer is one whose relation to the subject is not disinterested, detached, and remote but is, as it were, hermeneutically imagining a Piagetian "allocentric" perception that engages lovingly with difference, with what is being observed, rather than starting with the separation of subject from object.

If one begins investigation with a sense that the object is to be discovered and understood through an observation and experiment that is coolly indifferent to the conditions either of investigator or investigated, then that object is immediately transformed, in narrative, into alien and enemy—recognizable in the monster or alien of science fiction stories and movies. It is not enough, however, simply to step back and allow the object its utter incomprehensible difference. For if one begins by refusing the possibility of knowing the object and leaving it respectfully alien, there is left no possibility of connection between subject and object. The narrative that follows is of solipsistic isolation, which in turn can lead to narratives of insular combat—perhaps, one might say, ethnic wars. The narrative toward which the narrative of this book is groping is one that recognizes both difference and the possibility of moving beyond the limits of the self. Perhaps not quite dying to know, but living to know. Living entails work, however wary and distrustful along the way, toward a position beyond solipsism and perspectivalism into a condition in which one can share the authority of one's arguments with others. Radical skepticism is radically corrosive.

The story I am telling through these excurses into the various ways in which the narrative of epistemology has been constructed should lead—through a recognition of the inevitability of personal engagement—in the direction of this movement beyond the personal. The narrative of epistemology, as MacIntyre says, always also an ethical narrative, manifests the inescapable and valuable presence of human desire, human action, human will, human trust, and love in the enterprise of knowing. Although the explorations in this book will constitute in part a critique of that "magnificent edifice of the physical sciences," I expect that I am not being a "little sentimentalist." This is in part because I am moved, as James has figured it, by the stories "of disinterested moral lives . . . of patience and postponement" that went into the building of that edifice and, more important, because I believe it is extremely dangerous to think that the construction is not, in the end, invaluable. The question raised is not *whether* the ethical is inserted in epistemological enterprises but *at what point* the ethical is inserted.

While arguable for its own sake, knowledge is always inevitably for some human end, and the Harawayan question, inserted into the narrative, is, At what human cost? So we might say that underpinning traditional epistemology is the desire for human benefit, which produces a desire to efface desire because desire has proven to be counterproductive. The pressure to turn self-discipline into a radical severance of knowledge and desire is narratively impossible. Scientific epistemology, always based in narrative, cannot escape from it, or from the ethical conditions that necessarily frame it. Bringing narrative and ethics into epistemology is consistent with the history of epistemology. No culture can escape the need for a narrative that establishes the conditions by which knowledge can most effectively be pursued, the criteria by which those who can be trusted (and under what conditions) are identified. It will be necessary now to establish these arguments on stronger grounds, to examine various ways in which epistemological ideas flow into and out of narrative, and various ways in which the ethical import of questions of knowledge have been confronted by writers who have engaged significantly with the questions of what it takes to find things out, with what it means when one says, "I'm dying to know."

— 2 —
Dying to Know Descartes

Descartes has become the villain in the drama of Western epistemology. Most lay readers know him for the radical skepticism that precedes the famous "Cogito ergo sum," for his mechanistic view of the world, for his radical splitting of mind and body, and for his insistence on the priority of reason in establishing the criteria for valid knowledge in the "clear and distinct." Although recent scholars have well demonstrated that Descartes's epistemology, read in historical context, does not affirm the idea of the kind of aperspectival objectivity that modern readers have found in it,[1] for more than a century he has been taken as the patron saint of the epistemology of detachment and self-restraint and, as I am calling it, "dying to know." One great irony of this reading of Descartes is that he develops this epistemology of self-effacement through a new affirmation of the self by the self.

The charges, in the broad contemporary case against Descartes, are multiple. He is not likely to be forgiven for the Cartesian anxiety: the fear that once a transcendental ground for knowledge is removed, we are plunged into the contingencies of history and experience, and there is nothing left for us but a debilitating and incoherent relativism. As Susan Bordo has put it, "For Descartes there are only two possibilities: absolute certainty or epistemological chaos."[2] He is also one of the great modern fathers of what Richard Bernstein calls "objectivism"—that is, "the basic conviction that there is or must be some permanent, ahistorical matrix or framework to which we can ultimately appeal in determining the nature of rationality, knowledge, truth, reality, goodness, or rightness."[3] But the point of most importance for the argument of this book is that while Descartes's theory is certainly dualistic, his foundational formulation of the arguments in the *Discourse on Method* and *Meditations* depends heavily on narrative devices that implicitly call the dualism into question.

Of course, there are few real Cartesians among modern philosophers, although Thomas Nagel unashamedly sees himself as writing within a Cartesian tradition because of Descartes's unequivocal belief in rationality. But as bogey, Descartes is very much alive, to be constantly reassassinated by the hermeneutics of suspicion (itself, I believe, a direct descendant of Descartes's Great Doubt, but differing in that it is not intended to be displaced by a firmer and better method). Another irony of the Cartesian inheritance is that he gave renewed impetus to a tradition of radical skepticism that has often, as in recent years, issued in a revulsion from any notion of the "clear and distinct" as epistemologically authoritative. The problem Descartes addressed—the problem of certain knowledge—remains intensely alive, at least among the broad-thinking public that has not resolved to accept indeterminacy and skepticism as final. Addressing these issues, I want here to begin my "literary" readings not by looking for the philosophical coherence of his argument but for the ways in which the theory is embodied and the way it is driven by ethical energies. The literature of philosophical analysis and criticism of Descartes is, of course, enormous, and in what follows I will occasionally lean on some of it. But my concern here is not ultimately with the philosophical difficulties of Descartes's argument—the problem, for example, discussed by Bernard Williams, that Descartes's ultimate appeal to God simply does not work.[4] It is rather the question of the relation between the overt Cartesian claims and the rhetoric in which they are embodied.

Despite what I take to be a quite justified and widespread skepticism, both about Cartesian foundations and about the imperial claims that science regularly makes for itself as institution, there is plenty of evidence that on a broad range of matters society as a whole has a rather Cartesian confidence in the general validity of what it knows and is willing to take at face value the Cartesian claims for epistemological authority that science seems to assert. Looking at the Cartesian narrative might help clarify why a Cartesian stance —subject to a long history of detailed, brilliant, and critical philosophical argument—might seem so much like common sense, why it might have been felt to answer so immediately to human needs and interests.

On the margins of a broad, commonsense, epistemological confidence there are, however, enough threats to make people jumpy; there is enough evidence for possible alternative ways of thinking, alternative interpretation of the facts, and maybe even—at the extremes of doubt—alternative facts, or worse, no facts at all. Narrative analysis of Descartes's arguments may help suggest ways in which they threaten to undercut themselves and lie vulnerable to critique. Certainly, most of Descartes's epistemological ideas have not survived well against the critiques, but the question of how we know, and how we can trust the authority of people who claim to know, has not gone away. Modern skepticism lives close by the side of a culture that is passionate about knowledge, that has been transformed by it, and that believes that its future depends on extensions of knowledge. Society, that is, seems to be dying to know and to be very angry that some people in positions of pedagogical power are happy in the anguish of their aporias. In any case, a consideration of Descartes's version of the "dying-to-know" story can only be relevant to current conditions.

Descartes's name provides the label for a belief in the universal power of rationality to allow the human mind to get in touch with objects outside of itself and, simply, for the possibility of knowledge. Descartes's epistemological shoring up of his science became, at least retrospectively, deeply influential in the development of epistemology as a central activity of philosophy. Historically, it became a way of contesting traditional forms of authority and power, of establishing new forms of authority in new social organizations, of giving to the developing institution of science its centrality as an institution and as authority.

In his initiating moves, described personally and cautiously, with a self-conscious avoidance of any social or political implications, Descartes put into question not any particular knowledge but the possibility of knowledge itself. What, he famously put it, if we are dreaming? How do we know we're not? He steps back from his own (quite confident) discoveries to their justification and thus initiates the narrative that underlies every knowledge claim—how do we begin, how do we find out, how do we justify, how can we claim authority? Descartes struck a chord that has rung through virtually all of modern European thought, and that continues to ring in the ears of those who, rejecting his affirmation of an intrahuman and ultimately transhuman rationality, find the foundation of all knowledge in local and particular perspectives, and thus find all knowledge claims inadequate to the kinds of authority Descartes sought. Descartes's answer to the questions he posed were, however, entirely consonant with the ethical ideals of his culture (and they remain so). The self he evoked as foundation turned out immediately to detach itself from what it knew and to defer to and find *its* foundation in a much larger power, his God.

Stephen Gaukroger has argued that Descartes was invented primarily as an epistemologist and father of the mind/body problem by later generations, but while posterity's focus on Descartes as an epistemologist is a historical distortion both of his project and of his career, it is as an influential epistemologist that I want to consider him. Though his epistemology came late, its enormous influence is undebatable, and it is certainly representative of a significant stream of Western thought about knowledge since his time. The epistemology grows not out of some general and abstract need to produce a formally coherent method, but out of intense social and political pressure that made self-justification imperative. Gaukroger argues the following:

> [Descartes's] interest in skepticism was relatively late, and took shape in the context of providing a metaphysical legitimation of his natural philosophy, a task which he never even contemplated before the condemnation of Galileo in 1633.... Skepticism was simply a means to an end, and that end had nothing to do with certainty about the existence of the material world, but rather with establishing the metaphysical credentials of a mechanist natural philosophy, one of whose central tenets—the Earth's motion around the Sun—had been condemned by the Inquisition.[5]

This is to suggest that Descartes's attempt to find a "ground" for his science in a method that could be recognized as operating beyond the limits of the local was, as modern historical critics would not be surprised to discover, determined by local conditions. He was moving nervously toward the publication of large treatises (particularly, *The World*) and frightened at what had happened to Galileo, who certainly at that time was far more prestigious than he. Preparing the ground, gingerly, Descartes meant the *Discourse*, as George Heffernan puts it, "to probe carefully and cautiously the uncertain reaction of the rest of the world."[6] The strategy of this narrative, yet another aimed at discouraging all future ones, is to legitimate a local and personal condition by showing its general pertinence.

It is probably for that reason that perhaps the most famous "Discourse" in the history of Western literature is in the form of an autobiography. The form is personal, defensive, and impersonal at the same time, but the *Discourse on Method* is virtually the perfect text with which to initiate the argument that epistemology is always, implicitly at least, narrative. Here is a case where there can be no argument. At least in this instance, epistemology is born inside a narrative.

The *Discourse on Method* is first of all an intellectual autobiography, a story (*histoire*), a "fable," as he calls it in the first pages. It is precisely not a treatise. Rather, like Newman's *Apologia* 230 years later, it prefers to tell the story of the personal development of a creed to elaborating that creed and making the case for it. Rhetorically, the case is in the "fable" itself, but the overt claim of both texts is that the case is not being made for the ideas, but only to establish the authenticity of the writers' commitment to them. For Newman and Descartes, concern for their own situations as public figures, that is, personal and contingent conditions, provoked them to talk about themselves and their ideas publicly. Yet if along the way they also manage to persuade people of the validity of their ideas, that—the rhetoric goes—should be taken only as an almost accidental byproduct of the work. Descartes worried about what had happened to Galileo because he feared that his scientific ideas might be received in the same way. He was concerned that people had expectations about what he had been writing, and he decided at last not to publish the writing itself but rather a work that talks about writing it. So, he says, "I shall be very happy to reveal in this discourse the paths I have taken, and to present my life as in a picture [*comme en un tableau* (p. 4)], in order that each may judge it."[7]

In a move that anticipates the trope of modesty characteristic both of scientific epistemological humility and of many scientific autobiographies after Descartes, *The Discourse on Method* becomes autobiography rather than treatise because Descartes believed that, as he puts it, "I may be wrong" (p. 28) [il se peut faire que ie me trompe (p. 3)]. He writes not an argument for the system, but an explanation of how he arrived at it. He opens the fourth chapter wondering whether he should "tell" us about his first meditation because it is so abstract and unusual and will not be "to the taste of everyone" (p. 53) [*au gout de tout le monde* (p. 31)]. But he finds himself "to some extent forced to speak of them" (p. 53) [*en quelques façon contraint d'en parler* (p. 31)].

Such rhetorical manipulations make the *Discourse* an almost perfect exemplar of the motif of "dying to know." The expressed humility, utterly necessary for the argument and representative of the ways in which the work of scientific self-legitimation proceeded in the centuries that followed, also works within a frame of almost overweaning ambition. Disingenuously or not, the strategy of self-humiliation can also be a strategy of self-empowerment; the paradox is built into the very metaphor, "dying to know," since of course the desire to satisfy curiosity is driven by a more or less intense quest for authority even if it also always suggests delayed gratification[8]—much delayed. Recent critiques of scientific impersonality and universalizing often focus on the way in which the lust for power lurks within universalizing assertions of neutrality, which incorporate even the potentially most reluctant elements into themselves.[9]

Descartes's ambitions were not, of course, modest. He writes of himself as—like Victor Frankenstein, who followed in his footsteps—a benefactor of mankind. Moreover, he believed that he was indeed capable of accomplishing his project of building and restructuring *all* human knowledge on the foundation of his kernel of undeniable certainty, the "cogito." In a work written after the *Discourse*, he makes his own high self-estimate clear:

> I do not yet feel so old; I do not have so little trust in my strength; I do not judge myself so far from the knowledge of what remains; that I would not dare to undertake to complete this project if it were possible for me to perform all the experiments I would need in order to support and justify my reasonings. But, perceiving that to do this would require great expenditures which a private individual like myself could not meet without the help of the

public, and not perceiving that I should expect that help; I believe I must henceforth content myself with studying for my personal instruction and that posterity will pardon me if I cease henceforth to work in its behalf.[10]

He does not hesitate, however, to go on to describe the ways he has already served posterity.

But the rhetoric of the *Discourse* is modest, punctuated by moments of bluff honesty when he admits that he has done well at his project (how else account for his desire to tell its story?). Both the modesty and the honesty are characteristic strategies of tale-telling in the tradition of scientific epistemology. "The capacity to judge correctly and to distinguish the true from the false," says Descartes, "is naturally equal in all men [le bon sens ou la raison, est naturellement esgale en tous les hommes (p. 2)], and consequently . . . the diversity of our opinions does not spring [ne vient pas] from some of us being more able to reason than others, but only from our conducting our thoughts along different lines and not examining the same things" (p. 27).

Nominated as the father of T. H. Huxley's notion that true science entailed constant skepticism[11] and of the nineteenth century's scientific impasse in the transformation of the human into a mechanism whose connection with the spirit, with thought and feeling, was inexplicable, Descartes built his project with a rhetoric of what I want to call hubristic modesty. Like the pursuers of knowledge who followed him, Descartes established a rhetorical stance that expressed utter humility in the pursuit of Faustian goals. He wanted to do no less than describe the world with absolute authority. But he could do so only by announcing at the very opening of the *Discourse* that his mind is "mediocre" and that anybody can do it. That rhetoric is not merely a disingenuous preliminary to his epistemology: it is an aspect of it, as Darwin's (I believe more authentic) modesty was to be essential to his achievements more than two hundred years later.

Proceeding with the self-justification both for his total work and for the *Discourse* itself, Descartes claims that perhaps the general response to his little book will provide him with "a new means of instruction [m'instruire (p. 4)] which I shall add to those which I am in the habit of using" (p. 28). That he would later, in the same discourse, protest his distrust of pictures, and the necessity to release ourselves from reliance on the tangible, does not affect his strategies of disclosure, although it ought significantly to affect a reading of them.

From the start, the *Discourse* makes apparent compromises with clarity and distinctness that its arguments will not sanction. It describes a method of what Bernard Williams, in the very title of his study of Descartes, calls "Pure Enquiry," the initiating step of which is the "Great Doubt," entailing the immediate rejection of everything that one takes for granted, but particularly the evidence of the senses. While Descartes will eventually, circuitously, reinvest the senses with some kinds of authority and save appearances—as was his purpose from the outset—the procedures by which he gets to that point require resistance to appearances' potentially demonic force. Confronting the problem of how to convince the public and the authorities that his scientific conclusions are correct, though they may seem to be based on appearances, Descartes needs to produce an epistemology sufficiently defensive that the worst possible reading—that the material world is merely an illusion—could be accounted for and defeated.

So beyond the absolute epistemology for which it argues, the *Discourse* implies a contingent and human way of knowing. The strategy of mortality, of dying to know, implies here the necessity for narrative and for the contingent, the desirable, the humane. I want to argue, then, that despite Descartes's extreme articulation of the "dying" epistemology and the absolute rationalist foundationalism of his overall argument, a consideration of the narrative and metaphorical elements of his work forces qualification of his absolute commitment to the clear and distinct, the authoritative and the objective. I propose this way of reading not because of a confident or even tentative antibinary holism. Narrative allows no living separation of knowledge and desire, of mind and body, but it also dramatizes the conditions that make it hard to trust knowledge produced by desire. Much modern skepticism, rejecting the idea that people can achieve an affectless and abstract relation to knowledge, in effect reenacts the Cartesian Doubt. Intellectual authority shifts from trust of authority and the intrinsic power of reason—the possibility of distancing desire and of self-suppression—to distrust of those who claim to be capable of such distancing. It is no adequate answer to Descartes, or, to take a well-known example, to Matthew Arnold, to declaim that all knowledge is "interested"—of course it is. There is no view from nowhere, but there *are* views from a distance. Even if Richard Rorty is right in his persistent argument that "everything we say and do and believe is a matter of fulfilling human needs and interests,"[12] Thomas Nagel can still be right on this issue: "it is worth trying to bring one's beliefs, one's actions, and one's values more under the influence of an impersonal standpoint even without the assurance that this could not be revealed from a still more external standpoint as an illusion."[13]

In seeking a critique of Descartes's position, I want to recall the earlier argument I developed from a discussion of Rom Harré. The exposure of the interestedness of supposedly disinterested investigation is itself inspired by a Cartesian animus: that is, by the belief that if you can show that knowledge is interested, you can show that it is not valid. The very attack on disinterestedness is likely to depend on an aspiration to it. Once the philosopher or scientist is caught—as we have already caught Descartes—implicated in history, society, and politics and not really working detachedly with the structure of abstract thought and argument, his or her work is demystified and implicitly falsified.

But why should that be so? Interest being a condition of all knowledge, the exposure of interest tells nothing about the knowledge. Or does it? If Cartesian objectivity ultimately dehumanizes both subject and object, the alternative rejection of objectivity belies human practice. Descartes himself, in the tensions between his rhetoric and his argument, provides a remarkable commentary on the dilemma.

In his quest for certain knowledge and a unified science, Descartes seems indeed to have fragmented the world absolutely into spirit and body. Despite his rhetorical and strategic attempts to avoid the disapprobation of his church and despite what seems to have been his authentic need for God as a ground for truth and then as a bridge to the material world, Descartes's position is unquestionably secular.[14] He does not, after all, begin with faith, with a primary intuition of God—which, for example, Newman takes as a condition of serious religious thought—but relies on "natural reason" to lead him to God through the cogito, the intuition of his own existence. (One of the problems with his whole argument is that in the end it is not clear whether he is willing finally to adhere to this position because, as we shall see in discussing the Sixth Meditation, confidence in the clear and distinct may ultimately be dependent on God and the fact that God does not deceive.) The autobiography is not an obviously religious story. It begins its investigations by focusing on the self, not on Descartes's relation to God. That turn to natural reason is central to all of Descartes's intellectual preoccupations. "I have always thought," he says, in the dedicatory letter to the *Meditations*, that "demonstrative proofs" of God and the Soul "ought to be given with the aid of philosophy rather than theology."[15] The standpoint is secular even if the faith is not dead. As I have already suggested, philosophers, too, have debated the potential circularity of Descartes's argument: beginning with consciousness, he derives God; but then God, who does not deceive, becomes the indispensable sanction not only of personal existence but of the "clear and distinct" senses of the material world, which consciousness alone could not accept.

Despite this secular foundation, no religious order could have propagated a doctrine more skeptical about the body or the material world. It was Descartes who inspired the modern secular transformation of the body into a mechanism utterly separate in its nature from the spirit that seems to impel and animate it. The body, as the seat of the senses, cannot be trusted to produce reliable knowledge without a conception of a God who can be trusted. By transforming the body into mechanism, an object for study from the outside, Descartes opened the way, for better or worse, for the "human sciences," whose history is so grotesquely checkered by the social and political consequences of ostensible "objectivity." Certainly this was one of his primary objectives, for when he published the *Discourse,* he published with it a study of "vision," *"La dioptrique,"* and he understood certainly that he was opening the way to the kind of study of the human that could lead, both to the sort of experimental medicine that Claude Bernard was to do and, I would suggest, to the disciplines of sociology and anthropology and psychology. The irony of this development is important to an understanding of the ways in which the epistemology of "dying" does its work. Beginning in a kind of quasi-religious distrust of the body, it ends by opening the body (and the human) to self-scrutiny and scrutiny. To put it in an extreme form: study of the human begins as Descartes carries over a traditional distrust of the body into a new dehumanization and humiliation of the body. Scientific epistemology—and the modern transformation of the human into the subject—is linked indissolubly to a religious view of the world.

Descartes also anticipated the move toward the study of the human by way of his intuitional model of mathematics, which provided the clarity and distinctness he sought in all knowledge. For that model led him to the descent into the cogito (an inner world free of empirical obscurities) and thus also to a modern subjectivism that has provided a necessary antecedent to our turn toward self-reflexivity in philosophy, art, criticism, theory, and even anthropology. In the story of the modern development of objectivism, Descartes's narrative is particularly ironic, since for him objectivity is grounded in subjectivity. But the "subject" Descartes creates is narratively detached from most of what we might consider human. As Dalia Judovitz puts it, "the establishment of the philosophical subject, as the universal subject of truth, involves a negative procedure, one of emptying the subject of any content other than that of thought. This leads both to the disembodiment of the subject and to its isolation and autonomous constitution as an entity independent of historical or social context."[16] The empiricist tradition, which would seem more humanly, contingently engaged, shared with Descartes's rationalism primary responsibility for the intellectual development of modern science, and was equally preoccupied with consciousness and the conditions of knowing. Bacon, like Descartes, chose to begin with an exploration of the possibilities and limitations of consciousness. So, too, did modern positivism. The ostensible oddity of Descartes's point that the only thing that can be absolutely certain is the fact of consciousness itself is startlingly compatible with modern science, whose effectiveness has been established by means of complex strategies to overcome the limitations of individual human sensation and consciousness. The elimination of the "subject" from technical scientific writing is fully compatible with Descartes's subjective turn, which, with his Great Doubt, was central to the construction of epistemology ("of all things," laments Richard Rorty) as the great project of Western philosophy—more important, it seems, than morality or God.[17]

Rorty is at least a little disingenuous in his surprise at the event, for the very possibility of the new science depended upon "natural reason," and a gradual distancing of God, the ground of all knowledge, in pursuit of knowledge of God's second book, nature. We have seen that development in my earlier discussion of Bacon's clearing the mind of its idols. Neither Bacon nor Descartes would begin by relying on God's immediate revelation, and for Descartes the equivalent of clearing away the idols was a discourse on method.

Both insisted on the profound disparity between human intellectual capacity and human achievement. For Descartes the disparity is a result of the lack of an adequate and universally used method of acquiring knowledge. To establish a method that would bring all people of good sense, everyman, to knowledge, required for Descartes beginning from the beginning, moving into a radical, one might almost say deconstructive skepticism in which nothing given in language or experience is decidable, all could be a dream. Only by clearing the mind of the intellectual debris of history, by wrenching consciousness from its conditions, and by developing a coherent method that would release the mind from its limitations could real intellectual advancement and objective understanding of the natural world become possible. The descent into doubt and denial is an intellectual, but surely also, a spiritual pilgrimage, parallel to religious pilgrimages into the dark night of the soul, or to the stripping away of all worldly things on the way toward at least the possibility of salvation. The secular road is mapped immediately onto the religious one, and entails a radical distrust of the material. One consequence of the view that the path to true knowledge is through such a spiritual pilgrimage is that epistemology has become the focal philosophical subject; and that the endless debates about reality, representation, correspondence, and coherence go on even now is in no little way attributable to the Cartesian narrative.

The descent into consciousness, paring away the confusions and pluralities and contingencies of history, leads to the shift of focus from goodness to consciousness, and their virtual separation. In the famous stove-heated room, with "no cares or passions to disturb" him, Descartes quietly detached the pursuit of truth from the complexities of action and ethics, and helped initiate the potent dualistic distinction between "is" and "ought" that would have seemed utterly inappropriate in a fully religious world. For there, what is is right, because divinely created and sanctioned. Among the consequences of that shift is Descartes's famous mechanization of organisms, which itself is only part of the massive detachment he sustained. On his accounting, consciousness was entirely separate from the material: he could contemplate the material from Nowhere—or, to intimate the way his narrative tends to subvert his theory, from a stove-heated room during a winter hiatus in military action.

The "stove-heated room" is one of the more striking sources of ambiguity in Descartes's narrative. On the one hand, Descartes quite humanly emphasizes the importance of getting away from it all:

> I was, at that time, in Germany, whither the wars, which have not yet finished there, had called me, and as I was returning from the coronation of the Emperor to join the army, the onset of winter held me up in quarters to which, finding no company to distract me, and having, fortunately, no cares or passions to disturb me, I spent the whole day shut up in a room heated by an enclosed stove, where I had complete leisure to meditate on my own thoughts. (p. 35) [ou ne trouvant aucune conversation qui me divertist, & n'ayant d'ailleurs, par bonheur, aucuns soins ny passions qui me troublassent (p. 11)]

On the one hand, the passage enacts the movement away from full human engagement. Serious thought only happens away from amusements, diversions, social connections. In particular, knowledge comes when there are no passions. On the other hand, the passage (though not at all detailed) is thick with a sense of contingency and the body, of the winter out there, the warmth inside making work possible. And the universality of the "method" is a product of the chance break in a life of great, even violent activity, and the happy good fortune of Descartes's not having any anxieties at the moment. The passage doesn't even indicate whether Descartes sought out such a place or whether he merely stumbled upon it. Of course, this context does nothing to change the thoughts that he claims to have had in that room. But if the method depends on a clearing out of the merely material and the local, its development depends on locally contingent circumstances.

Nevertheless, post-Cartesian epistemology, characterized by a faith in the reason of humanity and the rationality of the natural world, detaches itself from any overt moral project: Descartes's morality of knowledge comes to depend on a characteristic Enlightenment assumption that the will acts according to what the understanding reveals. Thus, morality is a form of rationality. Knowing the truth is sufficient for acting on it.

There is a morality of epistemology, but in Descartes and his followers it is disguised in the method itself. The highest morality inside epistemology is the willingness to put away desire, pride, and the material contingencies of selfhood to become pure consciousness. Bacon says the researcher, even as he invades and tortures it, must finally submit himself to nature; Descartes requires the emotional austerity of absolute skepticism. The turn is from desire to method, and the values that might have predominated in philosophy preoccupied with virtue are no longer the explicit object of concern. Quietly, the virtues of ideal self-surrender, of endurance of pain and sacrifice, are absorbed into the procedures of the mind itself so that epistemological success becomes a morality of its own, without acknowledgment. This internal, self-denying morality is ostensibly unrelated to ethics, and yet the cultural force of epistemology surely resides in these ethical borrowings. Epistemology is of the mind. The test of truth is not its consistency with virtue, but the power to detect the truth is implicitly virtuous, and the rejection of objectivity is felt, culturally, to be immoral.

The stories we tell ourselves about knowing, our "fables" of consciousness, remain far more Cartesian than many would like to acknowledge. Descartes is still one of the great mythmakers of modern consciousness, and one whose inheritance we cannot afford to reject with the Cartesian skepticism we might play out in the solitude of some solar-heated room. It is important to recognize that the Cartesian project has through history often seemed a liberating move, not a mystification of present authority but a shift of authority to the common powers of individual human intelligence: Descartes's epistemology was an opening to that secular democratic respect for the individual that, I believe, lies behind our current outrage at its newly detected mystifications of power. These qualities, too, are part of the tradition of scientific objectivity that is often treated now as a radical falsification of the interested nature of all action and acts of knowing.

One needs to consider how intricately Cartesian theory is bound up with universalizing and essentializing abstractions and at the very same time with embodying metaphors and narrative. Cartesian epistemology always implies some narrative even in its most abstract and acontextualized logical moves. As Bernard Williams points out, the *Discourse* and, even more thoroughly, the *Meditations,* are "not a description but an enactment of philosophical thought, following what Descartes regarded as the only illuminating way of presenting philosophy, the order of discovery."[18] Descartes himself asked that the *Meditations* be read straight through like a novel (p. 181). Thus, despite the directions of his arguments toward an ideal, universalizing rationality, Descartes constructs his position from materials apparently inimical to it. Williams claims that the "I" of the *Meditations* is not really Descartes but some representative and idealized figure, and that the "order of discovery" is not "just arbitrarily individual, but idealized" as well.[19] But even if this is correct—and while it was surely Descartes's intention it is not at all clear that he achieved it—both of his major epistemological works remain "fables," "pictures," as it were, of the pilgrimage to truth rather than starkly austere arguments.

Descartes retreated from the tangible, but invoked it through narrative and metaphor. In his move from science to epistemology, he in effect asked to be judged not by his argument but by his picture of it. The method of the *Discourse*, then, is not the four-step way of skepticism, analysis, orderly system, and inclusiveness that he lays out as the basis of his arguments, but storytelling and connection-making and rhetoric. And this countermethod, as one might call it, may help provide a way to reimagine the Cartesian inheritance, to find accommodations that incorporate into the old idea of objectivity the historical and social location, the contingency and partiality that Haraway, to recall an earlier example, seeks as a way to make possible "a faithful account of the real world" that is not incompatible with human desire.

Descartes's "autobiography" implies in its forms and rhetorical strategies the epistemology it frankly offers, but with the difference from, say, a formal treatise, that the form itself compromises it. In chapters 4 through 6 I will turn directly to autobiography as a form, but it is useful to anticipate the argument here because Descartes's autobiography has in its broad dissemination and later influence an almost foundational quality. The form and its strategies recur with remarkable similarity in later narratives that imply a similar epistemology, strategies that I will be examining in the next chapters, in discussions of Carlyle, Darwin, and John Stuart Mill, among others. I would claim as well that the dominant narratives of Victorian novels echo this Cartesian pattern, for in their form they not only assert the power of the external world over the desires of the protagonists, but they tame their protagonists to a modesty and diminution of personal claims that bring them to the condition of humble scientists of the Cartesian narrative tradition.[20]

The story Descartes tells about himself is the story modern science has told about itself with remarkable consistency since the Renaissance. It is the story of a release from history by way of history, the establishment of a view from nowhere paradoxically in this time, in this place: the pattern entails a renunciation of old ideas and old desires, a descent into the self enacted as a form of self-alienation, an abasement of self before some larger system, of nature or of God, a recreation of self through a new harmony with the external world, which might well be figured as a marriage. Self-alienation and self-abasement are triumphant. The narrative of scientific objectivity, which

Descartes in effect tells in his *Discourse,* reintroduces into epistemology the very human questions of value and contingency that Descartes had, from another direction, banished. At the root of Descartes's epistemology, with its radical separation of truth from moral obligation, history, contingency, the body, and desire, there lies a narrative that—willy-nilly—reintroduces the body and all the muddied distinctions that Descartes's arguments are ostensibly designed to banish. Almost every step of the argument is couched in metaphor, dependent on images, far from the abstract distinctness and clarity Descartes requires for knowledge.

How, then, do we justify this dependence on pictures, images, and imagination? It would be foolish to assume that Descartes did not know what he was doing. The question of the imagination engages him directly, intensely, and often. And in the first pages of the *Discourse* he can confess that his book is a fable and then note of his own education that *(a)* fables awaken the mind; *(b)* he had given more than enough time to fables; and *(c)* fables make one imagine many events to be possible that are not. Fables, to say the least, are dangerous. Descartes entertains a similar ambivalence about the imagination. Imagination, he says, "is an obstacle to spiritual knowledge":

> what persuades many people that it is difficult to know [the certainty of God's existence], and even also to know what their soul is, is that they never lift their minds above tangible things, and that they are so accustomed not to think of anything except by imagining it, which is a mode of thinking peculiar to material objects, that everything which is not within the realm of imagination seems to them unintelligible. (p. 57)

> [ils n'eslevent jamais leur esprit au dela des choses sensibles, & qu'ils sont tellement accoustumez a ne rien considerer qu'en l'imaginant, qui est une façon de penser particuliere pour les choses materielles. (p. 37)]

Tangibility and corporeality are the realm of illusion, of the evil demon of the First Meditation. The corporeal is, in true Platonic tradition, dependent and therefore, as Descartes says, manifestly a defect. The corporeal is dependent on consciousness and not the other way around, and therefore the imagination, which deals in images of the corporeal, is dependent on the tangible. Thus, though it is a part of consciousness, it is utterly inferior to the consciousness of clear and distinct ideas.

There is nothing in the *Discourse* that would belie this view of imagination except the fact that it is so intensely imaginative a work. Many have quite reasonably argued that the metaphorical and fabulous texture of the *Discourse* is intended by Descartes for those who do not look higher than the tangible, who are not ready—as Descartes assumes at the beginning of the fourth chapter when he introduces the cogito. "I do not know," he says, "if I ought to tell you about the first meditations I pursued there, for they are so abstract and unusual that they will probably not be to the taste of everyone" (p. 53) [car elles sont si Metaphysiques & si peu communes, qu'elles ne feront peutetre pas au goust de tout le monde (p. 31)]. On this view, Descartes uses the resources of the imagination to seduce his unprepared readers away from the senses that the imagination always implicitly evokes. But that, I believe, is too simple. Imagination applies both to metaphor and to fable, to fiction and even to hypothesis. None of these is immediately if at all connected to that axiomatic world of the clear and distinct toward which Descartes attempts to persuade us. And surely, the view that he uses metaphor and fiction as a mere device in the rhetoric of his theory suggests a much more innocent view of the powers of the imagination than we can reasonably attribute to Descartes.

Knowledge in Descartes is always a process, a history of discovery, even if it is a discovery of the end of history. Knowledge becomes knowledge when it seems to be alienated from its narrative beginnings in the contingent and tangible, just as true knowledge only begins when Descartes can affirm his existence against the rich, contingent complexity of his fully human life. Yet Descartes's procedure makes full alienation impossible. Consciousness is a process, always linked to an "I." Somewhere, in the crevices of his argument, there seems to be space for the corporeal, the dependent, the contingent, and the perhaps circular conclusion, in which God finally confirms the validity of the imagination that ought not to be trusted. Some of Descartes's other comments about the imagination provide clues to its importance and to the way it works in constant tension with the overt arguments.

Commentators have noted that the "I" of the *Meditations* is not to be taken for Descartes's. Judovitz marks a distinction between the Cartesian "je," which is at most a thinking thing, and Montaigne's "moi," which always carries with it the weight of human complications. Bernard Williams suggests that the "I" who is having the thoughts may be the reader.[21] That the "I" of the *Meditations* is a stripped down, thinking thing, and that it is allowed no personal complications, is certainly right, although the "I" of the *Discourse* has a life and desires that are the perhaps accidental condition of his turning himself into a "je" rather than a "moi." But the strategy of temporal engagement, of struggle with the process, and the rhetorical energy implicit in the attempt to persuade by way of such personal engagement, keep Descartes from that radical objectivization of thought to which he ostensibly aspires.

This is particularly so in the *Meditations*. The preface to the *Meditations* adopts a rhetoric rather more confident and less humble than that of the *Discourse*, but it is explicit and demanding in its insistence that a serious reader must surrender his normal human ways: "I would not urge anyone to read this book except those who are able and willing to meditate seriously with me, and to withdraw their minds from the senses and from all preconceived opinions" (p. 8). He argues that the method of the Great Doubt is particularly useful for "providing the easiest route by which the mind may be led away from the senses" (p. 9). If the *Meditations* allow a particular location, they also require that annihilation of body that confirms once again Descartes's commitment to the narrative of dying to know. To be the Descartes of the *Meditations*, the reader is enjoined to die to his senses in order "to meditate seriously."

The *Meditations* do register with intensity the apparent clarity and distinctness of sensations, but also back away from accepting them as reliable. Yet in the Sixth Meditation, as we have already noted, Descartes reinstates sensations as reliable after all, though only after a difficult pilgrimage from the cogito through God. Descartes begins by relying on his own capacity to rid himself of his "senses," and thus to get at the clear and distinct; yet he ends by determining that it is only because God does not trick us, and God is the purveyor of all clearness and distinctness, that he can be confident that his sense of the clear and distinct leads him to the truth. The narrative of this question replicates the circularity of the narrative of the whole movement of the *Discourse* and the *Meditations:* that is, the story begins by stripping away all that has been learned or taken for granted, and ends by reaffirming what had been doubted.

Descartes's rhetorical methods allow him narratively to establish a nowhere from which to know, a distance that will keep the urgency of experience from overwhelming intelligence. Judovitz claims that Descartes's fable or "fiction of philosophy as an autonomous field of knowledge . . . fosters the illusion of truth, verisimilitude and objectivity—in other words, a truth beyond language, metaphor, style and rhetoric."[22] Yet Descartes's narrative strategy constantly reminds the reader of the difficulty of being nowhere, of the necessity for keeping alert to the movements from the tangible somewhere to the intangible nowhere, from the uncertain to the certain. Those movements leave space, in the crevices and interstices of his thought, for the corporeal, the dependent, the contingent.

The crevices are in the form of narrative itself, which allows a glimpse behind the illusion of the nowhere related to the omnipresence of the somewhere back in the language, in metaphor, style, and rhetoric. It is particularly striking, for someone who has read Descartes on the imagination in the *Discourse*, that *La dioptrique*, published with the *Discourse*, begins this way: "Toute la conduite de nostre vie depend de nos sens" (p. 81).

Descartes believes in the senses and commits himself to the "body"—and therefore must discuss the imagination in detail, as if to find a way to "disembody" himself. The *Meditations* are the embodiment of the story Descartes tells in the *Discourse*,[23] and the first *Meditation* begins with greater physical detail than the *Discourse* provided. Descartes tells his epistemological story *as* a story because he must work his way through the material and the imagination before he can get to the utopian space of nowhere—the narrative plausibility of these stories of the implausible abolition of the senses is partly dependent on the novelistic recognition of the immediacy of the physical and sensible. The epistemology only makes full sense as a narrative. Descartes begins, wonderfully,

> although the senses occasionally deceive us with respect to objects which are very small or in the distance, there are many other beliefs about which doubt is quite impossible, even though they are derived from the sense—for example, that I am here, sitting by the fire, wearing a winter dressing-gown, holding this piece of paper in my hands, and so on. Again, how could it be denied that these hands or this whole body are mine? Unless perhaps I were to liken myself to madmen, whose brains are so damaged by the persistent vapours of melancholia that they firmly maintain they are kings, when they are paupers; or say they are dressed in purple when they are naked; or that they are pumpkins or made of glass. But such people are insane, and I would be thought equally mad if I took anything from them as a model for myself. (p. 13)

The *Meditations* begin, then, in the material and in common sense, and with an effort to discriminate epistemology from lunacy. The narrative of the *Meditations* moves radically away from that embodied common sense, and from the imagination that expresses it, and then concludes in recuperation.

As the Sixth Meditation is preoccupied with the existence of material things, it is also, not surprisingly, initially preoccupied with the imagination. The imagination is somehow allowed greater importance here, although it remains confined to material things and is subsidiary to intellect. "The faculty of the imagination" is "capable of persuading" Descartes of the existence of the material,[24] whereas in the pre-cogito doubt, the imagination was not to be trusted. As he divides mind and body, so Descartes distinguishes intellect and imagination. The intellect, he says, can conceive of a thousand-sided figure, but the imagination cannot imagine one. One does not need to "imagine" a triangle, for example, to have a conception of one, but one *can* imagine the triangle. So the imagination is not necessary to the essence of his self, but the intellect is; the imagination is subsidiary to the intellect.

It is, finally, difficult to disentangle, even temporally, Descartes's relation to the imagination. Certainly, as the preface indicates, he believes that detachment from the sense is a condition of adequate inquiry. And yet the Sixth Meditation allows the imagination back in, at least with the authority of conceiving a triangle. The intellect turns to itself and "inspects one of the ideas which are within it." But the imagination "turns toward the body and looks at something in the body which conforms to an idea understood by the mind or perceived by the senses" (p. 51). This is, to be sure, only a conjecture, Descartes admits, and he is not yet fully satisfied that this provides evidence that the body really exists. Yet a final step is required to validate what the imagination teaches, and here Descartes turns to God as the bridge from consciousness to the material world. "There is no doubt," he says, "that everything taught by nature contains some truth," for Nature means God (p. 56). The nature Descartes will know is, in certain respects, as stripped down as his self—it is merely extension. He goes on to list those things that now, with God behind him, he can trust from nature's evidence. He trusts the fact of his own body but also now knows that "I am not merely present in my body as a sailor is present in a ship [note again the figurative language], but that I am very closely joined and, as it were, intermingled with it, so that I and the body form a unit" (p. 56). Moreover, the imagination convinces him that "various other bodies exist in the vicinity of my body" (p. 56).

By the end of the Sixth Meditation, having worked his way through the limitations of the senses and the tangible, Descartes concludes with a remarkable expression of faith in the body:

> I know that in matters regarding the well-being of the body, all my senses report the truth more frequently than not. Also, I can almost always make use of more than one sense to investigate the same thing; and in addition, I can use both my memory, which connects present experiences with preceding ones, and my intellect, which has by now examined all the causes of error. Accordingly, I should not have any further fears about the falsity of what my senses tell me every day; on the contrary, the exaggerated doubts of the last few days should be dismissed as laughable. (p. 61)

The end of the narrative of the *Meditations* is a curious restoration of something like traditional common sense, buttressed now by the recognition of the reality of consciousness, the reality of God, and the recognition that God is no deceiver: this fact means that there is no possibility "of there being any falsity in my opinions which cannot be corrected by some other faculty supplied by God" (pp. 55–56). All that Descartes requires for the full restoration of the validity of sensible experience is the rigorous—in other words, scientific—application of sense, memory, understanding, and the appropriate humility that comes with an acceptance of the inevitability of error because decisions must always be made "before we have had leisure to examine things so carefully." We must, so the *Meditations* conclude, "acknowledge the weakness of our nature" (p. 62).

Descartes here saves appearances and validates the sensible and the tangible (within a systematic study, of course). Despite his argument that everyone must go through the same experience as he, nobody will really have to go through the Great Doubt again because he has taken the risk for us, undertaken what turned out to be a "laughable" enterprise, and derived from it new confidence in what we already think we know. To be sure, appearances will never provide the kind of certainty that pure intellect can; but they are to be trusted. This method of pure enquiry manages, in the end, to validate the empirical study of nature, or, perhaps more precisely, to validate that study as long as it is understood that it can only be sustained through reliance on axiomatic thought and on an axiomatic nature, that gives to experience its ultimate authority. Imagination achieves an authority that Descartes had seemed to deny it.

The metaphors and fables of Descartes's narratives and arguments are, then, more than rhetorical devices to convince readers too dependent on the tangible. The tangible, in Descartes, is indispensable to establishing the authority of pure, transparent consciousness, so that it is literally necessary to move to epistemology through narrative. Descartes uses metaphors and fables because—despite the ostensible chronology of his revelation of progress through his method from cogito to nature—he already knows, before he proves it in the Sixth Meditation (the end of three days of thought), that they are legitimated by the theory that disallows their use.

While Descartes seeks a certainty beyond what is available to limited humanity in the historically contingent world of the senses, he can only do so by validating the senses after all. The stripped and drained "self" of the *Discourse* and the *Meditations* surreptitiously returns to its place in the world of human complexity. The ingenious contriver of the metaphors that drive the argument fills his world with analogical life, a procedure he can justify only if he is self-consciously falsifying in order to get at the lesser consciousness of his readers or if he has played some hanky-panky with the chronology and uses what he later finds out to strengthen his case. To make his points he fills his text with allusions to experiences of ordinary life: he invokes architects and city planners; he compares followers of Aristotle to ivy that does not seek to climb higher than the trees that support it; he talks of philosophers behaving "like a blind man who, in order to fight a person who can see, without disadvantage, brings him into the depths of a very dark cellar." Publishing his philosophy is like opening "a few windows" to "let the light of day into this cellar into which they have descended in order to fight" (p. 85).

The remarkable analogical energy of Descartes the writer and his ultimate reversion to modified trust of the senses and belief in the reality of the material world do not make him any less a dualist, a mechanist, a rationalist. In reading Descartes's metaphors and his story against the grain of his arguments, I have not meant to suggest that he was himself unselfconscious about the texture of his language or about the determination to turn philosophy into story, or, rather, to make over the story that determines the philosophy. Yet while Descartes is careful to validate and make coherent the reembodiment that follows from the determined disembodiment of self (and reader), he does not seem alert to the ways in which his subjective turn and his preoccupations with modes of persuasion can be taken to mediate his argument. He is not merely the sailor on the ship; the sailor and the ship are inextricably part of each other.

We, however, are not bound by his failure of acknowledgement. There is a great deal at stake here, and our ability to trace, in one of the most imaginative and powerful texts in the history of Western culture, the inevitability and consequences of narrative and metaphor, ought to help open that text to alternative possibilities. Tracing the moves by which Descartes establishes epistemology as the dominant mode of modern philosophy, one finds epistemology to be something rather different from what it is normally taken to be—never ethically neutral. Behind every epistemology, Descartes's and Bacon's most obviously, there is a narrative and a moral impulse, and built into every epistemology is a whole set of values imported from other realms. Epistemology will always be part, historically, of attempts to garner social and political power, and somewhere it will always turn on questions of trust, even matters of faith. It will always entail some kind of intense moral act, usually self-sacrifice—of desire, of pride (a favorite subject of Bacon), of one's instinctive trust in the senses. It will almost invariably imply the necessity of deferred gratification and of submission to the knowledge acquired. So, implicitly, it points to a force greater than the self—often "reality," usually in the form of nature or of God. These may be commonplaces, but the degree to which they impel Western scientific epistemology helps determine the way knowledge and its effects go in our culture.

The sin of epistemology in this respect is not the importation of values from other realms, but the denial of them. What epistemological work almost inevitably leaves out is the complex process, connected with contingency and history, by which consciousness makes its moves. It is here that Descartes's very work in establishing the illusion and effacing the process is particularly valuable. Even as he effaces narrative, value, and the moderating power of language and representation, he dramatizes them and forces them upon our attention. Reading Descartes as trapped in his own fables and metaphors allows the reimagination of knowledge, not as merely contingent but as imbued with the ethical, the human, the partial.

Descartes establishes the myth that knowledge is attainable only through shucking the senses, verging on what would seem, to the flawed perceptions of common sense, rather like death. We must die to know. The resurrection of the "I" in the cogito is a rather bloodless and scientific affair. As Marjorie Grene puts it, "it is the very alienation of mind from nature that," for Descartes, "liberates the attentive intellect for the clear and distinct apprehension of its objective."[25]

We mechanize our bodies and the material world and the consequence is a hardness of relationship between researcher and researched that can often seem a parody of that best ideal of distance that ought to characterize serious attempts to know. When Descartes describes the circulation of the blood in the *Discourse,* and particularly Harvey's theories, he begins by saying, "I would like those who are not versed in anatomy to take the trouble, before reading this, to have cut open in front of them the heart of some large animal which has lungs" (p. 67) [de faire couper devant eux le coeur de quelque grand animal qui ait des poumons (p. 47)]. Not many are going to run out and carve up their kitties or doggies for the sake of knowledge, but there is something of this attitude implicit in the Cartesian model of knowing.

But Claude Bernard has taught that the passageway to knowledge goes through the kitchen of slaughter,[26] and how can we get to know how the body works unless we find a way to step back from our feeling, overcome our love of mammals, transform them and other organisms into objects, and cut? My reading of Descartes has meant to imply no easy answers. The problem with Descartes is not the position of detachment, the commitment to get beyond personality and the obvious limits imposed by interest, desire, and ambition. The problem is that Descartes's sort of detachment seems to require, first, a commitment to the absolute; second, alienation from the body; and third, mystification of the narrative that produced it.

Exposing the ethical, the ideological, the human beneath the facade of Descartes's rationalizing theory does not get us far. Something like the distancing entailed in Cartesian objectivity is an indispensable, even a humanizing element in human intercourse. It made possible Bernard's important work and has made possible the whole "science" of medicine. It is striking that, for example, Bernard makes life out of death, by way of "experiment," or that Darwin's theory quite literally builds life in all its diversity out of selective death, and that Descartes struggles back to the body only by virtue of emptying it out, devaluing it, turning it into an object of epistemological discussion and ultimately into a mechanism. In keeping with a Christian sense of the fallen nature of humanity, Descartes worked into his final comments in the Sixth Meditation a recognition of how extraordinarily difficult is the detachment required to know. His myth of consciousness is indeed a fable, as extreme as Don Quixote's, but it does not obviously look like one. Nevertheless, the fact that Cartesian epistemology is a narrative is one of its greatest strengths, for narrative might be the first step out of the ideal of universal and statically ahistorical knowledge that can ultimately produce only one, and final narrative, the movement that makes the acquisition of knowledge a kind of death.

— 3 —
Carlyle, Descartes, and Objectivity: Lessen Thy Denominator

Here I make a long leap from the Renaissance to the nineteenth century, where this book will largely remain. In tracing the working out of an idea, I will not be discussing much the material conditions that connect Descartes with later thinkers. The history of epistemology between Descartes and the English nineteenth century is, of course, long, complex, and not at all entirely Cartesian. But my concern is not with the *history* of epistemology; it is with those basic narrative patterns I have already described in Descartes's *Discourse* as they emerge, however modified, in the thinking and storytelling of the great Victorians. Most of them seem to have lived with a set of assumptions and expectations about how one knows, how one articulates what one knows, and how one attains authority for knowledge that in many ways recapitulated the Cartesian story. The stories they found themselves able to tell were largely engaged with the same problems I have identified in my reading of Descartes: the implications of the quest for authority, universality, and truth. For them, as for Descartes, the pursuit of knowledge entailed a radical distrust of the body and a sacrifice, rather like death, of normal human desire.

I want to examine here what I take to be a singularly representative, powerful, and influential form of this dying-to-know story early in the nineteenth century—the story that Carlyle tells in *Sartor Resartus*. But I want first to suggest, through some significant examples from writers very different from Carlyle, the degree to which fear and hatred of the body, as a radical impediment to true knowing, dominates Victorian thought.

The problem of locating a kind of knowledge to be trusted has always been associated with the need to move beyond the personal. The potentially destabilizing idealism of Bishop Berkeley, like Descartes's, depended for its epistemological certainty on an all-perceiving and all-knowing God. Or, to take another example, the very lowest level of Kant's epistemological structure is the level of personal sensation. Indispensable it might be, but the sanction of ultimate rationality derived from a universal condition is unattainable by mere sense without the mediation of "judgment," or the aesthetic.

Or, to take a Victorian example, the radical empiricism espoused by John Stuart Mill, which imagined a structure of cumulative experience building to the level of deduction and general law, encountered the Kantian qualifications of the polymathic William Whewell because raw empiricism is so vulnerable to the Cartesian anxiety. Mill begins by denying the intuitive rightness of deduction, replaces it with induction, then gives induction the authority previously attributed to deduction, the impersonal and extrasensory certainty of which could then parallel Descartes's ideal. In effect refusing the Cartesian idea of necessary knowledge, and insisting that all knowledge, logical and mathematical included, depends on induction, Mill still needs to achieve that authoritative and universalizing power that Descartes claimed to have discovered. He tells the story of a process as he unfolds his theory of "induction," which then can become the basis for universalizing, vulnerable though it may have seemed to exceptions and an indeterminate future.

The narrative implications of Mill's empiricism end, then, by being not very different from Cartesian ones. Mill, too, has to get to the position of affirming general laws, and he must in the end disentangle "material circumstances" from "immaterial" ones.[1] Knowledge, even for Mill, is thus not quite compatible with embodiment. Into the realms of epistemological abstraction the Western and, of course, particularly Christian distrust of the body enters. Mill, who in his treatment of deductive logic constantly exposed axioms as founded on induction after all, had to have his own axiom—perhaps *the* indispensable axiom for the development of nineteenth-century science, "the uniformity of nature": "Whatever be the most proper mode of expressing it, the proposition that the course of nature is uniform, is the fundamental principle, or general axiom, of Induction" (p. 338). Mill quickly agrees that even *that* axiom is based on "induction," but in the end this leads to a *structure* of argument virtually identical to the structure of traditional logic. As Mill says, "no conclusion is proved for which there cannot be found a true major premise" (p. 339).

The long battle between Mill and William Whewell, one of the most interesting intellectual combats of the nineteenth century, had to do with how one moved from particulars to generality. How *did* science manage to establish universal truths if all knowledge derived from the particulars of experience? Mill makes his argument by establishing induction as the ground of authority, and the axiom of the uniformity of nature as deriving from (and serving as the foundation of) induction. In effect, Mill keeps separate those Cartesian poles, mind and matter. For Whewell, who had gone to school to Kant, this was a totally unsatisfactory way to think through the issue. For Whewell, too, and yet more directly, as I have suggested in my brief allusions to his portrait of Newton in chapter 2, the body is an impediment. If in Mill the mind, as it were, pulls itself up by its own bootstraps, dispensing with the body and its limits by using it, Whewell insists that the mind occupies the body from the start.

That is, as in perhaps the most famous line of his huge production, "there is a mask of theory over the whole face of Nature."[2] Unlike Mill, Whewell found it impossible to disentangle "facts" from "ideas." All fact for Whewell is "theory laden."[3] "In every act of my knowledge," he writes, "there must be concerned the things whereof I know, and thought of me who know. . . . No apprehension of things is purely ideal: no experience of external things is purely sensational."[4] Whewell's way of thinking, consonant in many ways with the directions of modern theory, might well have implied an anti-Cartesian narrative, but Whewell resolves to live with an irresolvable tension, and values science and knowledge not for its complex interweaving of body and mind but for the capacity of mind ultimately to transcend body. The image of Newton, totally abstracted from the life around him, fairly represents Whewell's sense of the possibilities and achievements of science. At the same time, however, he is committed to the view that a valid scientific epistemology can only be derived historically, through a *History of the Inductive Sciences* that must lay the ground for an adequate *Philosophy*.

The question of how to universalize knowledge, lift it from *mere* contingency and singularity, pervades almost all nineteenth-century thought about how we know. Raw fact is not knowledge at all.[5] The merely personal and local cannot be trusted. Like the religious traditions whose patterns it more or less self-consciously mimicked, epistemology sought something like transcendence.

The attempt to transcend the local and particular requires a narrative that includes the local and particular, and what I am looking for is the narrative, not the system. The narrative of the progress to transcendence (it should be noted that Mill and Whewell both believed in the progressive development of science extending into the triumphant study of the human) remains strikingly similar, and it can be read in an astonishing variety of places. Beyond any particular system of philosophy there remains the radical distrust of the corporeal, the contingent, the personal. Salvation—religious, aesthetic, or epistemological—depends on the capacity of the self to get outside of itself, to extend, as Schiller and Matthew Arnold after him insisted, from the individual to the species.

Arnold, like Descartes, invokes the "self" as the basis of knowledge and then salvation, but it is the "best self," which is also the "buried self," and this "self" has as little relation to the corporeal "selfish" self as Descartes's "ego." It is a representative of the species rather than of the local conditions of a private life, and lives beyond local politics and the practicalities of everyday life to attain to spiritual certitude that puts all individuals in common. The means is "right reason," a medium that transcends the personal—"the reconciling force."[6] And to take just one other of myriad examples, when John Ruskin tries to teach his audience how to "read," how to learn from great thinkers of the past, he insists on a Keatsian negative capability. Our responsibility is to put "ourselves always in the author's place, annihilating our own personality, and seeking to enter into his."[7]

The greatest obstacle, that is, to spiritual salvation, aesthetic value,[8] and epistemological authority is the self that seeks those things. Knowledge is only knowledge when it extends beyond the particular and can be shared. The crisis of knowledge comes when the self, in all of its inevitable situated uniqueness, needs to transcend its situatedness. The story of that crisis is the narrative of "dying to know."

The epistemological narrative, through all its variations among scientists and novelists and poets, is the story of the pursuit of something like the Arnoldian "best self." The pursuit requires paring away all local conditions of the living busy world—the numbing power "of all the thousand nothings of the hour."[9] It is a secular pilgrim's progress. The story is implied richly in Schiller's *Letters on Aesthetic Education,* one of the sources, surely, of Arnold's notion of the best self. Schiller puts it this way:

> Every individual human being . . . carries within him, potentially and prescriptively, an ideal man, the archetype of a human being, and it is his life's task to be, through all his changing manifestations, in harmony with the unchanging unity of this ideal. The archetype, which is to be discerned more or less clearly in every individual, is represented by the State, the objective and, as it were, canonical form in which all the diversity of individual subjects strive to unite.[10]

The story is complicated in Schiller by his rather un-English insistence on play and pleasure.[11] Schiller aims at the "sensuous drive" as Arnold never explicitly does, but Schiller, too, recognizes the fundamental paradox of the whole narrative: the absolute only manifests itself in the limits of the material, and "it is only through limits that we attain to reality, only through negation or exclusion that we arrive at a position of real affirmation, only through the surrender of the condition of unconditional determinability that we achieve determination."[12] Despite Schiller's enthusiasm for sensuous pleasure, the fundamental narrative is there—the denial of self to achieve self, the negation of embodiment for the sake of revelation.

If there be an archetypal Victorian version of this dying-to-know story, it would have to be Carlyle's. Nowhere does it emerge more clearly than in his *Sartor Resartus*, the work that raised him, if belatedly, from literary hack to the first and most influential of the Victorian sages. In Carlyle's model representation of the quest for knowledge, problems of embodiment dog every move. Writing out of two traditions deeply distrustful of the body—his father's rigid Presbyterianism and German romantic idealism—Carlyle is particularly sensitive to the claims of the body. He and his body wage a lifetime war against each other, and as writer Carlyle could never for a moment leap into the transcendent without being dragged back by his griping flesh.

Sartor Resartus is a gesture at establishing what might be called an adequate "method" of knowing—but it is cast in part as biography and enwound with the ethical and the material. It is not, then, a piece of philosophy but, if extravagantly, it is a Victorian version of Descartes's discourse, a "biography" of a philosopher. A Professor of Things in General as Descartes is a constructor of the world, Diogenes Teufelsdröckh undergoes a similar pilgrimage, which traces the same line of self-annihilation into revelation. *Sartor Resartus*, in its turn, becomes a kind of model and fountainhead for Victorian attitudes toward the self and toward knowledge. The strenuousness and deliberately contorted nature of its narrative reflect a quite literal battle with the body that is also an epistemological battle. For the body continuously reminded Carlyle of what he regarded as the dangerous threat of a worldview that was entirely dependent on its relation to the individual consciousnesses that perceived it. His story sets a pattern of tense sensitivity to the apparently inseparable presence of the body in the act of knowing.

No matter how much Fichte he read, and how little Kant, Carlyle was not a philosopher. His place in intellectual and cultural history—despite his frequent co-option of terms out of German romantic philosophy—has little to do with the sorts of abstract metaphysical speculation that he sought to replace with "Fact," usually with a capital F (giving it, then, a more than local resonance). But his way of thinking and arguing reflects and reflects back upon the great post-Cartesian tradition of epistemology. In what follows, I will be tracing the parallels in his narrative to those that developed on the Cartesian model that he explicitly resisted. Carlyle's strange book overtly enacts the way the ethical and the epistemological are, in the nineteenth century, sanctioned by the same values. Its narrative and its philosophy point toward that inaccessible view from nowhere that entails both the very disembodying Carlyle seemed bent upon resisting, and the annihilation of self. The critical word here is *inaccessible*. As his two essays on history—"History" and "On History Again"—demonstrate, Carlyle understood that a true "Universal History" is impossible. The limits of the human body and the flux and multiplicity of history make possible only the faintest glimmers of universal truth, the evocative Fact. Carlyle constructs his narrative out of doubt, limitation, and pain. When seen against Descartes's epistemological narrative, Carlyle's helps clarify the ways the most abstract of philosophical enterprises, ostensibly dependent on detachment and disinterest and distance from the constraints of the body, is in fact a deeply embodied and ethical enterprise.

The special virtue of Carlyle's thinking and writing in the context of this tradition is that they refuse to remain at the level of abstraction to which Descartes aspired. No prose can seem more un-Cartesian than Carlyle's, but the extreme difference does not include a narrative and formal difference about the process by which one moves to knowledge. As we have seen, Descartes tried to think of the metaphors and embodiments that figure so significantly in his *Discourse* as mere concessions to the limited consciousnesses of readers who had to be seduced into the truth because their minds stopped always at the level of the material, which they had to learn to transcend. Carlyle, on the other hand, whatever his painful struggles with the material, always had to begin with it, contend with it, even affirm it.

To be Carlylean and to love the world entailed something like madness. But it was not entirely mad, for his way of thinking and writing was marvelously contaminated by that world even if, in the end, his intensely puritanical consciousness and his sadly inadequate body required its ultimate sacrifice. Much of *Sartor Resartus* spins around the question of embodiment, however fancily and metaphysically it may play with the question. In effect, Carlyle's, or Teufelsdröckh's, vision is almost sacramental: that is, the embodied world constantly signifies, while it conceals, the sacred. "All visible things are emblems," Teufelsdröckh asserts.[13] To know "reality," one needs not only to recognize the Fact, but also to know what it "symbolizes." The fundamental paradox remains the same: the absolute only manifests itself in the limits of the material. For this reason, Carlyle and his surrogate take metaphors not as a concession to popular intellectual weakness, but as the fundamental condition of any adequate language. The world of *Sartor Resartus* is "The World in Clothes." Its subject is, "the Tissue of all Tissues, the only real Tissue . . . the vestural Tissue," and it is only by recognizing how fundamental this "Tissue" is that "Science" can attain its fullest knowledge (p. 5).

So Carlyle's starting point was very different from Descartes's: for him, the mind is not separable from the corporeal, and for him, Cartesian rationality and commitment to theory were representative of just that inhumane analytic mode that he saw as a fundamental failure of modernity. Contemptuously, Teufelsdröckh talks of how Cause-and-Effect philosophy appeals to "superior intelligences." But "for inferior Intelligences, like men, such Philosophies have always seemed to me uninstructive enough" (p. 36). Descartes may well be lumped among those "Cause-and-Effect" philosophers, one who was antithetical to Carlyle's argument that "[b]oundless as is the domain of man, it is but a small fractional proportion of it that he rules with Consciousness and by Forethought."[14] The mystery and the darkness cloaked by life are inaccessible to ordinary consciousness: "underneath the region of argument and conscious discourse, lies the region of meditation: here in its quiet mysterious depths, dwells what vital force is in us."[15] There is nothing much clear and distinct here. Beyond consciousness there lies the great Carlylean moral injunction, "Work while it is called Today; for the Night cometh, wherein no man can work."[16]

It would not be unreasonable to suggest that for Carlyle, "nescience" is at times more valuable than "science." With his emphasis on unquestioning work, he turns from a theory of knowledge to a theory of action. He even seems at times modern in his exploration of the indeterminacy of knowledge and of the conditions of life. Yet his response to that sense of the inability of the human to understand the meaning and conditions of its existence is an almost absolute affirmation of action without "knowledge." He always praises the "unconscious"; "self-consciousness" is the mark of ill health. As he claims in "Characteristics,"

> A region of doubt hovers forever in the background; in Action alone can we have certainty. Nay, properly speaking, Doubt is the indispensable inexhaustible material whereon Action works, which Action has to fashion into Certainty and Reality; only on a canvass of Darkness, such is man's way of being, could the many-coloured picture of our Life paint itself and shine.[17]

Without faith in the power of rationality (of course, for Carlyle "Reason" has the metaphysical meaning implicit in German romantic philosophy) he could not believe unequivocally in the possibility of complete and universal knowledge, as Descartes did. At the limits of the body, of perspective, of moral limitation, only unscientific intuition can open glimmers of the possibility of universal truth. Ironically, however, by rejecting the discourse of abstract metaphysics, by embodying his ideas in metaphors and narrative, by moving from philosophy to history, Carlyle exposes (and spells out some of the implications of) the narrative of heroic self-abnegation that underlies and validates Cartesian epistemology and links the quest for truth with a moral imperative that philosophy has traditionally evaded or denied. Epistemology, for Carlyle, is a form of ethics, after all; and the ethics tend to depend on a dualism between mind and body, in which he, too, however tortuously, acquiesced.

The difficult history of epistemology's deep distrust of the body and sometimes secret longing to satisfy it can be detected working its way through Carlyle's strange affirmation of the need to deny the self. As I have been arguing, largely by way of Descartes, epistemology tends to resist the slightest implication that desire is operative in the pursuit of truth—with the exception of passion for truth itself. Yet official dogma notwithstanding, the various idealisms and rationalist accountings, from the cogito to the anti-rationalist resistances of the late nineteenth century in the thought of philosophers like Nietzsche and James, depend largely on a sense of the enormous, the possibly unconquerable power of the body, with its accumulated heap of irrational desires. Carlyle, along with Nietzsche and, for example, a writer like D. H. Lawrence,[18] was greatly influential in calling into question the Cartesian tradition even if, against Nietzsche and Lawrence, he wanted to affirm a world touched by divine meaning and overseen by God.

Carlyle served his apprenticeship seeking to transcend the body. The body, which ought to have been recognized as the soul's clothing, was for him the seat of the ego, and thus the source of "desire," the great temptations that his world-resisting temperament of rigorous Presbyterianism needed to overcome in order to achieve at least moral salvation. His narrative, by confronting the body directly, attempts ultimately to conquer it. So his prose reflects an obsession with things bodily, a very powerful sense that the body is always there, always in danger of taking command. In his own Scots Presbyterian way, that is to say, Carlyle was a sensualist.

He was particularly obsessed with his bowels, though other bodily parts also bothered him.[19] He and his bowels had a long, largely unpleasant relationship. Bowels and digestion became for him a clue to the world's ills and they figure importantly in almost all of his major works. His bowels betrayed him early, and in 1819, at the age of 24, he was complaining noisily about them. J. A. Froude talks of his being "intolerably irritable." "Reticence about his personal sufferings was," Froude says, "at no time one of his virtues."[20] Some of his lamentations, from those youthful days, suggest the painful relation between body and spirit that would become a central topic of his mature work. Fred Kaplan quotes him as writing, "And do but think what a thing it is that the etherial spirit of a man should be overpowered and hag-ridden by what? by two or three feet of sorry tripe full of ———."[21] Castor oil and epsom salts were steady companions, as was pain. As with his impotence, Carlyle was humiliated by the urgency and failures of his bowels when his business was so passionately spiritual. His endless complaints about his digestion, his special orders for home-style oatmeal, his flatulence, his constipation, accompanied as they were by sleeplessness, excessive sensitivity to sound, and a perpetual sense of discomfort in the world, suggest that no account of Carlyle's thought can be adequate unless it takes the body into account. One must imagine him writing always with the rat gnawing at his stomach, and wonder what that might be doing to his prose.

That is to say that Carlyle, like the Victorians who followed him, even like the scientific naturalists who wrote so comfortably about the material world, had a mind/body problem. He had a strong suspicion that much of his pain was psychosomatic, but it lasted for a lifetime and it felt real enough. It seemed, then, to be further evidence that mind and body were inextricably entangled, the thinking self not separable from the dyspeptic one. The pain encouraged him to think of the relation between mind and body and to find the Enlightenment critique of traditional belief both persuasive and deeply unsatisfying. Its rational force was irresistible, and that force led Carlyle to his touching conversion away from the dogmatic religion of his beloved and perhaps austerely intimidating parents, particularly of his mother. But beyond the mind with its reasons lay the bowels with their violent and irrational demands. The systematic work of logic and even rationality, which could obliterate "systems" of religion and spirituality, was helpless against the raw power of the irrational in his bowels.

Still, while Carlyle recognized that his own way of thinking and feeling would be altered if his body were more friendly to him, he could not believe that simple satiation of the body, or of the material needs of society, would be adequate to overcome the spiritual malaise he felt so intensely. His works are peppered with contempt for the body and contempt for the need to satisfy it. The hostility toward what he called "mechanism," announced in his crucial and personally liberating essays, "Signs of the Times" and "Characteristics," is only another version of his embarrassment with and aversion to material, bodily things. It expresses itself as a rejection of the view that he took to be dominant in his culture, that technological and material advances can save the individual from spiritual malaise. Carlyle concludes "Signs of the Times" with a figure from astronomy: the earth's path at the moment, he claims, "lies towards *Hercules,* the constellation of *Physical Power:* but that is not our most pressing concern. Go where it will, the deep HEAVEN will be around it."[22] The emphasis on the physical and the pains the physical produces was not to be overcome by the perfect laxative, although the history of the world might have been transformed by it. The very intensity of his physical discomfort argued for some extramaterial significance. Indeed, he believed, and argued with characteristic bitter irony, that the material pain was a consequence of a greater spiritual pain and that the cure would be in the spirit, not in the body: "Art thou nothing other than a Vulture, then, that fliest through the Universe seeking after somewhat to *eat?*" [23]

"With Stupidity and sound Digestion man may front much," says Teufelsdröckh. "But what in these dull unimaginative days, are the terrors of Conscience to the diseases of the Liver. Not on Morality, but on Cookery, let us build our stronghold" (pp. 160–61). The passage sounds a bit different if one thinks of it in the literal context of Carlyle's dyspepsia. This is no abstract irony but a worrying through of the implications of deep physical discomfort. While it is serious enough, in its comic way, it implies Carlyle's ironic turn against himself and against the insistence of his own pain. But it also suggests a refusal to allow that pain the primacy he is deeply tempted to give it—his strategy is a rhetoric of prophetic self-contempt. The body will not do, but neither will the rational mind—or not quite. While it is easy, and important, to think of Carlyle as hostile to *mere* rationality, it would be a mistake to minimize his totally characteristic nineteenth-century belief in the possibility of a functional rationality that will—under the right circumstances, of course—allow for intelligent solutions to difficult problems. There is an ultimate rationality in the world: "Truly," says Teufelsdröckh, "a Thinking Man is the worst enemy the Prince of Darkness can have" (p. 118). But a "thinking man," needless to say, would not be a Cause-and-Effect philosopher, like Descartes or John Stuart Mill. So, in that strange twist in the middle of "Characteristics," an essay that celebrates the lost virtue of unconsciousness, Carlyle also recognizes that the self-consciousness that is a symptom of illness ("The healthy know not of their health, but only the sick") "is also the attempt towards cure." [24]

In his continuing preoccupation with the intersection of spirit and body Carlyle attempted to overcome what he argued was the narrowing effect of Enlightenment rationalism—its tendency to "mechanism." On the surface, at least, he is totally unsympathetic to Descartes, who comes in for criticism fairly early. After invoking Descartes's "cogito ergo sum," Teufelsdröckh immediately dismisses him: "Alas, poor Cogitator, this takes us but a little way." [25] But Carlyle's rejection of Descartes is more pervasively evident in the almost ferocious commitment of his prose to metaphors.

As I have already pointed out, Descartes rejected metaphors because they implied embodiment, because they rely upon "images" and therefore the senses. In his view of the world as emblematic (or even sacramental), Carlyle saw metaphor as the predominant vehicle for shadowing forth the meanings embedded in the material. But for Carlyle, metaphor was not even strictly a vehicle, because as a form metaphor is mimetically accurate.

It is not the yoking together of heterogeneous things, but the revelation of mutual spiritual sources. Metaphor is itself a perfect metaphor for the material, since the material world, too—even the bowels—is not detached from meaning but the divine and human way of articulating it. A hint at Carlyle's commitment to the meaningfulness of corporeality comes in his perhaps excessive etymologizing. That is, he was always eager to dig the metaphorical roots of common or abstract words, and when he found what he took to be the "origins" of ordinary language, he wrote as though he had discovered the extrapersonal and even divine sanction of meaning.[26] So, in one of his more famous etymologies, he finds in the word *king* the roots that mean "Able-man."[27] Real kings, benefiting from etymology, become worthy of the submission they demand.

The metaphor, a signal of embodiment, is also a signal of the way Carlyle's mind was constantly preoccupied with the body. It may be an exaggeration to talk of Carlyle as a sensualist, but it is consistent with his own sense of his work—insistent as he was on the way the mind and its moods were shaped by the physical. When Teufelsdröckh, committed to "Historical" not "Lyrical" epitaphs, writes the epitaph for the patriarch of the only important family with whom he had been able to become close, he uses a formal Latin for something that "will surprise an English reader." What, after all, is Count Zähdarm's monument but a heap of manure, converted, not without "a racket" (sine tumultu) from "a hundred million hundredweights of food," into ——— (IN STERCUS).[28]

Moreover, Carlyle was as interested in the physicality of sex as he was in the bowels, although much more discreet about it and much more embarrassed about his own failures. As his intestines failed him, so, notoriously, did his sexual organ. He wrote what reads now as a very sad postnuptial letter to his physician brother, which surely is a cry for help out of what must have been a very great ignorance and a deep shame.[29] Less than a year into his marriage, he copied into his notebooks this line of Teick: "My whole life has been a continual nightmare, and my awakening will be in hell." And it was then that he entered a line from Aristotle that recurs importantly through his early work: "The end of man is an action, not a thought."[30] Given his belief in the dominant power of the male, Carlyle could only have been humiliated by his failure to perform on his wedding night or, as far as we know, ever. His sensitivity, not explained as having to do with sexuality, kept him and his wife sleeping in separate beds all their long married life. The tone of his marriage to Jane Welsh was set by sexual failure, and that failure accompanied him at his work as persistently as did the rat in his belly. The physical failure threatened—as with his bowels—to be a spiritual one. But, as Kaplan describes it, Carlyle resisted that almost inevitable conclusion. "In his eyes,"

Kaplan writes, "it was a failure not of love but of a body that would not do love's bidding."[31]

Carlyle's demanding and unresponsive body became his enemy, especially in sex and in defecation. In 1848, Kaplan shows, Carlyle wrote an essay, to remain unpublished, attacking what he called phallus worship[32]—another, even more overt piece of evidence of the degree to which his social critiques and metaphysical ventures were partly built on the problems he had with his relationship with his own body. The satisfactions of sex, like the satisfactions of digestion, would never be available to him, and in his work he transformed these losses into an affirmation of spiritual value, a denigration of the material, but not, as he put it for Count Zähdarm, without a racket, not without continuing attention to the body's sensuous demands, to its grossness and absurdity, to its capacity to humiliate. And so it is that his characteristic rhetorical strategy, made most famous in the clothes metaphor of *Sartor Resartus*, is to develop his ironies and his critiques through focus not on the spiritual, moral issue that, he will argue, is the real one, but on physical objects that he means to make resonate—perhaps in more ways than one—with nonmaterial meaning. The awful attention to Quashee's "watermelons" in "The Nigger Question" is one of his more egregious but most characteristic moves of this kind.

There is no need to recapitulate here the famous progress of Diogenes Teufelsdröckh except to note briefly how this remarkable little fictional biography recapitulates the dying-to-know narrative—again, literally "embodying" it. For Teufelsdröckh, in the crisis that follows the failure of his "romance" with Blumine, loses belief in everything, sinks into what might be called a Cartesian Great Doubt. His first great affirmation in the "Everlasting No" is of absolute negativity and the consequent necessity to assert the self against the bewildering emptiness of the frightening universe: "then was it that my whole ME stood up, in native God-created majesty, and with emphasis recorded its Protest." With this profound negation Teufelsdröckh "began to be a man."[33] Here the analogue to Descartes's descent into the self for a foundation is remarkable except, of course, that Carlyle's "affirmation" is saturated with feeling, with heroic insistence that it can take it, that no worse can be done. Ironically, the Teufelsdröckhian "No" has some of the force of an existential "yes" and the "Yea" that will follow has the feel of suicidal self-surrender. Teufelsdröckh's move through the "Center of Indifference," when, with all old connections with the world severed, there seems no point in being, is followed by the "Everlasting Yea," which splendidly enacts the paradox of the dying-to-know narrative. That is, the affirmation of the "Yea" is a radical denial of the self, a refusal to satisfy the body or normal human desires. The moment of revelation (even if it points to the value of work over self-consciousness) is the recognition that the "Ideal is in thyself" (p. 96). It is a movement toward what Arnold was to call the "best self," that which connects the individual human being with the species and therefore obliterates all the singularity of the individual. If Carlyle's focus is on the eccentricity and singularity of his "protagonist," the eccentricity (like the particulars of the body) is to be taken as a disguise for that which is totally concentric.

Although I focus here on the epistemological implications of *Sartor Resartus*, the book is primarily a moral, quasi-spiritual narrative, and it acts out in a new secular form the narrative of religious pilgrimage that Carlyle would have known quite well through Bunyan. But I want to emphasize here how it duplicates the narrative by which the epistemologist reaches toward true knowledge, beginning with the impulse to know, which has the assertiveness of an everlasting "no," and with the purging away of the past, and concluding with radical self-denial. One achieves moral, spiritual, and intellectual salvation through *Selbstödung*.

It is nevertheless not quite true that for Carlyle the body and the physical world are imagined as utterly without value, as worse than valueless, except as they manifest energy, are alive, and are rightly perceived, that is, understood as metaphors. Much of the energy of his rebellious relation to things as they are was generated by his sense of physical contiguity and physical suffering. If he wanted always to minimize the body and emphasize the spirit, he was clearly outraged by the suffering and brutality he saw all about him. In the essay on "Chartism," which, it is true, waffles on the question of the New Poor Law, and demands "justice" for the poor before it demands physical assuagement, Carlyle is nevertheless sympathetic to the Chartists' anger. He is hostile to the *theory* that assumes the problems of the poor can be resolved by closing down outdoor relief. But most important for my purposes here, the essay is thick with the sense of unjust suffering, of physical deprivation that requires treatment. Talking of the failure of the New Poor Law, Carlyle complains:

> To say to the poor, Ye shall eat the bread of affliction and drink the water of affliction, and be very miserable while here, required not so much a stretch of heroic faculty in any sense, as due toughness of bowels. If paupers are made miserable, paupers will needs decline in multitude. It is a secret known to all rat-catchers: stop up the granary-crevices, afflict with continual mewing, alarm, and going-off of traps, your 'chargeable labourers' disappear and cease from the establishment. A still briefer method is that of arsenic; perhaps even a milder, where otherwise permissible. Rats and paupers can be abolished.[34]

The bitterness here is not abstract. It is a resentment of the government's failure, as it lapsed from the realities to the abstract theory, to feel the pain of hungry workers "while here." Carlyle's imagination of the body tends to imply a continuing battle, and at the same time a recognition that the body is always there, always indispensable, and always in quest of satisfactions that, somehow, it can never fulfill.

The importance of the body in Carlyle's engagement with himself and resistance to Enlightenment rationalism makes his ultimate choice of career, as historian, seem inevitable. It did not seem so for him, as he worried among the possibilities of mathematics teacher, lawyer, clergyman, poet, journalist, and asked himself that dominant Victorian question, What shall I do?[35] His apprenticeship was a very long one, and *Sartor Resartus* becomes his *Discourse on Method*. But clearly he could not be an abstract thinker or a poet. Ideas mattered forcefully to him, but only as they were embodied, and where else to find embodied idea but in history itself? History, he claimed, is the "Letter of Instructions, which the old generations write and posthumously transmit to the new."[36]

Descartes sought knowledge in ideas. Carlyle sought it in embodied life, in history. Struggling to avoid the reduction of life and meaning to sets of intellectual abstractions, he was drawn to meaning that emerged out of experience itself. Descartes sought to see through the corruptible body to the permanent essence of thought. Carlyle was impressed by the opacity of the body and, as John Rosenberg points out, consistently rejected the old saw of John Bolingbroke that "history is Philosophy teaching by examples."[37] "Truly," he says, "if History is Philosophy teaching by Experience, the writer fitted to compose History is hitherto an unknown man."[38]

And yet the past is, for Carlyle, "the true fountain of knowledge." And what he requires in relation to that past is "vigilance and reverent humility." It is what Bacon requires of the investigator of nature, and what Carlyle's contemporary, John Herschel, asked of the aspiring scientist in his *Preliminary Discourse,* a work that was written at about the time Carlyle was writing "Signs of the Times," *Sartor Resartus,* and "Characteristics." The true scientist, according to Herschel, is one whose powers of observation are always active, and who attends meticulously to the slightest detail, for the larger idea may be buried in the detail and the detail itself can only achieve the status of knowledge if it is in fact incorporated into larger clusters of detail. This is to say that Carlyle, in his ostensibly eccentric mysticism and his glorification of the Fact, participated in a great epistemological tradition, one that was critical to the development of science and that serves as science's most powerful justification. It begins and almost ends in the requirement of humility before the Fact, and the recognition of the inadequacy of human powers to comprehend the fact independently. Here, Descartes's project had moved ambitiously beyond Carlyle's romanticism. Unbefogged by romantic generations who had rediscovered the mystery in things, he could think such a project possible precisely because he eschewed embodiment in his method and relied on "pure thought."

While Descartes seems really to have believed that given a not unreasonable stretch of time he could have constructed the whole world out of the ground of the cogito, Carlyle insists on the inescapability of fragments and incompleteness. So in the essay "On History," he writes:

> For though the whole meaning lies far beyond our ken, yet in that complex Manuscript [of history], covered over with formless inextricably entangled unknown characters, nay, which is a *Palimpsest*, and had once been prophetic writing, still dimly legible there,—some letters, some words, may be deciphered; and if no complete Philosophy, here and there an intelligible precept, available in practice, be gathered; well understanding in the mean while, that it is only a little portion we have deciphered; that much still remains to be interpreted; that History is a real Prophetic Manuscript, and can be fully interpreted by no man.[39]

Materiality makes for mystery. And yet knowledge is *in* the material.

So, having rejected the Enlightenment sense of the powers of pure rationality, having complicated knowledge by requiring recognition of material incompleteness and groundedness in particular places and particular times, Carlyle nevertheless adumbrates a world largely consonant with that of the Enlightenment. Meaning really is there, behind the material—even if never to be wholly accessible to corporeal beings—and it requires precisely what his Enlightenment predecessors required, humility before the real, and the attempt (if not entirely successful) at the obliteration of the self in the representation of reality. Here is Carlyle's worship of the Fact, with a capital *F*.

But history, full of eccentricities, oddities, opacities, is the source of knowledge for Carlyle, and in his own romantic way he treats it as a sort of philosophy teaching by experience after all. Rosenberg points out that Carlyle's history requires recognition of the uniqueness of every historical moment and therefore the inadequacy of the notion that human nature is everywhere and always the same. And while this anti-Enlightenment element of Carlyle's work is very important, it is equally important to understand how much his work constitutes a struggle to discover in the fragments and eccentricities of experience the universal and the permanent. (The very form of the biographical sections of *Sartor Resartus* is a structure out of the shards and fragments contained in Teufelsdröckh's laundry bags.) Carlyle surely believed that the present and the past shared all but their "clothing." It would be a mistake to underestimate, even in the face of Carlyle's predilection for particulars, embodiment, and eccentricity, the degree to which he believed that a humble encounter with the Fact of the past would reveal its similarities to the present. At the end of book 1 of *Past and Present,* he notes: "The Past is a dim indubitable fact; the Future too is one, only dimmer; nay properly it is the *same* fact in new dress and development."[40]

Through a reading of several Victorian novels in later chapters, I will try to confirm that this view underlies even the realism of Victorian fiction. The "universality" of the particular, if observed correctly—an inheritance from Wordsworth's commitment to the language of common men, for example—is explicit in the Proem to George Eliot's *Romola.* There, in a language that for George Eliot is uncharacteristically Carlylean, she announces the deep similarities between the apparently alien life of "more than three centuries and a half ago" in the singular political imbroglios of Florence, and the life of our own time:

> the great river-courses which have shaped the lives of men have hardly changed, and those other streams, the life-currents that ebb and flow in human hearts, pulsate to the same great needs, the same great loves and terrors. As our thought follows close in the slow wake of the dawn, we are impressed with the broad sameness of the human lot, which never alters in the main headings of its history—hunger and labour, seed-time and harvest, love and death.[41]

For the Victorians, even the particular is general, even the embodied leans toward the disembodied. Value lies—even for the realist novelists who insist on the sacredness of the particular fact, as Carlyle himself did—in the permanent and the stable. The paradox of dying to know is pervasive (and it is no accident, I believe, that this passage, claiming the continuity of history, ends in "death").

Carlyle is, then, both resistant to and fully implicated in the dominant ideas of the Enlightenment. He became a figure attractive to late-century scientists like Huxley and John Tyndall[42]—who seemed as distant from him as any science might have seemed from such an outspoken antirationalist—in part because he did insist on materiality, did require radical self-effacement before the Fact, and did aspire to large general inferences from the Fact. Over these issues, the culture seemed to converge, all being ready to write the epistemological narrative of dying to know. (One might add that some of the sympathy of these ostensibly antithetical and materialist scientists with Carlyle might have had to do with the fact that they shared with him a chronic dyspepsia.) One way to think of Carlyle's place in this tradition is that he was, after all, a dyspeptic child of the Enlightenment.

Carlyle saw empiricism as a ghoulish, lifeless materialism, a failure to recognize the ideal in the real; and for the scientists Carlyle influenced the observable real was, precisely, the source of all value and meaning. Yet he was rather easily enlisted in the antiintuitive project he had intuitively affirmed, and provided for the scientific naturalists "the foundation for a new view of nature, religion, and society that allowed them to regard themselves as thoroughly scientific and naturalistic without becoming either materialistic or atheistic and to accept secular society with good conscience and the finite universe without spiritual regret."[43] As Frank Turner puts it, Carlyle and the naturalistic spokesmen "regarded loyalty to fact and veracity in thought, speech and action not only as good in themselves but as constituting the legitimate foundation to claims of social and intellectual leadership and moral superiority."[44]

Physical austerity and cold showers are the mark of the ostensibly unsensuous Carlyle, whose ethic of work and *Selbstödung* became the official core of the Victorianism against which several generations of ostensibly more sensuous cultures have rebelled. His lean body and long, gaunt face seem antithetical to the pleasures of the flesh. Yet the body leads him to history, not philosophy, and is a constant presence in his work, an inescapable condition of the spiritual well-being toward which he aspired. His organicist imagination, which led him to embody the transcendental in the material particular, never in the end gets beyond that Cartesian dualism. The body is quite literally on Carlyle's mind all the time, reminding him that he is not well, "for the healthy know not of their health, but only the sick."

Far from being an absence in Carlyle's work, then, the body and its detritus become the core of it and lead Carlyle back into the central Cartesian tradition of modern Western thought that might seem to be defined by the absence of body, that is, epistemology. Carlyle's thought suggests how deeply epistemology is enwound with the body in all its materiality, how much the intellectual project is an aspect of the physical and of the ethical, and how fundamentally difficult it is within Western culture to break loose from Cartesian dualism. His uneasy relationship to his body parallels the unease of a culture that was at once obsessed and driven by the body and determined not to let it get the upper hand: let the ape and tiger die.

The beginning of Inquiry is disease. The question of truth only gets raised at the point at which one recognizes that there are problems in attaining the truth. If we follow out this theme of Carlyle's "Characteristics," we find that the beginnings of epistemology are in the body: consciousness is a condition of illness, since only with discomfort do we begin to notice ourselves. Descartes's *Discourse on Method,* in its very distinctive way, seems to affirm Carlyle's aphorism, and traces a similar path. In effect, Descartes founds modern epistemology on the experience of doubt he suffered at the end of his formal education. Like Carlyle, primarily enamored (at least for a while) by mathematics, he turned away from the abstract speculations of men of letters to confront the firmer realities of experience. And while he was not so outspokenly skeptical (he was, indeed, more nervous than Carlyle about the consequences of disbelief), he clearly allowed himself to put in question, or at least in brackets, any dogmatic belief. The subdued intellectual narrative of Descartes, ending in a discovery of the self, is in fact very much like Teufelsdröckh's progress.

It would seem that in building epistemology upon the ego, the ego that thinks, Descartes was doing something quite different from what Carlyle struggled to achieve in Teufelsdröckh's narrative. Descartes's "ego" is obviously less connected with body than anything in Carlyle and is constituted precisely by its existence outside any bodily container. It does not suffer from indigestion. Carlyle struggled to achieve the sort of desireless certainty of ego that Descartes discovered through entirely abstract speculation, but he could only do so through narrative, through relentless historical and metaphorical embodiment. It is not unreasonable to consider the narrative he actually does finally contrive as an extended commentary on the unarticulated assumptions and implications of Descartes's great epistemological pilgrimage. The climactic "Everlasting Yea" for Teufelsdröckh is not an abstract conclusion based on a necessary, rational argument. Rather, it is what he calls the "first preliminary moral act." That act is, indeed, *Selbstödung,* "Annihilation of Self," and with it "my mind's eyes were now unsealed, and its hands ungyved."[45]

In *Sartor Resartus,* then, the cogito is not a speculative proposition, it is a moral action. The purging of body and desire and what in ordinary understanding would indeed be the "self," comes through acts of moral self-restraint and refusals of desire. The annihilated self of Carlyle's famous pilgrimage is the equivalent of Descartes's thinking ego. It is the foundation on which true knowledge can be built, and it is also the foundation for a whole way of acting and being in the world. But Carlyle dramatizes the price of building that foundation and the more violent language of what translated into "suicide." This and "annihilation" speak more convincingly to the drama than Descartes's comfortable reflections in the famous stove-heated room, which change thought, not action, and allow him to continue in the public world exactly as he had been before. Yet in both narratives, the liberating moment is the moment of the abolition of the self, or rather its transformation from the self of the body—driven by dyspepsia and desire—to the self of the spirit.

From a modern, secular perspective, one that begins with the assumption that the essential and the universal are precisely not human conditions, there is no escape, as Carlyle warned, from one's own shadow, from one's sheer physicality and the limitations of perspective physicality must bring. Carlyle's narrative dramatizes the act of epistemological purification as a quite literal human sacrifice: for the sake of certainty and clarity, the Western tradition requires suicide. It is striking that in *Sartor*, it is the moment of the No that affirms the self, in which "my whole ME stood up." The Yea, the supposed moment of affirmation, is the moment when the self is annihilated. Morality and knowledge seem, somehow, to require death as a condition of goodness and truth. "*The Fraction of Life can be increased in value not so much by increasing your Numerator as by lessening your Denominator. Nay, unless my Algebra deceive me, Unity itself divided by Zero will give infinity.*"[46] This is a perfect metaphor for the model of Cartesian objectivity. The "preliminary moral act" is reinscribed through a mathematical metaphor. Here the body, at last, will disappear behind a language self-consciously without affect (though again, the language describing the move to the affectless is riven with passion). The great and ambitious Baconian enterprise ultimately disappears behind the language of technical science. The capacity to contain the impulses of desire in the pursuit of truth is, surely, necessary. But when morality and epistemology come together by way of the absolute obliteration of desire, this formulation becomes an algorithm for suicide as well.

Carlyle was too much a man of the body to succumb to the pure disembodiment toward which *Selbstödtung* pointed. One might want to say, perhaps with Carlylean descendental irony, that his bowels almost saved him.

— 4 —
Autobiography As Epistemology: The Effacement of Self

While this book has begun with discussions of discursive texts tinctured with narrative, usually autobiographical, I want to move on to consider predominantly narrative texts tinctured with epistemology. Since modern epistemology entered the world by way of a quasi-autobiography, it makes sense to move now to real autobiographies in which epistemology is not a subject, but whose form and texture, I am claiming, is informed by views consonant with those of the Descartes of the *Discourse*. Exploring the ways the narrative of scientific epistemology informs these very different and even nonscientific stories will further develop my argument about how narrative forces a reimagination of epistemology. The central paradox of the narrative of "dying to know" evidently disturbs truly autobiographical narratives. It surely disturbs the autobiographies of John Stuart Mill, Charles Darwin, and Anthony Trollope, whose stories about themselves are both fascinating and banal, often fascinating in their banality. In their diversity these books come to look alike: records of selves made possible only by the denial of self.

The very qualities that mark the autobiographies of Darwin, Trollope, and Mill as unliterary—their casual and relatively unrhetorical styles, their tendency to formlessness, the registering of apparently unimportant details, and, particularly, their persistent self-deprecation—can be read as reflections of a pervasive scientific vision which gives the documents the authority the rhetoric seems to be disclaiming. It is not, of course, that the narratives—not even Darwin's—*claim* to be scientific, but as I have been suggesting, the narrative of scientific epistemology assimilates more traditional ones as it claims its superior authority.

When Matthew Arnold sought compensation in poetry because the "fact has failed us," he revealed how completely he ascribed authority to science over religion, even in his attempt to reconstruct religion.[1] The authority of science in matters of knowledge manifested itself in everything from the condition of the church and religion, to social programs, to narrative, to the textures of these three not terribly inspiring autobiographies.

Recent critics have noticed the way scientific or epistemological theory has informed the construction of some Victorian autobiographies,[2] and they frequently focus on the specifically "scientific attitudes" in Darwin's self-portrayal. Linda Peterson goes on to discuss a genre of "scientific" autobiographies. In Darwin's *Autobiography,* she finds the rudiments of a new hermeneutics governing the form of autobiography—a hermeneutics that grounds the mental and spiritual in the physical, which controls the construction of the life. She notes that the book is marked by an "undirected gathering of facts," and that the whole narrative has a quality of objective rendering rather than a sense of the "private-experiential."[3] She cites James Olney's comment that Darwin writes about himself "as if he were a coral reef in the South Seas." Christopher Herbert has called this "anti-anthropomophism," a self-conscious resistance to the phenomenon Ludwig Feuerbach describes in *The Essence of Christianity*, and regards it as a mistaken positivistic response to the inevitability of the imposition of human consciousness on the world.

It is the body that sets limits to what can be known and that threatens, by obscuring the realities that lie outside of it, to create great epistemological difficulties. The transcendent by definition lies beyond the reach of a world entirely embodied, and intellectuals like Mill and Darwin, as my primary examples here, are regularly confronted with obstacles to knowing and certitude that put to question *any* intellectual authority. Darwin, it will be remembered, in his reflections on religion, asks can "the mind of man, which has, as I fully believe, been developed from a mind as low as that possessed by the lowest animal, be trusted when it draws such grand conclusions?"[4] In *what*, more skeptical followers might ask, might it then be trusted? The relativism and agnosticism that marked a main stream of Victorian thought were tied to this embodiment, for the embodiment locks consciousness into particular conditioned and local spaces. Thus, the narrative of scientific epistemology, as it enters more fully into the problems of embodiment, becomes tenser and more difficult. How might it carve out stability and authority from the flux that Darwin was elsewhere describing and from the mere secularity of the human that—through thinkers like Feuerbach and Mill—was being broadly argued and accepted. The point at which the extracorporeal, the universal and the stable, might be known, is the point of death.

The transformation of philosophy into narrative was already evident in the strategies by which Descartes developed and propagated his epistemology. Jonathan Loesberg has noted "the voraciousness with which Victorian narrative, through its structural correspondence with definitions of consciousness, devoured Victorian philosophies of the most diverse kinds."[5] Science, in effect, assimilated many of these "philosophies," and public justifications of the new power of science were offered regularly in periodicals, while T. H. Huxley was supplanting religious with "lay sermons." He adopted the language of religion to displace religion, or rather to allow for an undogmatic religion that would be entirely compatible with scientific epistemology.[6] Darwin might have been an anti-Christ, but he was comfortable in the church and was buried in Westminster Abbey. Positivism, taking science as the apotheosis of human thought, tried to transform itself into a church and George Eliot prayed to be allowed to join "the Choir Invisible." One can feel the presence of "science" in discourse even when it is not itself the subject. "There is but one kind of knowledge and but one method of acquiring it," Huxley proclaimed.[7] "For cultured early Victorians," writes Susan Cannon, "natural science provided a norm of truth. There cannot, as Victorians were fond of saying, be two truths, and the norm by which proposed truths were judged was, explicitly or often implicitly, the norm of natural science."[8] Beatrice Webb, looking back on the years of her intellectual apprenticeship, emphasized the dominance among the Victorians of "the current belief in the scientific method, in that intellectual synthesis of observation and experiment, hypothesis and verification, by means of which alone all mundane problems were to be solved."[9]

The narrative, or narratives, of scientific epistemology fit so neatly into the dominant narratives of the culture, so closely echo the patterns of ethical, religious, and aesthetic assumptions about value and human possibility, that they make it difficult to determine which sets of values have priority. Religious salvation comes only at the end of a life lived well; the good taste that leads to aesthetic satisfaction comes only when the vulgarities of immediate sensual pleasure are displaced by the more subtle perceptions that come with discipline, training, self-containment, impersonality. Ethical distinction follows the same pattern, for it can be achieved only through surrender for others of personal desire. The true scientist, as Galison has described the nineteenth-century condition, defers absolutely to the demands of the external world and refuses to impose his interpretations upon it.

The narrative of scientific epistemology thus implies a narrative of disciplined, self-denying progress through enormous difficulties toward a highly valued end, marked by constant repression of desire. This narrative gets written into real autobiographies and, in defining both the limits of what is knowable and the psychological and ethical conditions necessary to reach those limits, describes the constraints on self-definition that govern the way nineteenth-century writers could imagine the lives of their protagonists. Thus, in what follows I will be reading these narratives, normally taken in quite different ways,[10] as themselves epistemological narratives.

To start with, however, I want to look briefly at a roughly contemporaneous autobiographical narrative that does not fit the dying-to-know pattern, John Henry Newman's *Apologia pro vita sua*. It is important to mark qualities that, even if they seem to replicate the emphasis on modesty one might expect, distinguish themselves from that pattern, in that they are inconsistent with a scientific perspective. Such qualities might go far in helping to account for the strange raggedness and nondramatic quality of these other autobiographies.

In the *Apologia* Newman writes with conscious gentlemanly modesty. But he is not backward about the authoritativeness of his writing or the strength of his self. Authority derives not from precision of disinterested observation but from intuition, or revelation, and he thus speaks of "two and two only distinct and luminously self-evident beings, myself and my creator."[11] No "fact" of nature could contravene his primary intuition of God, and his style reflects that certainty. Modesty here does not mean self-effacement in the prose.

While Newman frequently reminds his readers of how awkward and embarrassing it is to have to talk about himself, he never for a clause or a phrase gives evidence of self-doubt. Nor does he ever show himself to be uncertain about his past actions. If he claims to be reluctant to speak about himself, he does so with an all-enveloping sense of gentlemanly propriety, not because he seems to thinks of himself as undeserving of attention. He is embarrassed by the need and thus the claims become an effective strategy to evoke sympathy for himself: How unfortunate that he should have been pressed to unmanly self-advertising by Charles Kingsley's even unmanlier poisoning.

Newman's deep intuition of self and creator allows him to speak with stunning confidence. That confidence is connected with a providential vision that allows him to read every natural phenomenon as related to divine intention. But providentialism was not a science, or even a quasi science, like "natural theology." Science, for Newman, *could* tell us legitimately about the natural world, but he did not need it, and his rhetoric does not reflect its values. For Newman, the attempt to "prove" the existence of God by reference to the observable world was to turn things on their head. One needs not empirical proof but the initial intuition. The world is "providential" and beyond understanding by the "wild, living intellect of man." Science could not explain things in any of the ways that matter humanly, and he rejected the developing mid-Victorian assumption that scientific method was the means to progressive improvement of the human lot.[12] He also rejected natural theology's conclusion that the world gives evidence of design: rather, it "is a vision to dizzy and appall; and inflicts upon the mind the sense of a profound mystery, which is absolutely beyond human solution."[13]

Stylistically, the difference manifests itself in that tone of absolute spiritual certitude. The secular writers are filled with doubt, but Newman "never doubted that in my hour, in God's hour, my avenger will appear, and the world will acquit me of untruthfulness, even though it be not while I live" (p. 8). Characteristically, near the start of the *Apologia*, he claims in a sentence of declarative briefness about the slanders against him, "They will fall to the ground in their season" (p. 7). Arguments against him he will "easily crumble into dust" (p. 10). These pronouncements, like many others that follow them, have a visionary intensity in their simplicity utterly distinct from the casual realism of the other autobiographies.

Newman takes Kingsley's attack as part of the providential plan to acquit him, and thus his narrative never deflects to focus more on others than on himself. In his sharp and concise discussions of important people in his life, virtually every detail works to establish his own authenticity, no matter how loving or deferential he may be. He does feel obliged to explain, in keeping with conventions of gentlemanly reticence, "how I came to write a whole book about myself" (p. 1). But the book *is* about himself, an "apologia," and Newman confesses that he had waited many years for the opportunity to write it: "I had long a tacit understanding with myself, that, in the improbable event of a challenge being formally made to me, by a person of name, it would be my duty to meet it" (p. 2). Self-abnegation would—in Newman's strategy—be an immoral default of duty, to himself and particularly to his church.

In contrasting the *Apologia* with the other autobiographies, I am not suggesting that the three others are somehow identical in rhetoric and form. But they do have common differences from Newman's in their almost unselfconscious secularity, their bias toward naturalistic explanations, and their minimizing of their own powers, with a totally shared sense that anyone might have done it. Only hard work makes the difference. The point can be made through a comment Darwin made in a letter of 1877: "Trollope in one of his novels *[The Last Chronicle of Barset]* gives as a maxim of constant use by a brickmaker—'It is dogged as does it'—and I have often and often thought that this is the motto for every scientific worker."[14] Odd though it may appear at first to find Trollope providing a maxim to guide all scientists, it makes sense within a uniformitarian context inimical to Newman's. (Ironically, moreover, within Trollope's narrative the advice is somewhat inappropriate to Crawley's already destructively dogged nature.) Hard work, that is, a certain kind of willed and hence moral commitment, is the key to what is often thought of as genius, which is available to anyone who puts his or her mind to it. Such strategy is most notorious in Mill's *Autobiography*. Darwin certainly indulges it, calling himself a poor critic with a weak memory and "no great quickness of apprehension or wit."[15]

Most famously, of course, we have that painful sequence in Mill's autobiography in which, after discussing his terrifying and dehumanizing education, he comments about his early achievements: "What I could do, could assuredly be done by any boy or girl of average capacity and healthy physical constitution: and if I have accomplished anything, I owe it, among other fortunate circumstances, to the fact that through the early training bestowed on me by my father, I started, I may fairly say, with an advantage of a quarter of a century over my contemporaries."[16] The governing moral assumption of Mill's education thus became part of his belief about himself and is reflected in the bathetic impersonality of the whole book. To the modern reader, the "fortunate circumstances" are appalling, and it seems clear that they almost annihilated Mill's sense of himself.

The assumptions that govern Mill's rhetorical strategies, and Darwin's as well, are exposed in a letter Darwin wrote to Francis Galton, whose book *Hereditary Genius* was to argue for the *inheritance* of the qualities necessary for success. "I have always maintained," Darwin says, "that, except fools, men did not differ much in intellect, only in zeal and hard work"[17] "It is dogged as does it." Galton did, apparently, "convert" Darwin, but the autobiography does not reflect the conversion.[18] There is a striking moment in Darwin's *Descent of Man* that suggests how Darwin's view moralizes genius and eliminates it as a distinctive category: "Genius," he says, "has been declared by a great authority to be patience; and patience, in this sense, means unflinching, undaunted perseverance."[19]

This mode of denying the self's genius might be taken as a form of uniformitarianism, which constitutes a central aspect of the narrative of scientific epistemology as it works itself out in the nineteenth century. The strong self, the inspired self, the conventional hero—these are consonant with what geologists might have called a catastrophic imagination of the world, or with transcendental intuition. Nineteenth-century realism displaces traditional narrative protagonists with a new domestic heroism that is embroiled in the secular quotidian and cannot call itself heroic. The realist hero belongs to a world in which gradualism reigns.

Ironically, as the culture's new heroes emerged from science, science was itself participating in the deprecation of heroism. For one thing, there is nothing mysterious about science. Available to everyone, it is simply, as Huxley was to claim, "organized common sense."[20] One of the phenomena characteristic of the new propaganda for science, initiated in a way by Bacon, is that the scientist becomes a moral hero by virtue of his powers of self-renunciation and dedication to his subject and to the "race" that his discoveries will redeem. "Sedulous attention and pains-taking industry," claimed Samuel Smiles, writing about scientists, "always mark the true worker. The greatest men are not those who 'despise the day of small things,' but those who improve them most carefully."[21]

As against the Carlylean and Hegelian narrative of the great man, which was also current and which, surprisingly, was often made to overlap with a uniformitarian reading,[22] the great achievements of Mill, Darwin, and Trollope are to be taken as the result of "causes now in operation," that have *always* operated and forever will. The guiding hand of God disappears from nature, even if Charles Lyell was reluctant to drop it, and in his place myriads of small natural causes shape the world. The world is not planned but the result of mere increment—and this is the world of the autobiographies. As within stories by Samuel Smiles, that which produces what looks like greatness is only persistence in minute particulars: Darwin's thorough and minute observations of the natural world; Trollope's unremitting commitment to write an allotted number of words every morning, barely pausing from one novel to another; Mill's steady work under his father's direction. The authority of uniformitarian science entailed—if it were to be consistent—a fundamental denial of genius and heroism, the traditional models for virtue and knowledge. With the death of the authority of God and the authority of the text, science became or at least aspired to be the new moral exemplar, asserting its authority through a nature doggedly uniform; and its most effective mediators were the keenly trained hardworking scientific observers, who, like the nature they studied, were products of gradual and persistent development.

Although these uniformitarian assumptions were not universally held by Victorians, their pervasiveness was remarkable—especially given the power of alternative and popular non- or antiscientific positions like those of Carlyle. Nevertheless, when uniformitarianism appears blatantly, amidst the various pyrotechnics of Victorian rhetoric, it seems odd, flat, and banal, seems to be lacking precisely that quality it was invented to assert—intellectual authority. The genius of the realists and of the autobiographers I am talking about here can be understood in the context of this new assertion of secular authority through uniformitarianism. Even Mr. Micawber's apparently catastrophic view of the world is built on gradualism. His claims that sixpence makes the difference between happiness and misery is exactly parallel to the geological claim that, for example, the great mountains of the world, some sublime and astonishing in their mass, were the result of millions of years of events still going on, lifting the land sometimes a few inches, sometimes a few feet.

The border between catastrophe and uniformity can be difficult to detect. The mountains, after all, are vast, and uniformitarianism and catastrophism are simply two different ways to account for an agreed-upon phenomenon. Much depends on what one defines as gradual—is it an inch, is it a foot, or is it a mile? And does it make sense to think of an earthquake as "gradual," when it seems in the experience so clearly catastrophic? The self-deprecation of the autobiographers might, analogically, be taken as a rhetorical device meant to imply the specialness it literally denies. But in their lives and writings, self-effacement again shows itself to have both moral and epistemological bearings: its effectiveness as a means to authority depends on a culturewide assumption that knowledge is attainable only as investigators are willing and able to control and limit the self, to refuse it its tendency to impose itself on nature and the world. If the Victorians were increasingly alert to the ways in which knowledge was contingent on perception, they were—sometimes almost in compensation for that contingency—dogged in their determination to minimize that contingency by repressing the self.

The authority of the scientific model was built on faith in the possibility of secular disinterest and in the uniformity of nature. I use the word *faith* advisedly, for, as Stephen Jay Gould has shown, Lyell's uniformitarian geology, which was so important to Darwin's intellectual development and to his theory, affirmed certain unprovable assumptions that in effect accorded to science ultimate authority in matters of secular knowledge. For Lyell and most working scientists in the nineteenth century, natural laws were constant through space and time, and all phenomena could be accounted for by causes now in operation. The view that they could be so accounted for—"actualism"—made part of a concentrated quasiprofessional attempt to exclude religious or spiritual intervention in matters of secondary causes, that is, of the natural world. The assumptions were heuristic. As Gould has put it, "to proceed as a scientist, you assume that nature's laws are invariant and you decide to exhaust the range of familiar causes before inventing any unknown mechanisms."[23]

While the idea of the uniformity of nature was often taken as support for the view that the world was "designed," it also made explanation of nature independent of the world of spirit, and thus eroded the need to see God's hand everywhere. It became less necessary to invoke God to account for anything that science might account for; the moral "scientific" thing to do was to postpone invoking God until all other possibilities had been tried. In the world of secular knowledge, uniformitarianism led to the dominance of scientific method that also seems to dominate the modes of self-representation in these autobiographies.

Lyell's uniformitarianism smuggled in other meanings, some not universally accepted by other scientists and, according to Gould, not legitimately inferable from the assumptions of the uniformity of natural law or from the empirical evidence. Lyell, like Darwin after him, claimed that nature does not make leaps: it moves gradually and at constant pace. This assumption excludes miracles, but excludes also *any* radical or abrupt changes. Thus, uniformitarianism, establishing the authority of science by excluding divine intervention, also by implication excludes romantic or heroic narratives. Great voluntary changes are not possible in a world where the mechanisms of change are ineluctable and unintentional. In George Eliot, this view of history explicitly denies the possibility of revolution, as it justifies the narrative patterns of these autobiographies, where causal explanation dominates.

Uniformitarianism is part of a cluster of ideas and attitudes that evokes in narrative a realism intent on the minutiae of domestic life and on the possibility of detached and therefore disinterested registration of phenomena, and in politics a (slow but inevitable) movement toward democracy, or at least meritocracy. For contemporary theory and criticism it is a commonplace that all of these movements are linked to the development of bourgeois individualism, and certainly these three autobiographies are, each in its own way, a kind of celebration of hard work, industry, individual effort, while they imply a bemusement at the fact that these virtues *are* rewarded. All three writers affirm that there is nothing that any of them achieved that anybody else, sufficiently dogged, could not duplicate. But the most effective mediators of these values and these ways of working were the keenly trained, hardworking scientific observers, who, like the nature they studied, were products of gradual and persistent development.

The scientist's doggedness requires self-effacement, and the scientific autobiography demonstrates the intellectual and moral necessity of eliminating the self in a way that extends beyond mere modesty or a sense of social propriety.[24] But then, how write an autobiography that eliminates the self describing itself? To write one's life is either to believe in its importance or to recognize that others believe in it. Fame and achievement are the occasion and justification for dwelling long upon oneself in public—that, or narcissism. Yet all these writers act at least a little surprised at the attention they have attracted in their lives.

The most literal example of self-effacement comes in a passage that first impelled me to write this book. At the very end of the very first paragraph of his autobiography, Darwin explains, "I have attempted to write the following account of myself as if I were a dead man in another world looking back at my own life. Nor have I found this difficult, for life is nearly over with me."[25] Here Darwin turns scientific objectivity into literal narrative. In so doing, he implicitly raises the kind of human and moral questions that do not emerge immediately from formulations couched in the impersonal language of epistemology. An autobiography, even if it claims disinterest, can never be entirely impersonal; and even this purportedly "styleless" assertion, affirming the affectlessness of the argument, abrupt and no nonsense and therefore "scientific," implies a sadness it does not express. As a consequence, the passage both embodies scientific epistemology—I can get at the true story now because I am virtually dead anyway—and implies a critique of it.

A perfect example of the "dying-to-know" narrative, it suggests that only in death can one discover what life has meant. The *Autobiography* is not theoretical, but whatever Darwin might really have had in mind when he set out to tell his story to his family, the narrative style and structure (or structurelessness) are conditioned by fundamental epistemological assumptions. Epistemology and narrative inform each other and thus embody a literal impossibility: Darwin speaks from "nowhere," as a dead man. The banishment of self, which is affectless in philosophical discourse, emerges here—even in language almost flatly matter-of-factual—charged with feelings and experiences, expressed or unexpressed, giving authority to (and putting to question) Darwin's self-description.

Not each book begins with death, but each writer chooses to look at himself with scientific detachment, so that all three books echo with a sense of self-alienation similar to what we have already seen in Descartes's excavation of the "cogito." It is there in Mill's rationalist austerity, in Darwin's almost neurotic humility, and in Trollope's self-deprecation. As they try to explain themselves, each becomes a distanced narrator of his own life (though Darwin often assumes the voice of a father talking to his family) and separates himself from himself just as social scientists must avoid any personal sympathy with their human subjects. Indeed, only Darwin's autobiography is consistently about himself—written, as he says, to "amuse me" and because it "might possibly interest my children or their children."[26] Possibly! The tone is characteristic: not only does Darwin write as though there is some question whether he is an interesting subject for posterity, but he seems uncertain about whether his own children would be interested. Mill does not even allow that the book is about himself: he claims to be testing out a theory of education. "It may be useful," he says, "that there should be some record of an education which was unusual and remarkable." The tone and meaning of the book's second sentence are touching in the conviction of Mill's lack of respect for himself: "I do not for a moment imagine that any part of what I have to relate can be interesting to the public as a narrative, or as being connected with myself."[27] Note, too, that Trollope's book is called *An Autobiography*, the indefinite article suggesting no significant identity at all, just one thing among many. And he begins by making it clear that he writes not of himself but "of what I, and perhaps others round me, have done in literature; of my failures and successes such as they have been, and their causes; and of the opening which a literary career offers to men and women for the earning of their bread."[28]

Trollope's self-dismissal begins on the first page as well. He explains that he feels uneasy even about writing of "so insignificant a person as myself," and he claims that he calls the book an autobiography "for want of a better name." "The garrulity of old age, and the aptitude of a man's mind to recur to the passages of his own life, will, I know, tempt me to say something of myself,"[29] he apologetically notes—as though an autobiography ought to contain anything *but* a discussion of the self writing it. All three self-portrayals virtually disintegrate into collections of facts (Trollope sometimes sounds like an accountant), records of books written, or, occasionally, acquaintances made. Mill's leaves the drama of early education and crisis, but proceeds merely to tick off his intellectual activities so that the last twenty-nine years before the writing can be put in only very small compass—all, Mill claims, that "is worth relating of my life."[30] Although Mill's autobiography has been assimilated into the typological tradition of the spiritual journey, that assimilation is made possible by ignoring the way the book dwindles into not much more than a list. The work replaces the man who writes, as Mill seems to ignore the very culture of feeling that his crisis taught him was essential to his life.[31]

"Truth" trumps art and the display of self-concern art might imply. "Nothing I say shall be untrue," Trollope affirms.[32] Against the romantic celebration of uniqueness and of feeling, Darwin, Mill, and Trollope return to the ordinariness of uniformitarianism. Uniformitarianism explains great achievements as an accumulation of ordinary ones. Not holier than thou, these great writers and thinkers are perhaps barely as holy; Trollope implies his superiority only in his willingness to be honest about his banality. Their virtues, then, are even less in their claims for themselves or in their living achievements than in the manner they have found for talking about themselves.

Of the three biographies, however, only Mill's confronts directly and self-consciously the problem of the relation of scientific knowledge to the condition of being a self who is something other than the passion to know. His famous "crisis" might be taken as a quite literal consequence of the attempt to embody epistemological pursuits, for Mill's notorious childhood education was, as he explains it himself, anesthetic.[33] No need here to recapitulate yet once more the most famous nervous breakdown in (at least intellectual) history. But as a metaphor for what happens when epistemological theory drops into narrative it is worth reconsideration.

The problem for Mill, framed in a sadly austere and rational language, is that having developed an understanding of the world through his amazing training in classical, mathematical, poetical, historical, and Benthamite materials, he suddenly became aware that none of this knowledge has any affect for him. If, he asked himself, everything that he knew he should want, on the basis of Benthamite-James Millian thinking, were to come to pass, would he be happy? And the answer came back, no. If I may push my own metaphor, it is as if he were a dead man looking back on his own life—the knowledge is there, but it does not matter. In this deeply personal and troubled form, Mill reproduces the romantic critique of the narrative of dying to know that is developed throughout the nineteenth century, not least by the novelists I will be discussing in future chapters. That is, the detachment required to achieve it is inhuman—the scientist would peep and botanize on his mother's grave. But when the activity of peeping is put to question, then the very pursuit of knowledge becomes not only inhuman but pointless.

Mill's narrative, then, raises the critical question inherent in the paradox of "dying to know": there is a passion requisite to knowledge, but to acquire the knowledge one must resist the passion, one must dehumanize oneself. Can the feeling and the knowledge coexist? This, as I will show in chapter 8, is the crisis of George Eliot's *Daniel Deronda*. In narrative, and at the foundations of epistemological theory, it does coexist. But to imagine a life in which knowledge just *is*, in which the affect is left out, is to imagine a kind of death. Mill's problem, then, becomes the problem of this book: how to refill knowledge with feeling, how to make knowledge matter, how to acquire it without either dying or becoming monstrous and inhuman like the intellectual villains of much science fiction.

The upshot of Mill's narrative remains strikingly ambivalent. He treats with flat, uninflected, and disinterested prose the virtually inhuman demands his father made on him, and what is moving about his description of his breakdown is the sheer rationality of the description. He turns back to Coleridge, but in no simple swerve from empiricism to idealism. For Coleridge, knowledge becomes historical:

[the] Germano-Coleridgean doctrine expresses the revolt of the human mind against the philosophy of the eighteenth century. It was ontological, because that was experimental; conservative, because that was innovative; religious, because so much of that was infidel; concrete and historical, because that was abstract and metaphysical; poetical, because that was matter-of-fact and prosaic.[34]

Coleridge was important to the empiricist Mill because he helped Mill understand that ideas survive only through history: "the long duration of a belief," Coleridge thought, "is at least proof of an adaptation in it to some portion or other of the human mind" (p. 261). To fill knowledge with feeling, Mill had to turn not merely to ideas, but to the *story* of ideas as they work through real minds of real people. His well-known turn to Wordsworth and poetry then helped him humanize the affectless, purely rational aspects of the utilitarianism he had been taught.

But "to know that a feeling would make me happy if I had it," Mill famously writes, "did not give me the feeling" (pp. 83–84). And the prose of the autobiography is a symptom of the doubts that survived past his ostensible recuperation into feeling. While the autobiography affirms a new conjunction of thought and feeling that romantic thought provided, its manner and its argument seem still to require repression of feeling for the sake of knowledge. Or perhaps they reflect his continuing incapacity to feel what he thought he should be feeling. Mill's refusal to complain about his education and his dispassionate critique of his father suggest that his efforts to bring thought and feeling together were neither complete nor entirely coherent. The Mill of the *Autobiography* comes through his crisis still burdened by the model of self-annihilation.

The autobiography embodies the idea that intellect is powerless without feeling. Latent in his therapeutic insistence on the compatibility of feeling and thought is a dualism that ironically ends by reenforcing the amazingly persistent objectivist scientific ideal. Mill claims that the emotion "an idea when vividly conceived excites in us, is not an illusion but a fact, as real as any of the other qualities of objects" (p. 91); and, he argued, "The intensest feeling of the beauty of a cloud lighted by the setting sun, is no hindrance to my knowing that the cloud is vapour or water, subject to all the laws of vapours in a state of suspension" (p. 92). But even as he attempts to save the feeling, the rendering of the fusion as a problem keeps the two perceptions utterly distinct. Feeling is preserved only because it is made consonant with scientific naturalism.

So Mill's story pathetically and sometimes tediously becomes evidence for the ultimate severance of feeling and knowledge. It is not only that he argues—against the particularity of pain and loss caused by his own education—that his education might, with some qualifications, be taken as a model; but also half the book steers pretty consistently away from feeling and from the personal. His written life becomes his public achievements.

Yet there is a tentative and unfulfilled movement toward a revision of the narrative of scientific epistemology latent in Mill's discussion of Harriet

Taylor. Mill's is the only one of the three autobiographies that connects a woman with his *public* life. He attributes to Harriet not only superiority of moral power, but superiority of intellect. The third stage of his mental development "was essentially characterized by the predominating influence of my wife's intellect and character."[35] None of the other autobiographies describe women as anything but ornaments or supporters. Yet Mill's unusual move only strengthens the pattern of scientific knowing that characterizes the narrative as a whole.[36] In a remarkable reversal that many later, usually male commentators have refused to credit, Harriet Taylor is described as having what were usually thought of as the "masculine" virtues of intellectual discipline and personal modesty (so modest, indeed, that the world "had no opportunity of knowing"[37] her). Harriet, then, becomes the ideal knower—appropriately, I would argue, a woman since in the mythos of the culture, women had all the moral qualities the tradition takes as specially scientific.

But the ideal simply repeats the dying-to-know narrative, with feeling, for Harriet's superiority allows Mill to disappear once again. He dies, as it were, to Harriet, finding in her knowledge what he could not achieve himself. Yet I am sure that this hyperbolic minimizing of self was no mere rhetorical gesture but deeply felt and believed. What Mill found in Taylor reflects what he felt to be absent in himself. So he describes her remarkable intellectual gifts but adds:

> [these] did but minister to a moral character at once the noblest and the best balanced which I have ever met with in life. Her unselfishness was not that of a taught system of duties, but of a heart which thoroughly identified itself with the feelings of others, and often went to excess in consideration for them, by imaginatively investing their feeling with the intensity of its own.[38]

Admiring intensely the naturalness of Taylor's feelings, Mill nevertheless maintains that radical dichotomy between thought and feeling that he, the romantics, and his Victorian followers were ostensibly trying to break down.

Mill's autobiography is a dramatization, perhaps suffering from the fallacy of imitative form, of the inhumanity of the paradox at the center of the epistemology behind the narrative of dying to know. The Cartesian alienation of self from self, of mind from body, is deeply embedded in the mythos of science that emerges in Mill and that had come to dominate in the nineteenth century. As of the Newton described by William Whewell, the nineteenth-century scientist might be said to have "existed only to calculate and to think."[39] The abstraction of mind from body that Descartes describes as method, that Whewell ascribes to Newton, is in these autobiographies made an aspect of self-description. It suggests, too, if we take Darwin's self-underestimation as seriously as he seemed to do, that the ideal scientific stance is "irreconcilable with the common conditions of human life."[40]

Trollope's autobiography provides considerable evidence of the ways this sort of dehumanization in fact assists the novelist in getting at the truth: "I can tell the story of my own father and mother, of my own brother and sister, almost as coldly as I have often done some scene of intended pathos in fiction"[41]—that is, almost as coolly as if they were not human at all. Ironically, the deliberately philistine Trollope takes a position here consistent with the ideal of the dispassionate artist. Trollope dissociates the man who suffers from the artist who creates. The aesthetic ideal is the empiricist ideal, and both point toward a repression if not an annihilation of the author's self. While contemporary critics complained of Trollope's apparently unfeeling narration, the effect of reality depended upon it.

Part of the strain of these autobiographies is that they are ungenerous to themselves; and that of course is part of what the alienation of the dying-to-know narrative entails. T. H. Huxley treats himself in the same way,[42] doggedly unsentimental about himself and refusing inwardness, even when he describe states of feeling: "there is no alleviation for the sufferings of mankind," he toughly claims, " except veracity of thought and of action, and the resolute facing of the world as it is, when the garment of make believe, by which pious hands have hidden its uglier features, is stripped off."[43] It is all here: the absolute, almost religious commitment to secular truth, the sense of the impiety of piety as it disguises ugliness rather than exposes it, the moral vocation of knowledge, and the unnamed (but heroic) voice of the morally passionate, morally indignant writer.

"Resolute facing of the world as it is" characterizes the stance of much late-century narrative and, certainly, this tradition of autobiography. One must not shrink from the unpleasant: this is a feature of Darwin's science and, it would seem, of any honest conception of the self. If one makes one's numerator zero, pain can have no effect and one dares take a good look at the worst. Thus, built into these narratives there is a division that makes them remarkable exemplifications of that doubleness that Feuerbach describes as characteristic of human consciousness. How can they describe themselves honestly, without the corruption of Rousseau's commitment to himself? They need to speak from a perspective that allows them to take impersonally personal pain and that rejects the happy ideal and dream for the sake of the truth that derives from dispassionate observation. In effect, then, they all speak "looking back as if I were a dead man."

As they pursue the implications of the narrative of scientific epistemology, all three autobiographies fail to explain either themselves or their works. What is importantly missing from both Mill's and Darwin's consideration of the powers of the self is any Newmanian or Carlylean deep intuition of the self or of divine connection. Darwin's uniformitarianism would allow for differences, but it would tend to emphasize minute rather than large ones. Intuiting no genius in himself, he maintains that he belongs to the large mass of essentially undifferentiated members of the species. He can and does enumerate the qualities that might, or might not, lead to great scientific achievement like his, but as Olney points out, the dominant tone of the *Autobiography* is "bewilderment."[44]

In the effort to be transparent, mere vehicles for the transmission of truth (about the world, about their own lives), the books create a kind of clandestine drama out of what is not being recognized by their authors, and out of the dissonance between what they can say and what the authors' own stature in the world would seem to suggest. They are thus exemplars as well of a certain fundamental strategy of realism, which has evoked significant criticism from contemporary theorists.

The critique of realism that has dominated modern criticism focuses on the way realism effaces the writer in its concentration on and naturalizing of the "real world." It makes the reader forget that what is being described is being perceived in a singular way; its constructed and local nature is disguised by the illusion of objective transcription. Autobiography almost inevitably brings this problem to the surface because it does not submit to the fiction of an invisible narrator. While the narrators may try to disappear, their names are the very reason for the books' existence. The critique of scientific method similarly emphasizes the way objectivity actually obscures the work of the "observing" scientist in producing a knowledge that aspires to universal validity. This kind of critique issues from the most skeptical of thinkers as well as from those who do not give up on the possibilities of a knowledge that transcends the limits of singular perspective. Evelyn Fox Keller, arguing for recognition of the way science has disguised its particular, and highly gendered, way of knowing, insists that "without mediation, commonality, or intercourse between subject and object, knowledge is not possible."[45] A very different sort of thinker, Michael Polanyi, seeks a "greater objectivity." But he, too, argues that there is no escaping the "anthropocentrism" of our perspectives: the greater objectivity would not counsel "self-effacement, but the very reverse—a call to the Pygmalion in the mind of man."[46] And, to revert to an earlier realist example, Thomas Nagel argues that objectivity by itself can never "provide a complete view of the world on its own." "It is necessary," he says, "to combine the recognition of our contingency, our finitude, and our containment in the world with an ambition of transcendence, however limited may be our success in achieving it."[47]

Such critiques are to the point of these autobiographies. For in my description of them thus far, in my insistence on their self-deprecation, I have lost the paradoxical aspect of the ideal of knowledge I have been discussing. It is clear, for example, that Darwin's actual practice as a scientist only partially corresponds to his self-description as a Baconian fact collector. Far from being passive in the face of a real nature, Darwin collected his facts with a purpose. He claims, for example, that in his first notebook, opened in July 1837, he "worked on true Baconian principles, and without any theory collected facts on a wholesale scale, more especially with respect to domesticated productions, by printed enquiries, by conversation with skillful breeders and gardeners, and by extensive reading."[48] But by March of 1837, as David Kohn and Sandra Herbert show, Darwin "was ready to take up the transmutationist hypothesis."[49] Darwin was able to work out the theory before he had Baconianly sufficient evidence; it might be fair to say that although he in the end had enough evidence to make for probability, he *never* had evidence sufficient to satisfy the Baconian formula. No mere passive collector of what befell, he sought materials that would help him with the theory. His pride in his powers of observation is not, after all, negligible, but more interesting is his quiet but clearly disturbed resistance to the argument that he "has no power of reasoning." "I do not think this can be true," he responds, for the *Origin of Species* is "one long argument from the beginning to the end, and it has convinced not a few able men. No one could have written it without having some power of reasoning."[50]

From Darwin, although even the least self-assertion is qualified and softened, it is difficult not to infer what is also clear from Mill's autobiography, a steely confidence in the validity of his own work and ideas. With Mill, it is not necessary to question the authenticity of his neurotic self-distrust; he is not being merely rhetorical or polite, as a Victorian gentleman should be. But as with the other authors, he writes with a sense of virtuous truthfulness; there is a quietly proud frankness in his rhetoric. The voice of Rousseau can still be heard dimly, affirming the supreme value of the individual, precisely in his individuality. Mill concludes his introductory passage by saying, "The reader whom these things do not interest, has only himself to blame if he reads farther, and I do not desire any more indulgence from him than that of bearing in mind, that for him these pages were not written."[51] Self-deprecation almost slips over into arrogance. And this, too, is the dominant tone of Trollope's autobiography:

> That I, or any man, should tell everything of himself, I hold to be impossible. Who could endure to own the doing of a mean thing? Who is there that has done none? But this I protest;—that nothing that I say shall be untrue. I will set down naught in malice; nor will I give to myself, or others, honour which I do not believe to have been fairly won.[52]

The tone of virtuous frankness expressing humility captures precisely that quality of authoritative self-assertion that can be implicit in the strategies of self-effacement. Certainly, one hears echoes of the opening of Rousseau's *Confessions*. As it turns out there are grounds for self-effacement, for it allows utter confidence in one's power to face the hard facts. Trollope's autobiography can be read not only as an explicit expression of his "ordinariness" and perseverance into the writer's craft, but, from another perspective, as a brilliantly contrived triumph over the humiliations of his own miserable childhood and painful vulnerability. Walter Kendrick argues that Trollope's case for realism entails perhaps the most literary of achievements, the creation of the illusion of reality. There is a paradoxical exploitation of artistic genius in the art that denies genius. In *An Autobiography*, Trollope produces a similar sort of paradox—a forceful denial of aesthetic genius that is made possible only by the presence of a rhetorically forceful artist.[53] The triumph is not in the text but in the fact that Trollope had become so important that such a text could be justified.

The other side of this almost self-destructive self-deprecation, then, is a style that insists without stating it that the author has courage superior to most others. And the courage consists in a dogged commitment to speak truthfully even where the self is alienated in the process. In Darwin, of course, the tone is never arrogant. He takes the middle road, but the note is clear in, for example, a brief passage about his Cambridge life when he says, "Looking back, I infer that there must have been something in me a little superior to the common run of youths, otherwise the above-mentioned men, so much older than me and higher in academical position, would never have allowed me to associate with them."[54] Quietly, Darwin's narrative, too, emerges as a vindication:

> When I left the school I was for my age neither high nor low in it; and I believe that I was considered by all my masters and by my Father as a very ordinary boy, rather below the common standard in intellect. To my deep mortification my father once said to me, "You care for nothing but shooting, dogs, and rat-catching and you will be a disgrace to yourself and all your family." But my father, who was the kindest man I ever knew, and whose

memory I love with all my heart, must have been angry and somewhat unjust when he used such words.[55]

The Darwin who tells his self-deprecating tale triumphs over this judgment. Convinced of his ordinariness, he can look back at his own achievements, perhaps bewildered, and recognize that they were not ordinary. His father was, indeed, "somewhat unjust," but Darwin is too modest and unassuming to put it any more strongly.

There is, then, in the raggedness and understatement of these autobiographies a kind of failure to engage with the reality that they *are* autobiographies, and of extraordinarily distinguished and powerful men. Ironically, they gather their power from their very failures, from the way the form of autobiography itself—the necessary embodiment, or at least voicing, of the selves who write—contradicts or at least puts to question the narrators' self-conception. The self that claims its own annihilation—or abasement—evokes sympathy. Its rigorous observation of the way the world is turns out, perhaps against the explicit intention of the narrative, to be an imposition on the world of a detectable way of feeling.

And the genuine self-effacements in these texts are the instruments of empowerment. Here, too, they run parallel to the self-effacement required in the scientific ideal of objectivity. All of these writers, strong men who achieved celebrity because of extraordinary work, can share a self-confidence that is inextricably part of their self-deprecation. The patterns in any case are similar, and the autobiographies all exploit the ultimate authority of disinterested observation of particulars. Together they help dramatize the connections between the scientific uniformitarian ideal, which is the new condition for authority, and an ideal of self-annihilation. Nevertheless, they also imply not only a latent self-contradiction in the scientific ethos that persists into our time, but an imperial assertion of a new secular authority. The self-annihilation in the ideal of scientific epistemology becomes a kind of ceremony of resurrection, and the implications for the epistemology itself are complicated. If the self is reasserting itself secretly, its duplicity has to be engaged. But one moral of the story is that the self can't do without self-denial; and self-denial can't do without the self. So self-deprecation lapses over into arrogance; the plain style becomes rhetorical after all. The implicit hubris of scientific intellectual imperialism shyly reasserts itself. The complications of the passion to know, dying to know, are inextricably entangled in the life it must renounce.

~ 5 ~
My Life As a Machine: Francis Galton, with Some Reflections on A. R. Wallace

One need look only at Richard Feynman's autobiographical work to realize that not all scientific self-portrayals have been so personally self-effacing as Darwin's,[1] so thick with insecurities as Mill's. I make no claims for the universality of the condition of self-effacement, but only that it marks a very important tradition within the history of scientific autobiographies—particularly in the nineteenth century—and that its strategies are informed by the assumptions of scientific epistemology. The moral, the personal, and the epistemological are unselfconsciously interfused in the ways Darwin and Mill describe themselves. And however much one can hear the notes of an affirming and even imperial self behind the self-deprecations, it is difficult to resist the sense their narratives convey of genuine self-doubt, of charming and generous openness to others. Even if one doesn't much like Darwin's autobiography as a work of literature, it is difficult to come away from it disliking "Darwin," the great man who denies his greatness. The implicit connection between epistemological and humane deference is central to the experience of his narrative.

That connection is not assured. Self-effacement does not inevitably imply moral—as opposed to epistemological—generosity and openness. Another and now more popular notion of the scientific stance—one broached at the end of the last chapter—treats the effacement of self as an arrogant authoritarian assertion, utterly resistant to the appeals of ordinary human feeling. The pathos of Darwinian self-deprecation becomes, in the triumph of science and technology, the cold aloofness of moral and intellectual superiority. That superiority is manifest—not so paradoxically, as I have already tried to suggest—in the strength required to efface self, the willingness to risk something like a personal death for the sake of the larger value of truth. One of the implications of much dying-to-know rhetoric, particularly aggressive in today's new Darwinian discourse—in the writings, say, of Daniel Dennett or Stephen Pinker—is that the power to face the bad news, the willingness to recognize the merely animal and minuscule nature of the human, is an indispensable and haughty virtue for the scientist. Readers who see things differently are, implicitly, frightened, weak, and superstitious. The moral and the epistemological, beginning together, end by threatening to become antithetical.

I want to look briefly at two later scientific autobiographies as further evidence of these contrasting styles of self-effacement. First, Alfred Russel Wallace's *My Life* (1905) has many of the qualities of Darwin's autobiography. If it is more verbose (two hefty volumes' worth) and perhaps less overtly self-deprecating than Darwin, it describes a career that defers with full sincerity to Darwin's intellectual priority and superiority.[2] It is hard to imagine anyone as different from Darwin as was Wallace, in social class and personal style if not in scientific focus. Wallace's work took him into the heart of politics, and his voluminous publications indicate an eagerness to engage openly and aggressively many of the great issues of the time. He was a more stalwart and thoroughgoing defender of natural selection than Darwin himself, yet he broke from Darwin on the fundamental issue of whether this process extended to the development of humans. His deference to Darwin does not preclude a narrative that traces the development of his own ideas about natural selection and shows that he arrived at his conclusions without knowledge of Darwin's work. But Wallace describes natural selection as "the new idea or principle which Darwin had arrived at twenty years before, and which occurred to me at this time" (1 : 361). And he talks of the *Origin* in this way:

> I do honestly believe that with however much patience I had worked and experimented on the subject, I could *never have approached* the completeness of his book, its vast accumulation of evidence, its overwhelming argument, and its admirable tone and spirit. I really feel thankful that it has *not* been left to me to give the theory to the world. Mr. Darwin has created a new science and a new philosophy; and I believe that never has such a complete illustration of a new branch of human knowledge been due to the labours and researches of a single man. (1 : 374)

At best Wallace's claims are Trollopean—that is, a candid facing of achievements it would be mere disingenuousness to deny.

Wallace's "preface" to *My Life* virtually repeats Darwin's opening gestures in his *Autobiography*, indicating that he would not have thought of writing an autobiography if his publishers had not "assured me that [it] would probably interest a large numbers of readers." He confesses that he had "promised to write some account of my early life for the information of my son and daughter, but this would have been of very limited scope, and would probably not have been printed" (1:v). Moreover, the "preface" is full of familiar self-abnegating gestures: like Mill, Wallace tries to generalize the significance of the book, allowing that "a true record of a life, especially if sufficiently full as to illustrate development of character so far as that is due to environment, would be extremely interesting" (1:vi). Again, he realizes that he will be subject to "the charge of diffuseness or egotism," but he trusts that "the frequent change of scene and of occupation, together with the diversity of my interests and of the persons with whom I have been associated, may render this story of my life less tedious than might have been anticipated" (1:v–vi). But perhaps the most convincing rhetoric of self-abnegation comes almost casually much later, as when he confesses, "It never occurs to me at any time to talk about myself, even my own children say that they know nothing about my early life; but if anyone asks me and wishes to know, I am willing to tell all that I know or remember" (1:410–11). James Marchant talks of his friend Wallace's "innate reticence and shyness of manner," which was "noticeable throughout his life."[3]

While Wallace charmingly and ingenuously notes that "it never occurs to me at any time to talk about myself," in Francis Galton's *Memories of My Life* (1908) the Darwinian charm is almost entirely missing, while the sense of intellectual and moral superiority is pervasive even if never directly announced. Galton's narrative represents another version of the tradition of dying to know, for there is little evidence of the modesty Wallace exhibits. He registers no surprise that he has managed to accomplish so much; while he does not claim for himself the title of genius, he takes no pains to minimize his powers. Like Wallace, Galton evades the personal and the subjective as much as possible. Wallace does so by writing what amount to extensive essays on dozens of subjects brought to attention by his activities, Galton by almost entirely refusing to do more than note some few of the larger events of his life. As D. W. Forrest notes, "Galton was extremely reticent about his personal affairs."[4] I will be concentrating here on Galton because he so clearly manifests the other side of the trope of modesty: the cool, impersonal affirmation of intellectual authority.

The diagnostic mark of the scientific autobiography, as Linda Peterson notes, is that it discusses the self as an object, from an external perspective. Wallace's is relatively more lively than this. He is pleased to be criticized as an "enthusiast." But the curve of his own career is traced characteristically as though it were largely out of his hands and is noted with something like dispassion. Galton's, on the other hand, is determinedly "scientific." His refusal of subjectivity should by now seem familiar. In contrast with Darwin's and Wallace's and yet as self-evidently related to them, Galton's *Memories of My Life* reveals more clearly how the governing assumptions of the late nineteenth-century "scientific" viewpoint encourage the idea of self-effacement, and it suggests how this self-effacement can manifest itself as something other than ingratiating charm—and even become the implicit justification of power and violence. In these respects, a consideration of the assumptions underlying Galton's otherwise pedestrian and unrevealing autobiography can help illuminate my arguments about the scientific ideal of objectivity.

Galton discusses his life so as to imply that the denial of self makes strict scientific sense. His recollections are not driven by desire, like Darwin's or Wallace's, to "interest" his children and their descendants, nor by Mill's ambition to use his education as an example to others. Galton plunges immediately into his narrative, without self-consciousness about the possible vanity of writing about the self, and without apology or expression of purpose. But from the start, with its curious little biographies of ancestors on both sides of his family, the narrative is governed by the naturalistic and hereditarian assumptions and conclusions of Galton's "scientific" work. Wallace, too, imagines himself within the framework of hereditarian and environmental determinism, but as Wallace turned from Darwin on the matter of human evolution, and therefore on the issue of whether human consciousness could be accounted for on strictly material terms, he implicitly rejected a "pure" scientific materialism. The increased geniality and looseness of his narrative is consistent with this difference. Here is how he puts it:

> On this great problem the belief and teaching of Darwin was, that man's whole nature—physical, mental, intellectual, and moral—was developed from the lower animals by means of the same laws of variation and survival; and, as a consequence of this belief, that there was no difference in *kind* between man's nature and animal nature, but only one of degree. My view, on the other hand, was, and is, that there is a difference in kind, intellectually and morally, between man and other animals; and that while his body was undoubtedly developed by the continuous modification of some ancestral animal form, some different agency, analogous to that which first produced organic *life*, and then originated *consciousness*, came into play in order to develop the higher intellectual and spiritual nature of man.[5]

Although Wallace is unembarrassed about his class origins and will discuss connections between his class and his work, for the most part the autobiography does not explore the relation between his science and his politics. His focus is primarily on the ideas, and he is not eager to press conflicts, certainly not with Darwin. In Galton's case, however, there is explicit connection between work and social objective, the strong utopian program that came to be known as eugenics. As Ruth Cowan Schwartz has noted, Galton had decided, by 1865, that "what is wrong with contemporary society . . . is that the evolution of man has not kept pace with the evolution of technology and social institutions."[6] Eugenics was to be the means to cure the world's ills. Behind the detached and fact-laden narrative of *Memories of My Life* lies a strong moral commitment that turns Galton's life into an example of his own scientific discoveries and evidence for the validity of his eugenic theory. Thus the narrative, like Darwin's and Mill's, is connected with personal humility and high moral intention, while its shape and texture derive logically from the faith of late-Victorian science in regularity of causal sequence, the explanatory powers of materialism (even when practitioners are not, strictly, materialists), the "geological" vastness of time, and the possibility of sciences of the human.

The faith of scientists both in causal sequence and in human as subject provoked major moral and intellectual tensions. It is true that by the end of the century, a tradition of questioning cause-and-effect philosophy, as Carlyle had dubbed it, was strong enough to culminate in Einstein's theories. But the dominant scientific style continued almost to the end of the century to resist the idea that regarded causation as far too complex to be reduced to isolable streams of cause and effect. Debates over the mind/body problem were only one of the manifestations of those tensions. Wallace's decision that human consciousness fell outside the processes of natural selection was another. Scientific materialism—apparently dependent on the cause-and-effect pattern that, for example, Charles Lyell's insistence on "causes now in operation" supported—had profound consequences for religion, morality, and social science. Here epistemology and ethics become inextricable again. As we have seen, from the perspectives of scientific naturalism, the highest morality lay in being firm and unflinching in one's willingness to accept the bad news that all things might be accounted for in material terms, by causes now in operation, without recourse to the spiritual or the divine. Ironically, of course, this view was also a "faith."

The textures of the "scientific autobiographies" I am considering depend on how each writer works out the moral problem of determinism that follows from the conjunction of faith in causal sequence and in the scientific accessibility of the human subject. Each narrative reflects the degree to which the writer accepts a strictly biological explanation of human behavior as opposed to a cultural and social one. Darwin's vision of the world almost required such confrontation, which he evades in the *Origin* only to face it squarely—if perhaps not too effectually—in *The Descent of Man*. But well before Darwin had thrust the human into uniformitarian nature, naturalistic determinism had become a powerful intellectual force. As we shall see in chapter 6, determinism became a liberating faith for Harriet Martineau. In the novels of George Eliot, particularly *The Mill on the Floss*, faith in causal sequence and the power to learn from it is a condition of moral strength and is possible only through annihilation of self. In her essay on Riehl, George Eliot makes the point clearly and representatively:

> The great conception of universal regular sequence, without partiality and without caprice—the conception which is the most potent force at work in the modification of our faith, and of the practical form given to our sentiments—could only grow out of that patient watching of external fact and that silencing of preconceived notions, which are urged upon the mind by the problems of physical science.[7]

This formulation, so entirely unproblematic in its affirmation, is an almost perfect statement of the conjunction of epistemology and ethics in the reigning scientific ethos. The deterministic vision derives from silencing the desiring self and issues in a new faith in which self-effacement will be validated. But determinism was unproblematic for few, and George Eliot, who found it philosophically convincing, found it also emotionally disturbing.

For Mill, of course, it began as imprisoning:

> During the later returns of my dejection, the doctrine of what is called Philosophical Necessity weighed on my existence like an incubus. I felt as if I was scientifically proved to be the helpless slave of antecedent circumstances; as if my character and that of all others had been formed for us by agencies beyond our control, and was wholly out of our own power.[8]

But he is finally relieved from that incubus by perceiving that "what is really inspiring and ennobling in the doctrine of freewill, is the conviction that we have real power over the formation of our own character; that our will, by influencing some of our circumstances, can modify our future habits or capabilities of willing." This, he discovers, is "entirely consistent with the doctrine of circumstances."[9] The belief in that consistency led to Mill's interest in the project of August Comte's positivism, and to the fascinating Sixth Book of *The System of Logic*, in which he argues for a science of man that, extending the methods of the physical sciences and even exploiting the principles of determinism, will allow Mill to complete the project of his father, who, it is said, had determined to make the workings of the mind as clear as the road to Charing Cross. Even the new freedom, for Mill, is only possible within a scientific framework of determinist theory—a kind of Cartesian consequence of the epistemology of self-effacement.

Mill's touchingly abstract liberation has its parallels in Martineau and Wallace. Determinism, remarkably enough, because it allowed for the possibility of control and predictability—scientific regularity—was morally preferable to a conception of free will that entailed unpredictability, and thus the impossibility of science. Free will in that sense would be, as Darwin might have said, "fatal to my theory."

Wallace never seems to have suffered from the abstract doctrine of Philosophical Necessity. Charting his intellectual progress toward science, he notes that Robert Owen provided his introduction to "advanced views." Owen's "fundamental principle, on which all his teaching and all his practice were founded, was that the character of every individual is formed *for* and not by himself, first by heredity . . . and second by environment." Here, as with Martineau, Mill, Galton, and Darwin, philosophy intersects moral vocation, for this view requires restructuring the moral and legal system, which is based on the view that "all men *could* be good if they liked." In a determinist system, people cannot be "*deterred* from future aggression"[10] unless the conditions in which they develop are changed. Hence Owen's "successful" New Lanark. For Wallace, implicitly, the vocation of science and, one might hazard, the nature of his theory, grow from this insight into hereditary and environmental determinism.

What, then, of free will? For Wallace, scientifically unflustered by the need for philosophical consistency, it isn't a problem. "In view of such astounding success as this, what is the use of quibbling over the exact amount of free-will human beings possess":

> Owen contended, and proved by a grand experiment, that environment greatly modifies character, that no character is so bad that it may not be improved by a really good environment acting upon it from early infancy, and that society has the power of creating such an environment. Now, the will is undoubtedly a function of the character of which it is the active and outward expression; and if the character is enormously improved, the *will*, resulting in actions whether mental or physical, is necessarily improved with it. To urge that the will is, and remains through life, absolutely uninfluenced by character, environment, or education, or to claim, on the other hand, that it is wholly and absolutely determined by them—seem to me to be propositions which are alike essentially unthinkable and also entirely opposed to experience. To my mind both factors enter into the determination of conduct, as well as into the development of character, and for the purposes of social

> life and happiness, a partial determinism, as developed and practised by Owen, is the only safe guide to action, because over it alone have we complete control. (1:91)

Part of the resolution here is like Mill's: the will is involved in the causal sequence that determines us. But Wallace also talks of a "partial determinism" (which is consistent with the stochastic model of natural selection), a theory that might seem oxymoronic. Any chink in the causal armor makes it vulnerable. Darwin's science could not live, he believed, if the naturalistic regularity of sequence were violated *anywhere*. (It is just this chink in the determinist armor that keeps Wallace from eugenics and—in its absence—drives Galton to it. To this moment, the question of whether environmental conditions or human choice can influence behavior remains controversial, scientifically and politically. Eugenics implies a powerful material determinism, a "natural selection" that leaves no space, as Wallace believed it did, for human choice.)

While Wallace defers to chance in amusing ways[11] and exhibits a Darwinian modesty in relation to his career, he sees a pattern of happy accident that implies something other than mere material causation. For himself, he can argue that

> many of the conditions and circumstances that constitute our environment, though at the time they may seem unfortunate or even unjust, yet are often more truly beneficial than those which we should consider more favourable. Sometimes they only aid in the formation of character; sometimes they also lead to action which gives scope for the use of what might have been dormant or unused faculties (as, I think, has occurred in my own case). (1:198)

But often, he says, those circumstances are not favorable, and if they consistently lead to bad consequences, "the system of society" is at fault. Wallace's willingness to accept inconsistency, to refuse the totalizations of system making, marks his autobiography and his scientific life, and surely was consistent with—either as cause or effect—his own strong leanings toward socialism. Galton's eagerness to accept totalization is similarly evident. For him, determinism was not partial. He sought no loopholes and accepted unequivocally the scientific naturalists' faith.

John Tyndall and T. H. Huxley were the clearest and most pugnacious propagandizers for the faith of scientific naturalism.[12] From their formulations—with which Galton seemed to identify himself—it is possible to infer how these "scientific" assumptions feed into and derive from the tradition of selflessness or self-effacement. Tyndall, in his notorious "Belfast Address," in which he went as confrontationally far in materialism as any of the scientific naturalists would ever go, finds language that is designed to evoke the tragic—a move that often follows from materialism. In making the case for causal inevitability, he claims that the doctrine of the conservation of energy

> 'binds nature fast in fate' to an extent not hitherto recognised, exacting from every antecedent its equivalent consequent, from every consequent its equivalent antecedent, and bringing vital as well as physical phenomena under the dominion of that law of causal connection which, so far as human understanding has yet pierced, asserts itself everywhere in nature.[13]

Galton, unlike Wallace, includes "vital" as well as "physical" phenomena under this severe regimen. Tyndall and Huxley were both careful not to call themselves materialists, arguing that materialism is as much a metaphysical conception as idealism. But in their efforts to explain the world and scientific method, they either accept a material explanation or halt where, they believed, human consciousness can penetrate no further.

Thus Huxley was comfortable with the idea that material explanations of natural phenomena—and this includes all vital phenomena—are all we need. And materialism—challenging as it does the efficacy if not the reality of the "spirit," raises crucial questions about the possibilities of selfhood and agency. The scientific laws of nature, extending to the human, insert the human into a system in which no volitions can bring about material changes. Huxley explicitly argues that "consciousness" cannot be the cause of physical action. The individual human is only part of a larger system of natural relations (Huxley abjured the idea of "natural laws"), in which that (human) part is often falsely perceived to be an active agent.[14]

"The problem of the connexion of body and soul," says Tyndall, "is as insoluble in its modern form as it was in prescientific ages." But he also argues the "extreme probability" that "for every fact of consciousness, whether in the domain of sense, of thought, or of emotion, a certain definite molecular condition is set up in the brain."[15] Huxley, insisting on the primacy of scientific over all other forms of knowledge, also recognizes the cost to the scientist of that primacy. Although he rarely uses a rhetoric of sacrifice, even at his most dogmatically exuberant he affirms the delusiveness of our sense of ourselves as separate spirits:

> We are conscious automata, endowed with free will in the only intelligible sense of the much-abused term—inasmuch as in many respects we are able to do as we like—but none the less parts of the great series of causes and

effects which, in unbroken continuity, composes that which is, and has been, and shall be—the sum of existence.[16]

Our individual selves passively symbolize the material conditions of the brain and body, but do not change the material world in any way. The self becomes a fantasy of the brain, which is driven by the laws of nature, not by our own agency. As happens frequently, and nowhere more clearly than in Darwin's theory of natural selection, the ideal of objective and selfless observation replicates itself in its discoveries. For Huxley, the disinterested observing self sees a world in which self is merely an illusion and from which it is banished. The self, on this materialist account, is no more a real existence and intentional agent than it was to become for poststructuralist theorists. This strain of thought became prominent among later positivists, such as Karl Pearson, for whom reality itself was only a human construct.

Galton's self-abnegation takes the shape, then, not of an instinctive modesty or a humane deference to others, like Wallace's, but from the other side of the dying-to-know pattern. That is, he takes it as a strong responsibility to face the worst; and he is particularly strong on the idea that human nature is what it is by virtue of inheritance. In a letter of 1869, he makes his position clear in a language familiar these days in evolutionary biology:

> I have no patience with the hypothesis occasionally expressed, and often implied, especially in tales written to teach children to be good, that babies are born pretty much alike, and that the sole agencies in creating differences between boy and boy, man and man, are steady application and moral effort. It is in the most unqualified manner that I object to pretensions of natural equality.[17]

While Darwin, as we have seen, eventually agreed with his cousin, he always believed that it's dogged as does it. Galton unequivocally discounts that view and thus believes very strongly in himself. After all, he and his cousin Charles were both grandchildren of Erasmus. Galton's first major work, *Hereditary Genius*, made a strong case for the way genius runs in families. The kind of humility his self-narrative shows, therefore, is the broader, almost cosmic humility that minimizes environmental and spiritual explanation of behavior. Galton's voice firmly represents that impatience he describes in the above quotation. He rejects the mere sentimentality of the childish worldview that assumes that being moral has any consequences for the nature of one's genius.

This perspective is implicit in *Memories of My Life*. Like Huxley, however,

Galton sees the materialist/determinist conditions of existence optimistically, and it is important to bear in mind that unattractive as Galton's vision may be to the modern reader, it was held at the time as part of a liberal and even utopian aspiration. Materialist determinism—as distinguished from but not entirely unrelated to Calvinist fatalism—was consistently associated in the nineteenth century with liberal and reformist positions, from Robert Owen to John Stuart Mill to Harriet Martineau to Beatrice Webb. The denial of self is the means to a greater good. Galton's autobiography makes it clear that scientific self-denial is connected to the determinism that issues from several "scientific" ideas that we have already encountered: (1) the uniformitarian insistence on regularity of sequence, which eases into determinism; (2) a naturalistic materialism, excluding from its explanations anything that does not issue from the pressures of environment or the effects of heredity—nature or nurture; (3) the view that, as Huxley had put it, "our method must be exactly the same [for vital phenomena] as that which is pursued in any other scientific inquiry, the method of scientific investigation being the same for all orders of fact and phenomena whatever";[18] thus Galton comfortably treats humans—himself included—as objects in the world, available for study like any other parts of nature; (4) a rejection of the idea that chance plays a role in nature: all phenomena and actions are explicable by causes now in operation; and (5) a recognition that, in wider human terms, the world is, like Darwin's, significantly shaped by chance, in the sense that what happens occurs regardless of human intention or agency, but because of forces in nature, not all of which can be understood.

Galton's obsession with statistics allows him to transform individual aberrations into consistent parts of a larger system and thus to reconsider all chance events so that there will be nothing unpredictable or inexplicable. Wallace's response to this crisis of understanding and individual aberration is to allow in something extrascientific, "some different agency," and thus to surrender on the idea that scientific thinking was the only true thinking and the disappearing scientific self the only true authority.

Galton will surrender no scientific authority, and all the views listed in the preceding paragraph are implicit in that dominant trope of scientific self-effacement and of Galton's narrative, which holds together the "scientific" assumptions within a utopian moralistic scheme of things: the unflinching confrontation of pain, the willingness to take a good look at the worst. Such confrontation implies or makes explicit the heroic powers of the observer and makes possible an understanding of the "worst" that allows transcendence of it. Knowing the laws of nature becomes the only means to avoid their worst consequences. But knowing them means knowing not only the insignificance of the self, as in Hardy's *Two on a Tower*, but the disassembly of the self, virtually into a mere machine, or "automaton." Some people might call it being dead.

The prose of *Memories of My Life* is almost entirely unimpassioned, even when it registers what certainly were for Galton very strong feelings, like his affection for his sister, Adele. He allows himself no "enthusiasm" of the sort Wallace defends in himself. Ostensibly, Galton's narrative is a record of facts, and these often tumble over one another without apparent formal or logical relations. The opening chapter is a model of "scientific" narration in its provision of what seems to be a collection of data about Galton's kin and their characteristics. None of them is presented feelingly, and for most there is some cool indication of their accomplishments (oddly selected and enumerated), of their skills, and of their claims to fame. The paragraph about his father's elder sister, for example, ends with an abrupt sentence: "For more, see Dict. Nat. Biog."[19] For an uncle who died young, Galton concludes: "There is a touching notice of him in the *Annual Register*" (p. 11).

The principle of selection according to which these various "lives" are presented becomes evident in the chapter's final paragraph when Galton at last allows himself to become the subject: all the relatives are sources of his inherited characteristics. For the most part, the Galtons and Darwins are presented as talented, successful, prosperous, or cut short before their full promise was realized. Samuel John Galton, grandfather, "prospered in his business as a 'Captain of Industry,' to use the phrase applied to him in a book treating of Birmingham" (p. 4). The grandmother's half-brother, Robert Barclay Allardice, "was a noted athlete and pedestrian, and in later years an active agriculturalist" (p. 5). The list includes his grandmother's half-sister, who was married to Hudson Gurney, antiquary, verse writer, man of large fortune. The maternal grandmother lived to the age of eighty-five and probably bequeathed her longevity to Francis's immediate family. With the distance of the scientific autobiographer, Galton then notes: "My own age is now only eighty-six but may possibly be prolonged another year or more" (p. 7). Darwin's *Autobiography* has at least the charm of family intimacy, but Galton's seems a memoir by the dead.

As he moves to self-description, Galton works more systematically, as he does in his *Heriditary Genius*, or *English Men of Science*. The self-representation is almost statistical: "The general result of the foregoing is that I acknowledge the debt to my progenitors of a considerable taste for science, for poetry, and for statistics; also that I seem to have received, partly through the Barclay blood, a rather unusual power of enduring physical fatigue without harmful results" (p. 11). Perhaps the most striking aspect of Galton's hereditarianism is that all personal qualities are accounted for by biological inheritance. Inheritance determines not only general proclivities of mind or tendencies of the body: an interest in statistics can be physiologically inherited as directly as the shape of one's nose. Each individual becomes a congeries of inheritances rather than a fully constituted self and agent.

Confronted by the question of determinism, Galton makes "a series of observations on the actions of my own mind in relation to Free Will." As Darwin used observation of his children for materials for his scientific arguments, so Galton uses himself. If there is something a little unnerving about Darwin's observation of his children, his manner is marked by an obvious love of his "subject," an entirely human excitement about having a baby and watching it grow. But in Galton's prose, there is no residual sense of a full life. The "self" becomes the object of investigation, evidence for a dispassionate scientific theory. And the result is rather drearily predictable:

> The general result of the inquiry was to support the views of those who hold that man is little more than a conscious machine, the slave of heredity and environment, the larger part, perhaps all, of whose actions are therefore predictable. As regards such residuum as may not be automatic but creative, and which a Being, however wise and well-informed, could not possibly foresee, I have nothing to say, but I found that the more carefully I inquired, whether it was into hereditary similarities of conduct, into the life-histories of twins, or introspectively into the actions of my own mind, the smaller seemed the room left for this possible residuum. (p. 295)

The "conscious machine" echoes Huxley's "conscious automata" and obviously carries with it the freight of Huxley's scientific determinism, but it emerges not in argument but in narrative—and the life and ethical implication go out of it as a result. The displacement of self in the act of observing it produces, not surprisingly, a "scientific" conclusion that the self does not exist as it has been imagined traditionally. One of the consequences of the scientific effacement of the self is that scientific investigations will almost invariably exclude the self from the natural world.

The autobiography treats Galton with the sort of detachment such a view would be likely to produce. Nothing in the prose personalizes this treatment. Everything is discussed with an abrupt, staccato, fact-asserting economy. The prose is clinical. "I think," he says of his early childhood, "I can revive my principal feelings at that early age with fair correctness, their change during growth seeming to have been chiefly due to increased range of mental prospect." The "feelings," if noted at all, are abstractly noted. And Galton glides easily from first to third person. So the passage immediately continues: "The horizon of a child is very narrow and his sky very near. His father is the supreme of beings. He has to learn by slow degrees that there are more and more appreciable stages between the highest and the lowest, and the number of such stages that he can discriminate affords a good measure of his mental calibre at the time" (p. 16). So, both in the autobiography and in his more "scientific" studies of heredity, Galton tends to move from a rigorous concern with statistical method and verification to ostensibly unsubstantiated generalizations. But the generalizations are inseparable from the statistical orientation: statistics are designed to enable generalization, and they do so by blurring individual difference. (Despite the numbing inhumanity of this prose, it will be important to show that Galton wants to avoid the inhuman consequences—already evident in his prose—of transforming individuals into numbers; in fact, he believes in statistics because they can help mediate between a cruel disregard for the individual and a sentimental preoccupation with the individual's fate.) The individual tends to get lost even as Galton attempts to protect the individual by understanding the natural system and to write an autobiography that registers the distinct life and career of Francis Galton without touching on the personal.

His vision is shaped by a transpersonal understanding of natural law. As he puts it on the final page:

> Individuals appear to me as partial detachments from the infinite ocean of Being, and this world is a stage on which Evolution takes place, principally hitherto by means of Natural Selection, which achieves the good of the whole with scant regard to that of the individual.
>
> Man is gifted with pity and other kindly feelings; he has also the power of preventing many kinds of suffering. I conceive it to fall well within his province to replace Natural Selection by other processes that are more merciful and not less effective. (p. 323)

Galton's project is to improve on natural selection. The project of making life less painful to the individual inspires the science. But, he says, "if a man devotes himself solely to the good of the nation as a whole, his tastes must be impersonal and his conclusions so far heartless, deserving the ill title of 'dismal' with which Carlyle labelled statistics." Recognizing the threat of heartlessness, Galton rejects what he takes to be its at least equally vicious opposite: "If, on the other hand, he attends only to certain individuals in whom he happens to take an interest, he becomes guided by favouritism and is oblivious of the rights of others and of the futurity of the race" (p. 322). This is simply to reassert the paradox (and the perils) of the "dying to know" metaphor. That is, despite the disturbing impersonality of his writing and of his scientific project, all of it is driven by the utopianism I mentioned at the start of this chapter, just as modern science was driven by a commitment to improve the conditions of individual humans everywhere. There is a touching disconnect, another version of the epistemological paradox, in the way Galton can think through his attempt both to recognize the need of the general and to save the particular. And this disconnect easily transforms itself into the terrible project of eugenics as it actually would unfold in the twentieth century (partly under the tutelage of his polymathic disciple, Karl Pearson) because the enterprise assumes intellectual authority in the heroic "objectivity" of its procedures. Such authority might easily be used (and was to be used) as a universalizing cover for genocide.

Galton's way of thinking about the good of humankind participates in the perspective and project of natural selection itself. Despite his desire to find a process of human development more merciful to the individual, Galton's utopianism is like natural selection in its diminishment of the individual: a conscious machine, a partial detachment from Being. The project of eugenics is to make a better natural selection by checking the birthrate of the "Unfit" and increasing that of the "Fit"—in effect, reducing the number of individuals that natural selection will have to eliminate. The last sentence of the autobiography contrasts natural selection and eugenics: "Natural Selection rests upon excessive production and wholesale destruction; Eugenics on bringing no more individuals into the world than can be properly cared for, and those only of the best stock" (p. 322). The statistical perspective wins out in a way that makes mercy to the weak individual immoral; true mercy to weakness can only come from the elimination of weakness entirely. Galton does not transcend natural selection so much as try to implement it by decreasing the number of unfit individuals whom it would have to annihilate.

Galton's attempt to find a path between the general good and the individual is the fundamental epistemological (and moral) problem with which this book is concerned. It depends, finally, on his commitment to statistics, which views the individual in relation to the larger group. Galton's great contribution to the social sciences was his development of modes of "anthropometry": the science that measures individuals within the context of larger systems and views their peculiarities as part of larger systems of peculiarities. Many of the pages of the latter part of *Memories of My Life* are given to brief descriptions of some of his anthropometrical methods. His eugenic program depended upon his ability to see peculiarities—Darwin's "variations"—as generalizable: "whatever quality had appeared in man and in whatever intensity, it admitted of being bred for and reproduced on a large scale. Consequently a new race might be created possessing on the *average* an equal degree of quality and intensity as in the exceptional case" (p. 266). Galton thus went on to develop "the method of centiles," a statistical procedure that treats all characteristics to be examined relative to a larger population.

The effect is to diminish the significance of the individual, as scientific method diminished the significance of the individual fact unconnected to some larger system. It is also to disguise, even from the scientist himself, the degree to which the objectively indisputable evidence is shaped by those very passions and desires the system is meant to minimize. In effect, the perspective entails resistance to the sentimental, while it assumes that eugenic transformation requires precisely the qualities of good science: a willingness to face the facts, an understanding of the participation of all individuals in larger systems—facts in laws of nature, humans in the "ocean of Being." The detached, alienated, staccato, matter-of-fact voice of the narrative is the condition of true observation, true understanding, and eugenic reform, as that voice quietly reveals both the degree of self-alienation required by the scientific perspective and the way self-alienation replicates itself in its perception of the world.

It takes a while, in a first casual reading of *Memories of My Life*, to become aware of what a painful world it projects. Most of the narrative proceeds before the theoretical underpinning becomes explicit, as it does in the last chapters. Rather, it is one damned thing after another, gradually unfolding how, as Galton puts it, "the growth of a mind has been affected by circumstances" (p. 45). The mind, of course, is constituted of inherited characteristics; the "circumstances" are the worldly chances that elicit from the mind its potentialities. It is not difficult, upon reflection, to infer the kind of "scientific" theory Galton would articulate, a theory that treats all individuals as observational objects in larger systems of which they are likely unaware, that insists on causal regularity and material determinism. Circumstances conspire to obstruct or, on other occasions, facilitate the natural tendencies of the mind. Galton the object of study is a conscious machine, whose mechanical properties need to be accurately understood if it is to fulfill its highest potential for production. Galton the writing subject is, too, a conscious machine, sufficiently alienated from the immediate experience to write about it objectively. So, pushing himself too hard at Cambridge, his health finally breaks down:

> I had been much too zealous, had worked too irregularly and in too many directions, and done myself serious harm. It was as though I had tried to make a steam-engine perform more work than it was constructed for, by tampering with its safety valve and thereby straining its mechanism. (p. 79)

But the alienation implicit in treating the self as machine is less disturbing than the alienation from feeling that insinuates itself into the extraordinary compounding of physical pain and grotesque suffering throughout the narrative. The reduction of the individual to its nature and circumstances reduces its capacity or need for personal feeling. The world perceived is as insensate as the perceiver. In a way, virtue inheres in the unwillingness to be swayed by feeling—the accidents of inheritance, the pains of circumstance—so that the reality of nature can be perceived and, if necessary, manipulated.

In any obvious sense, there is nothing violent about Galton's life. Indeed, much of the narrative is a rather tedious recounting of acquaintances' lives and scientific work accomplished. In place of intimate descriptions of private life come descriptions of work and travels.

Juxtaposed to Wallace's more abundant autobiography, it seems austere. And as Galton focuses on his travels and works, Wallace seems to focus on everything. His book is packed with information: on the enclosure of land and its consequences, on geology, surveying, local customs, architecture, poetry. Wallace shows himself to have an endless curiosity, and his own history becomes an occasion for discussing all sorts of interesting and miscellaneous facts.[20] But there is a generous and heated ebullience about Wallace's evasions of self-representation, and most of the facts are presented rather warmly, in order to argue for political or moral reform.

Galton's austerity ends in his presentation of moments of rather unpleasant violence. He finds a way to squeeze in more violence and death than seem necessary for his story. It is impressively full of gory details—mutilated slaves, a dead monkey, oystercatchers devoured by a fox as they stand trapped with legs frozen in the ground, slaughtered seals, medical operations. There is no flinching from the most sordid of facts. At one point, discussing "the horrors of savagedom," he describes in a short space a "wretched native whose muscles along the back of his neck had been severed to the bone," the clubbing of an ox thief, the whipping of a culprit (on Galton's command), and he concludes by exclaiming, "enough of these gruesome stories."[21] This is the first indication that "gruesome stories" have been told, and it is no accident, I think, that the recognition of their gruesomeness coincides with Galton's revulsion from the "savage." In the implicit treatment of non-Europeans and women (the cause, in savagedom, of most of the problems), Galton's sense of the superiority of scientific culture is most obviously to be felt, and perhaps most particularly, in its power to be unsentimental.

But violence is part of "civilization" too. The first corpse appears in the second chapter. Galton claims that the momentous step of his life was his father's decision to withdraw him from school so that he might pursue medical studies. And the first event discussed after this momentous step is his observation of a postmortem examination. Here at least the emotional possibilities are not entirely ignored, for the point is clearly the excitement of the event: "Oh, the mixture of revulsion, wonder, interest, and excitement" (p. 23). While there is some restraint in the description, as Galton concedes that the details would be "unfitting for these pages," he does not hesitate to register the swollen features, the "perforated portion of the stomach." And then he dispassionately notes: "Mr. P. pricked his finger while sewing up the abdomen. A dissection wound when death has followed peritonitis is proverbially dangerous. It was so in this case, for Mr. P. nearly died of it" (p. 24).

Later he does not restrain himself from observation of the morning rounds at a hospital. The prose steadfastly resists affect, and the narrator's voice is clinical. He notes, for example, that "[b]roken heads from brawls were common accidents at night; then it was my part to shave the head, using the blood as lather, which makes a far better preparation for shaving than soap. The wounds were stitched together with a three-cornered 'glove needle,' which cuts its way through the skin" (p. 30). I want to emphasize here how scrupulously precise Galton tries to be, how at no point does he indulge the feelings of revulsion that the descriptions might be expected to provoke. It is the voice of a writer committed to "truth," someone who has learned to dismiss the claims of the feeling self. These passages are surely not there to teach the reader about surgery; they are, rather, about Francis Galton and his capacity to submit himself to the pain of the outside world; they are thus about Galton's capacity to efface himself. And their effect, in the self-effacement of the prose, is the paradoxical one of powerful self-assertion.

Why, for example, does he provide quite so much detail about the hospital experience? Why does it make so great an impression on him? Here is a sequence that evokes for me Thomas Pynchon's brutally lucid and affectless description of Esther's nose job in *V*:

> One of these events made a great impression on me. It was that of a man, a small piece of whose skull had been depressed by something falling on his head and stunning him. He was brought in utterly unconscious, with the 'stertorous' or snoring respiration characteristic of such cases. The man had to be trepanned, so the surgeon was sent for. In the meantime everything was prepared for his arrival. The trepan is a hollow steel cylinder with teeth cut out of its lower rim, used to saw a circular wad out of the sound bone nearest to the fracture. A miniature steel crowbar is used to raise the depressed fragment, and a rod to lay across the sound bone as a fulcrum for

the crowbar. I seem to see it all before me as I write. The brightly lighted room, the apparatus in order, the surgeon at work, the eager faces of the bystanders, and the utterly unconscious patient. The wad was cut out, the crowbar adjusted, and still the monotonous snore continued unchanged. Then pressure was put on the free end of the crowbar, the broken bit of skull was raised, and instantly life rushed back. (pp. 31–32)

(It is difficult, in transcribing such passages, not to be aware of one's own—my own—fascination with the brutal details, which I recognize *not* as self-abnegation, but as its opposite. But part of the excitement of registering such horrors is the recognition of the cold dispassion with which Galton seems to claim to have experienced it.) The moral of such passages is clear: only by confronting directly the necessity for unpleasant and violent action can one act mercifully. Natural selection plays itself out in the operating room and the medical professions, and only the dispassionate observer can make creative use of its processes.

The catalogue of illness and pain continues for more than ten pages. Yet it would be difficult to claim anything particularly unusual about the accumulation of such details in discussing medical training. It is only after similarly violent details emerge with equal frequency throughout the first half of the narrative that the pattern emerges. Galton's imagination of himself, as of his world, entails a stony acceptance of the fact of the unpleasant, and a feeling for the moral power and authority to be derived from that acceptance. His description of an escape from accidental death provides another example of the developing pattern. Thrown from an old steamboat and battered by the revolving paddles, Galton found himself underwater, "held back by projecting nails which had hooked into my clothes." With a coolness of prose mimicking the coolness of behavior under life-threatening pressure, Galton writes, "My breath was becoming exhausted, so I passed my hand quickly but steadily all over myself, disentangling nails in two or three places, and then made my last dive for life. I fortunately rose clear, and utilised my former enemy the mass of wood as a raft" (p. 46). Galton rather proudly but in the matter-of-fact voice of the entire narrative describes his rowing to the ship: "My face was, I understood, a spectacle, being painted with blood that had flowed freely from a few scratches." The cool control, refusal to be upset by danger, is the active equivalent of scientific detachment. Through submission of self the self is saved.

I do not mean to suggest that the moments of violence make up more than a small part of the overall text, but throughout the early chapters Galton finds occasion to attend to matters that require a tough stomach. In a chapter largely devoted to well-known acquaintances he made during his

Cambridge years, he tells the fate of his friend Strickland, who died "demented" by hardship during a risky winter walk through the hills and was found almost wholly undressed. Later he discusses how long the dueling spirit survived at Cambridge. But it is in the sections devoted to his travels before he could focus on a vocation that Galton's indulgence in the painful is most prominent.

Cool detachment governs the examples as it governs the entire narrative. "I had shot a desert partridge, but not killed it, so, taking it up, I knocked its head with the stock of my gun. I ought to have cut its throat with my knife, while repeating the Moslem formula" (p. 108). "Youths are murderous by instinct," he claims, "and so was I," and then proceeds to describe seal killing, "the oozing blubber of the animal" making "a circular calm around the spot where he is shot, with the bloodstain in the middle" (p. 113). It is interesting to compare these sequences with Darwin's brief and undetailed admission that as a young man he loved hunting, although he hated the sight of blood and later came to displace hunting with the intellectual life of science. Galton's narrative voice, making no fuss, has the literal and affectless descriptiveness of the more extended and more tedious passages.

Memories of My Life, in its focus on public activities rather than private life, provides a clear example of what it means to embody objectivist methodology and theory in narrative. Galton's commitment to anthropometric study—a commitment that became so valuable to modern psychology and social science—follows from the materialism, determinism, and objectivism of post-Darwinian nineteenth-century science. Similarly, the peculiarities of the narrative—its cool registration of violence, its abruptness of representation of facts, its staccato style and its movement from subject to subject—issue from the same ideal of effacement of the personal. The willingness to face the violence of nature, the commitment to understand nature by means of statistical generalization, the calculations dependent on a recognition of regularity of sequence—all of these govern Galton's science and his life and are part of that stance of objectivity that squelches the sentimental and implies the ultimate virtue of truth seeking and observational accuracy leading to useful prediction. The intrusion of feeling threatens to distort perception and invalidate reasoning. One characteristic passage brings all these qualities together in a way that shows how the scientific vision is indispensable to practical life and to survival. On a caravan across Africa, Galton is concerned with the cattle on which the travelers' survival depends:

> One sheep—they were very lean—afforded twenty meals, and I found that men on full work required two meals daily. An ox was reckoned equal to 7 sheep, and would therefore feed twenty-four people for three days. The

> gross total of oxen, cows, and calves in the caravan was ninety-four; that of sheep was 24. Seventy five of the oxen were broken in; nine of these as ride-oxen and a few others as pack-oxen, the remainder only for draught. I considered myself to be provided for ten weeks, exclusive of game, while still preserving a sufficency of trained oxen. (p. 140)

This is Galton's voice—firm, calculating, scientific, and self-confidently masculine.

In my discussion of Galton and of the other autobiographies, I have emphasized the mutual dependence of self-effacement and scientism. But this peculiar dependence is characteristic of books written by men, and in these books, modesty, self-effacement, and commitment to truth are ultimately also likely to be part of an assertion of authority. The particular insistence on the necessity to remain cool in the face of suffering, and the determination to represent moments of violence and aggression, turns the passive, self-abnegating aspects of scientific epistemology back on the world it seeks to represent. There is a paradox of gender in the scientific ideal of objectivity: the effacement of self, the submission to reality, is a requisite for scientific practice, which was defined almost universally as masculine (there are, of course, exceptions, such as Caroline Herschel or Mary Somerville. The latter, widely admired by male scientists, carefully excluded herself from British Association for the Advancement of Science meetings in order to assure that women did not trivialize the proceedings).[22]

I have throughout been emphasizing the oddity that the proper relation of the male scientist to nature is passive: the objective observer accepts what nature offers, receives what is already out there waiting to be discovered. But the masculine is traditionally associated with aggression and the assertion of self: success in the world depends on a willingness to insist on one's own self-interest. Women are understood to be incapable of success in the world because they lack the power to fight adequately for their self-interest and, perhaps, the endurance to withstand the sorts of violence that Galton describes so matter-of-factly. Oddly, then, in the world of business and making one's way, men are self-interested and empowered, women are self-submissive and suited only for the domestic. In the world of science, men are disinterested and empowered, women are self-preoccupied and ineffectual. There is also, of course, the question of strength: Galton's is the book of an overtly strong man, submitting by virtue of a willingness to think and act unsentimentally in relation to the moral violence of the world.

This is not to suggest that the self-effacement in these male autobiographies is merely disingenuous. Rather, I would argue, it is precisely because people like Galton and Darwin, even, to a lesser extent, Trollope and Mill, are empowered by virtue of their masculinity and their social positions and professions that their intense self-effacement becomes rhetorically effective. Each begins from a position of strength and must train himself to restrain natural powers, desire, aggression. When Darwin denies his powers, he speaks as a male child of the successful Darwins and Wedgewoods, husband of a Wedgewood, from a position of power that allows his readers to admire his modesty without at all diminishing the sense of his authority. The voice emerges from a context that is already empowering.

Mill's autobiography recognizes the incompleteness of the austere self-effacement of scientific objectivism, and struggles, in its early chapters, to incorporate feeling—and women—into the life of the mind. Wallace's autobiography, charmingly and ramblingly self-effacing, resists the coldness of much scientific autobiography by rejecting the totalization of epistemological system that Galton accepted. Galton's autobiography can serve as a strong example of the unselfconscious implications of the unsentimental (inhumane?) rigors of scientific discipline, restraint, self-effacement, and the commitment to a true social science and the doctrine of circumstances. All provide examples of what happens when scientific epistemology is overtly dramatized in narrative; and clearly, the consequences are different, but the fate of the self remains similar. Yet any adequate consideration of the way scientific epistemology has impinged on how people have told the stories of their lives requires some attention to the voices of women caught up in the ethos of a culture so strenuously dying to know.

— 6 —
Self-Effacement Revisited: Women and Scientific Autobiography

If self-effacement is a condition of knowing, then why were women, understood by the Victorians to be the self-effacing "sex," excluded from science? Women were maternal, admired for their capacity for self-surrender, expected to make their lives secondary. It is obvious that in the kingdom of gender, epistemology is not determining, and the question, asked from that perspective, seems naive.[1] The narrative of scientific epistemology, I want to argue, worked consistently across gender, and yet its implications are different as the realities of embodiment, and gendered embodiment, play out the implications of the austerities of dying to know. Galton's version of the problem provides a simple and unequivocal reason women were not going to make good scientists.

As he looks back, wealthy and fulfilled, from his old age, Galton writes with imperial and overtly masculine modesty: "The virile, independent cast of mind is more suitable to scientific research than the feminine, which is apt to be biased by the emotions and to obey authority."[2] Women, then, are slaves to their feelings and thus can't be trusted (although, as the most famous Victorian narratives tended to dramatize, it is just their capacity to feel that makes them capable of self-surrender).

There is such a thing as too much self-effacement, it seems. Efface yourself enough and you give in to "authority," and as Huxley had taught, "The improver of natural knowledge absolutely refuses to acknowledge authority, as such."[3] Does this mean that women are barred from becoming "improvers of natural knowledge"? It seems so. Their self-annihilation comes too early. In the heroic narrative of scientific epistemology in the nineteenth century the improver of natural knowledge is going to be a man, for although the scientist may have to efface himself before the "object," he must refuse to efface himself before the authority of tradition. Women would be incapable of Descartes's great doubt or the Huxleyan ideal of resistance to all mere authority. Thus, the paradox of the inadequacy of women's self-effacement is resolved.

But in the narrative of scientific epistemology the division between observation of the real and acceptance of inherited tradition was never as clear for the scientist as Huxley's ringing, heroic formulation would imply. When John Herschel described the traditional empirical ideal of scientific method, he asserted that "Experience" was the source of all knowledge. It was not, however, the "experience of one man only, or of one generation, but the accumulated experience of all mankind in all ages, registered in books or recorded by tradition."[4] What happens, then, to the Great Doubt, which sweeps away all prior intellectual authority?

The "authority" that the Cartesian adventure rejected sneaks back by way of a history of science. There *is* a trustworthy past, and the development of science depends upon it. The good scientist needs to accede to it or be caught up forever in the limitations of merely personal experience—the Great Doubt can happen only once. If *Pilgrim's Progress* makes a perfect fictional narrative translation (not accidentally religious) of the epistemological venture described by Descartes, a modern version of the narrative might be detected in George Eliot's treatment of Mr. Casaubon's failures in *Middlemarch*. That this is a woman's story is, I believe, important, for it raises immediately the question not of whether one should trust one's own experiences—though that is part of it—but which if any external authority one should trust. That is a woman's crisis in *Middlemarch*, and the growing good of the world depends upon the answer.

A timorously ambitious investigator, Casaubon austerely gathers the facts of ancient myths fastidiously and meticulously, accumulating them, Baconian style, in pigeonholes—and looking like a "hero" in the eyes of innocent Dorothea. His problem is not in the fact gathering but in his failure to know about, no less defer to the authority of German investigators who had gone way beyond him in range of knowledge and interpretive power. Dorothea's problem is to learn whom to trust, and her narrative describes a tension between her superior understanding of how to make that choice and her sense of responsibility to an authority she knows is untrustworthy. Casaubon, for his part, is a bad scientist because he does *not* defer to authority. But the important thing to note here is that lack of knowledge of past knowledge dooms the empirical investigator: mere openness to the facts is a waste of time.

Part of women's responsibility is to mediate the right authority, and sacrifice the self—in the great tradition of self-annihilation that marks the narrative of epistemology. The woman then is both chooser and mediator, still secondary, but indispensable. Dorothea's story can suggest some of the kinds of complications that emerge when the narrative of epistemology is embodied. Bacon had long before lamented that science was not yet incremental, and he sought a way through system and prosthetics to improve upon a method that otherwise depended on the scientist starting from scratch. The fuller epistemological narrative, then, would treat of a protagonist heroic not simply through fearless dismissal of traditional knowledge, but through relentless struggle to determine what traditions to trust. Trust, entering the hitherto heroic narrative of scientific epistemology, changes it; and trust, as Steven Shapin argues, is a condition of the work of every scientist. The very quality that Galton identifies as disqualifying women—their tendency to submit to authority—turns out to be indispensable for qualification as scientists, although this story rarely got told by any but women.

The various Victorian myths of gender (and many of our contemporary interpretations of them) break down under any serious examination. Throughout this chapter, I intend my attribution of "masculinity" or "femininity" to particular modes of thinking and action to be reflections of other writers' views. It is Galton, for example, who talks of the scientist's "virile, independent cast of mind," which would seem also to place the masculine above the call of desire and feeling. To pursue this line of reasoning about women is to end hopelessly entangled in contradictions. If, for example, women are morally superior in their powers of self-sacrifice, it can only be because *they* have Galton's "virile" power to resist the anguish in confrontations with individual suffering. But the question of generality and particularity does become a gender question in the nineteenth century. Women, as Galton would have argued, are too weak to be scientists, too warm to be statisticians: they have not the strength of self-effacement before reality because they are so immediately attracted to and absorbed by the particular person to save, the particular pain to cure. In this mythology, effacement of self in science means the death of feeling, and effacement of self in women means the triumph of feeling.

Women writing about themselves often reaffirm the mythology, but more often make it seem merely silly. Deborah Nord argues that women memoirists tend to fall into two camps: those public women who are "loath to emphasize" their public lives "or to make claims for [their] importance," and those who resist discussing their "emotional and physical experiences that are peculiarly female" and who try to transcend their femininity. But the "tone of self-effacement and embarrassment"[5] that Nord finds in much female autobiography is rather like the self-effacement we have seen, for example, in Darwin's autobiography. Certainly, the traditions of reticence overlap. But the "self-effacement and embarrassment" in women's memoirs suggest forcefully that women *ought* to have been seen as potentially strong scientists—they, too, were dying to know.

Since the ideal of scientific objectivity developed a strain in Western thought that had always valued spirit over body, women were going to be a problem, or at least it would be a problem *for* women. The culture of science seemed never to be able to imagine women free of the constraints of the body. Even in the throes of self-sacrifice, women were "alive," while self-abnegating men could be "dead," dead to the pain of others, dead to their own feelings. Again, within this mythology, study of the material object, "nature" (gendered feminine), required ideal (masculine) detachment.

Do women's autobiographies imply some alternative hermeneutic or antiobjectivist epistemology? I don't think "gender is the hermeneutic key to authorial intention and textual production"; there may be, as Linda Peterson argues, quite other considerations, and "gender need not be the crucial factor."[6] The autobiographies of Mary Somerville, Harriet Martineau, and Beatrice Webb—whom I am here taking as counterbalances to Mill, Darwin and Trollope—in many ways repeat the patterns observable in the men's autobiographies. These women, too, subscribed to the scientific ideal of self-effacement; they, too, valued mind over body. All of them, however, violate the standard view that women can handle particulars but not generalizations. All three, in distinctive ways, seek in their narratives to submerge the singular and the concrete in the general and abstract. This, in the messy mythologies of gender, might mark their narratives as "masculine." Their stories follow within the Carlylean tradition that foregrounds the work and theory of self-submission. The moments equivalent to Teufelsdröckh's "Everlasting Yea" or Mill's "anti-self-consciousness theory" emphasize the liberating energies that come from suppression of self and the possibility thus of recognizing the world as fully knowable.

Although there seems to be no generic difference between male and female scientific autobiographies, the trope of modesty tends to work differently across gender lines. Everyone would have agreed that true scientific work requires selflessness, and these women chose vocations that required of them extraordinary efforts of self-discipline—if not self-sacrifice. Florence Nightingale, in her passionate, visionary, and angry *Cassandra*, confronts the question of how she might explain "the many cases of women who have distinguished themselves in classics, mathematics, even in politics?" Her answer is prophetic: "Widowhood, ill-health, or want of bread." She allows the rare possibility of "an indomitable force of character,"[7] but for the most part women—unlike the men—cannot count on their own deliberate choices for important careers.

Nightingale's categories work. Mary Somerville's success as a mathematician depended primarily on widowhood—the death of a first husband unsympathetic to her mathematical interests. Harriet Martineau's success depended on "want of bread," her responsibility to her family to provide economic support, and her sometimes convenient deafness, which allowed her to fail to hear uncomfortable demands. Nightingale exploited ill health to sustain her post-Crimean work.[8] But it is true of all these women that "indomitable force of character" was also essential. Circumstances alone could not account for their capacity to enter the man's world of Victorian intellectual life.

All three had internalized the cultural demand for self-abnegation, but as Nightingale's argument strongly implies, the act of affirming the importance of self-abnegation required powerful self-assertion. Nightingale talks of how women "are accompanied by a phantom—the phantom of sympathy, guiding, lighting the way—even if they do not marry. Some few sacrifice marriage, because they must sacrifice all other life if they accept that . . . if she has any destiny, any vocation of her own, she must renounce" marriage.[9]

Obviously men do no such thing and count on their wives for support. Darwin makes his way through self-abnegation and self-effacement, and manages a strong patriarchal marriage while establishing extraordinary professional authority. Martineau, Somerville, and Nightingale begin by making claims for the self that require a renunciation of "normal" women's satisfactions in order even to *do* the work to which they aspired. Self-abnegation and self-denial are not then options, as they seem to be for men, but virtually the only mode of self-affirmation available. Ironically, in their work, women must readapt that ideal of self-renunciation that they had renounced in order to work at all. Renouncers of renunciation, they have to develop a new kind of renunciation in a pattern that reemerges in many women's narratives of the period, not least those of Charlotte Brontë (and in the story of *Aurora Leigh*).[10]

The texture of the women's autobiographies tends, consequently, to be somewhat different, if only subtly so, from that of the men. Martineau, Somerville, and Webb do not seem diffident, as Mill, Darwin, and Trollope claimed to be, about making the self their books' subject. Like Mill's—with the possible exception of Somerville, whose autobiography is, to say the least, incomplete—they trace a movement from unhappy childhood to intellectual freedom, which becomes a moral and emotional freedom as well. Somerville's, which we have only as edited and published by her daughter, is rather a set of memoirs with the scattered quality that Deborah Nord describes as one mode of women's self-writing. Martineau's has the confidence of someone who believes she is important enough to write about herself, but discreetly avoids much interiority. And Webb's, as Nord points out, has the shaping, fictional, almost teleological shape of the main stream of "masculine" autobiographical writing. Yet for the most part, these are "private" autobiographies in the great Victorian anti-Rousseauvian mode. They casually evade melodrama as they incorporate into their textures resistance to self and merely personal desire. Webb and Martineau relatively boldly direct their narratives toward intellectual discoveries. Martineau prefaces her story with a claim to the preeminence of the right to privacy, explaining why she will not publish personal letters; and in the course of her narrative, she only mentions without explanation the breakup with her famous brother James of the most emotionally intense relationship of her life. But self as a burden to be shed is urgently present in the early sections, as she recollects a Trollopean childhood of "misery, isolation, neglect, physical weakness,"[11] but there is a quiet boastfulness in Trollope's admission of humiliation. Mill describes his crisis so as to make its rejection of romanticism romantic.

But Martineau's narrative unromantically elaborates on her unhappiness as a child and traces a path to resolution without drama, as a kind of Comtean progression from her self-oriented religion, which made her hate and despise herself every day, to her prosaic adoption of a positivist, rationally grounded view of "scientific" and secular determinism.[12] Where Mill moves from a sternly intellectual childhood toward the adoption of a (highly qualified) romantic position, Martineau moves from a sternly religious and romantic childhood to the adoption of a rigorously unromantic and unsentimental position.

Similarly, Beatrice Webb (née Potter) tells her story rather impersonally, leaving out or diminishing merely personal aspects of her life. Like Mill, she traces the narrative of her development inside the elaboration of general ideas. Like Martineau, she omits much about the most passionate relationship of her life, with the charismatic Joseph Chamberlain, whom she loved but was too strong to marry. And like Martineau, she emphasizes the unhappiness of her childhood. But that unhappiness is itself a mixed blessing—as Martineau said of her deafness that it was "about the best thing that ever happened to me."[13] Webb dryly notes that she "was neither ill-treated nor oppressed: I was merely ignored. For good or for evil I was left free to live my own little life within the large and loose framework of family circumstance."[14] Even more than Martineau's autobiography, *My Apprenticeship* is a narrative structured by intellectual concerns and developments. Rather than linger on the pains of romantic disappointment and personal aimlessness, Webb traces a slow, logical progress from the religion of science she learned from Herbert Spencer toward her ultimate vocation in the systematic study of social structures.

It is rather more difficult to discuss Mary Somerville's autobiography, since what we have are simply her "recollections," edited by her daughter with a great many omissions and interpolations. While it is a book that one must distrust as an accurate representation of what Somerville wrote and thought, it convincingly reveals a Somerville who made a brilliant career in science out of the spare time of her domestic obligations, and leaves all emotional difficulties in the interstices of her recollections. She does not describe an unhappy childhood or dwell much at all on her "feelings." Yet it is not difficult to infer important things about how Somerville managed to become so distinguished a mathematician and interpreter of advanced science. She describes no "crisis," but she does make it clear that her capacity to learn and work as a scientist was made possible by her strong will and her determination not to be thwarted by the constraints of women's life. While in fact a strong supporter of women's causes, she is represented in the autobiography as an acquiescently "womanly" figure. Yet her great success in life was built on her working against family and cultural presumptions, "like a man." Her lifelong interest in the cause of women's education and franchise reflects the forces against which she had to work. But Somerville gives no sign that she was interested in rejecting or even modifying the objectifying and universalizing ideals of science that modern critics like Evelyn Fox Keller and Jane Duran have marked as masculine.[15]

Together, these three autobiographies suggest that the woman's path to male vocation, no matter how fully women are accepted into their "professions" by men, intensifies the implications of the dying-to-know narrative. Collectively, these self-portrayals make much clearer the central arguments of this book and what is only implicit in the men's stories: scientific ideals of impersonality are entwined with the ethical; objectivity and dispassion are saturated with feeling. Denying the self, in all these narratives, becomes a means to a freedom the men never had to seek. Rejecting the self for science meant finding it for life. Harriet Martineau, who of the three turned most self-consciously to a theory of scientific epistemology, formulates the point most clearly and directly. When she looks back on her life near the end of her *Autobiography*, she speaks exuberantly of her liberation:

> I had got out of the prison of my self, wherein I had formerly sat trying to interpret life and the world,—much as a captive might undertake to paint the aspect of Nature from the gleams and shadows and faint colour reflected on his dungeon walls. I had learned that, to form any true notion whatever of any of the affairs of the universe, we must take our stand in the external world—regarding man as one of the products and subjects of the everlasting laws of the universe, and as the favourite of its Maker; a favourite to whom it is rendered subservient by divine partiality.[16]

For each of the women, the study of science was a religious or moral vocation. Somerville taught herself to be one of the most astute mathematicians of her time, and like most of her early-century contemporaries—like her very good friend and scientific confidante, John Herschel, for example—she thought of science as entirely compatible with religion. It reveals "the goodness of the great First Cause in having endowed man with faculties by which he can not only appreciate the magnificence of his works, but trace, with precision, the operation of his laws."[17] In her most famous book, *The Connection of the Physical Sciences*, she argues that there is a force "which is mixed up with everything that exists in the heavens or on earth."[18] She traces that force in a detailed review of all the physical sciences and concludes by seeing the formulae of science, which can describe that force, as "emblematic of Omniscience":

> [The sciences, which are] united still more closely by the common bond of analysis condense into a few symbols the immutable laws of the universe. This mighty instrument of human power itself originates in the primitive constitution of the human mind, and rests upon a few fundamental axioms, which have eternally existed in Him who implanted them in the breast of man when He created him after His own image.[19]

So the book ends, with human consciousness, through science, achieving a simulacrum of Omniscience. Scientific method becomes an aspect of religion, for not only does it reveal the natural world to be fully coherent and designed, but the scientific investigator comes as close as humanly possible to the universal consciousness of God.

Behind trust in objectivity is the faith that it will reveal a world designed, meaningful, and intelligible. In a pattern that became common among the Victorians and that, I believe, survives today, Martineau risks the pain of loss of traditional Christianity because the world revealed by science is more just, more meaningful, better designed, than the Christian God's world. As R. K. Webb has shown, her vision of the world was deeply informed by the Unitarian thought of Joseph Priestley, whose necessarianism established the basis of her own "conversion." Through Priestley she found Hartley and Southwood Smith's *Illustrations of the Divine Government*.[20] She rejected dogmatic religion for the purer and unsuperstitious religion of a world rationally ordered by invariable laws of cause and effect. Thus, science became for Martineau "the sole and eternal basis of wisdom—and therefore of morality and peace."[21]

Like George Eliot after her, Martineau rejected conventional religion because of its built-in *self*ishness: the end is not goodness but personal salvation. As she puts it near the end of her narrative, "dogmatic faith compels the best minds and hearts to narrowness and insolence" (2:109). She, in contrast, turns her attention entirely from the self, to the ideal of a community in which selfishness (usually religion-inspired) is banished for the sake of the group. By the end of her life, Martineau describes something of that ideal "view from nowhere" that is the ultimate goal of the narrative of scientific epistemology.

Now, she claims, when she is "at full leisure to see things as they are," and when she is "placed apart where the relation of the past and future become clear," she is able to "discern the working of that great law of Progress." Her self, which will, she knows, disappear as she moves beyond the liminal state between life and death, is absorbed already in her prose in the work of imagining the future and caring about it. Her conclusion is relentlessly opposed to the selfishness of Christianity. So she writes:

> Now, while the state of our race is such as to need all our mutual devotedness, all our aspiration, all our resources of courage, hope, faith and good cheer, the disciples of the Christian creed and morality are called upon, day by day, to "work out *their own* salvation with fear and trembling." (2:123)

The language of plurals ("our") and the broad concern for "the state of our race," in contrast with the italicized "*their own*," marks the moral superiority of Martineau's positivism to the dogmatic myths of Christianity—a secular science that opens itself to the realities of the natural world.

Beatrice Webb, a later child of the same tradition who was deeply influenced by Herbert Spencer, shared for a while his belief in "The Religion of Science."[22] In fact, she never seems to have lost her feeling for "the supreme value, in all social activity, of the scientific method" (p. 391). Like Martineau she found in science a sense of liberation from the self and of creative action. At the age of nineteen, she wrote in her diary:

> Whatever may be the faults of the new religion, it accomplishes one thing: it removes the thoughts from that wee bit of the world called self to the great whole—the individual has no part in it; it is more than silent as to his future existence. Man sinks down to comparative insignificance; he is removed in degree but not in kind from the animal and vegetable. In truth, it requires a noble nature to profess with cheerfulness this religion; and the ideal it presents to us is far higher than any presented by the great religions of the world. (p. 118)

The hangover of religion in liberated secularity is not uncommon, of course, and it almost always plays the role of outreligioning religion—science truly purges an inevitably contaminated self in a way that messily affective religion cannot do. Here again is the heroic narrative we have seen reiterated by the scientific naturalists with whom Webb would have had much contact, and by the free-thinking avant-garde of positivism in the early twentieth century. In chapters 10 and 11, I will discuss this narrative as it reappears in the work of Karl Pearson, providing the austere satisfactions that derive from recognition that the world is meaningful, though "Man" himself is diminished. To "believe" in this science it is necessary to be "noble," to fight one's way through ostensible pain and ugliness toward a higher and less selfish belief than Christianity can inspire.

While the autobiographies of Somerville, Martineau, and Webb are spiritually charged, they are—with the exception of Somerville—at least equally concerned to avoid making the religious, moral, or social objective *determine* their scientific knowledge. (Somerville's extraordinarily popular work is unrelentingly technical, offering only a page or two at the beginning or end of her books to anyone not trained or training in mathematics or science; there is barely any pious exclamation or natural theological rhetoric. Her work requires no religious commitment for understanding.) Religious energies and beliefs precede and follow from each narrative, but each writer attempts to keep those energies from contaminating it. Webb records her passage through the religion of science to a conclusion that "science is bankrupt in deciding the destiny of man" (p. 127); Martineau sees the world as driven by forces that are beyond the material, but in the end benevolent; Somerville does her science from within a natural theological frame, taking the order she discovers as evidence of a divine hand. Yet all of them are bound, in discussion of the natural and social world, by a fundamentally materialist reading of the secular. Webb's view probably adequately represents the others: "any avoidance of the scientific method in disentangling 'the order of things,' any reliance on magic or on mystical intuition in selecting the process by which to reach the chosen end, spells superstition and usually results in disaster" (p. 127).

The narrative of scientific epistemology is built from this materialist reading of the natural world, of the world as shaped by universal laws; as thus "determined" always by antecedent causes; and as increasingly susceptible to precise abstract, systematic, and statistical analysis. Along the way, then, the narrative entailed a rejection of the kind of feeling that belongs to ordinary human relationships. Where Galton spent no energy avoiding the demoralizing implication of this scientific worldview, each of the women refuses total materialism and finds the objective truth inspiring. However true to the technical demands of their science, each believes that what scientific study reveals will be the divine source *behind* nature. But for the culture at large, liberation into true knowledge as opposed to "superstition" had frightening potential: it meant a rejection of the millennia-long faith in the spiritual source of the material, in the sacramental nature of the visible world, in the idea of individual agency, in free will, and particularly in personal responsibility. Only a cool rationality—so these narratives tend to imply—could resist the temptations of religion and self-importance. While this ideal seemed entirely consistent with the culture's professed imagination of the masculine, it seemed utterly antithetical to its imagination of women.

Martineau's commitment to an ideal of a human science (political economy, in particular), offended many. Scientific epistemology can lead not only to a science of the human, but to a *physical* science of the human—a positivist anticipation of sociobiology. Comte invents the word *sociology,* and Mill at the end of his *System of Logic* attempts to lay the groundwork for such a science. Martineau closes her autobiography, foreseeing the end of religious disputes and superstition:

> By the verification and spread of the science of human nature, the conflict which has hitherto attended such attainment as this will be spared to our successors. When scientific facts are established, and self-evident truths are brought out of them, there is an end of conflict;—or it passes on to administer discipline to adventurers in fresh fields of knowledge. About this matter, of the extinction of theology by a true science of human nature, I cannot but say that my expectation amounts to absolute assurance.[23]

Feeling none of Mill's despondency before a science that seemed to provide no emotional satisfaction, Martineau speaks with the optimistic clarity of the Enlightenment. When Webb addresses the same problems some years later, such optimism about pure knowledge is gone. The narrative of scientific epistemology gets bleaker as the world seems to offer unhappy information, but even in the face of such bleakness, the two women share a commitment to knowledge and the necessity of self-sacrifice to achieve it.

In pursuing knowledge in this way, these writers further disrupted the conventions that understood the feminine as capable of responding only to the particular and collapsing before the power of masculine rational generalization. Victorian narratives are full of such representations—the famous Sissy Jupe of Dickens's *Hard Times,* or the only less famous Glencora Palliser of Trollope's *Can You Forgive Her,* who disrupts the worldly innocence of the abstracting Plantaganet with her incorrigible focus on particulars. But Webb, who found a kind of vocational salvation in statistics, and Somerville, who mastered the most arcane realm of masculine abstractness, mathematics, write narratives that belie the gendered myth. If numbers were, theoretically, the province of men, these "scientific" women found them a perfect means to knowledge, to social action, and to a self-denial that satisfied the desires of self for power and control. Reginia Gagnier takes *My Apprenticeship* precisely as a denial of the stereotype of "woman" as preoccupied with the private, the particular, the individual.

Webb, she says, "has been marginalized" as an autobiographer because she succeeded "throughout a lifetime of struggle in overcoming 'the significant selfhood' (upper class individualism and 'autonomy') and 'individual detail' (women's sensitivity to the special case and minutiae) that her class and gender had provided for her."[24] Science lifts the scientist out of herself and joins her to a larger project of social and moral action. Redemption from domesticity though this be, it follows the pattern of the male autobiographies treated in chapters 4 and 5.

In committing themselves to the general, Somerville and Martineau persist in seeing the world feelingly. But again, this was the crux of the crisis Mill describes in his autobiography. Webb finds statistics a way to overcome the inhumanity of the narrow perspective. In an appendix to *My Apprenticeship*, she discusses the damage done by those who generalize from their own narrow experience. To avoid that damage, "We must," she argues,

> force those who present these striking statements to enumerate exactly the number of persons possessing the characteristics or living within the conditions described: we must compel them to compare this number with that of the whole population. And until the artist consents to bring his picture within a statistical framework, we may admire it as a work of art, but we cannot accept it as a verified statement of fact.[25]

As so often with Webb, who, it is claimed, never suffered in her work because she was a woman, this is fully consistent with arguments notoriously marked as "masculine." Webb sounds a bit like Romney as he tries to dissuade Aurora Leigh from her poetry—impatient with the aesthetic, distrusting feeling, denying the validity of poetry as knowledge.

Webb participated in and anticipated the development of statistics as a method of the sciences of the human. The supposedly masculine mode of mathematics and statistics pointed toward a way to translate the epistemology of self-effacement into an efficient tool that could avoid narrative (except for those who, like Webb, participated in the interviewing that produced the statistics). Webb becomes, in a way, a sort of transitional woman, finding the satisfactions of generality (a displacement from the generalities of ethics and religion) in an activity that responds directly to the particular conditions she encountered daily in the streets of London.

Theoretically, statistics allow an overview in which all individuals are ultimately taken into account. Perhaps the closest analogy to the view from nowhere, their vision transcends that of the individual human, the mere self. Like Somerville's scientific "laws," they become emblematic of omniscience: God sees all and cares for the individual. The generalizing—statistical— force of science is an attempt to see all, and through that larger vision it does more for more individuals than a merely personal or limited attempt (like that of art and the novel) could achieve. Statistics, then, are another effort at establishing a method of knowing that will break through the limits of the particular and the contingent. In this respect, science, indeed, was not much different from the religion whose blandishments women and scientists needed to avoid in doing their work. In this respect, too, religion continues to lurk behind the material and offers its spiritual satisfaction by indirection.

Materialism, determinism, and statistical system may seem odd allies in a spiritual and moral enterprise; but they were so for Descartes and they became more overtly so for these writers. They are the means by which the "self" and its interests are dispelled and truth emerges untainted by desire. They require toughness, a willingness not to be swayed by a momentary feeling, a determination to take a good look at the worst. It is *because* they are ostensibly incompatible with immediate gratification that they can provide satisfaction. In the end, they offer a truer compassion than religion or art, and Webb rejected the individualist means of art for the more communal means of science. She denigrated social studies tinged with personal compassion. The religious resonance of the objectivist spirit of science saved it, for these writers, from the cruelty conventional romantic critique attributed to it. "To me," says Webb, "'a million sick' have always seemed actually more worthy of self-sacrificing devotion than the 'child sick in a fever,' preferred by Mrs. Browning's *Aurora Leigh:* . . . The medical officer of health, who, made aware by statistical investigation of the presence of malaria in his district, spends toilsome days and troubled nights in devising schemes for draining stagnant pools and providing for the wholesale distribution of quinine, has a compassion for human misery as deep-rooted as, and certainly more effective than that of the devoted nurse who soothes the fever-stricken patient in the last hours of life" (p. 307). So in effect, Webb denigrates the traditional nursing function of women for the less immediately compassionate activity of the implicitly male doctor. Like the scientific epistemologist, the doctor seeks not the particular, but the general, and like the scientific epistemologist finds truth and salvation (for his patients, at least) in his capacity to minimize the individual. Statistics are not dehumanizing. They are the way to knowledge and to *effective* feeling. The drive to impersonality is the drive for truth, and it is a moral impetus.[26]

The culture's contradictory reading of women—defining them by their powers of self-sacrifice and, at the same time, by their incapacity to contain their emotions and desires—was, thus, contradictorily confirmed by the careers of these three remarkable, ambitious, and self-denying figures. Estelle Jelinek has noted that women's autobiographies tend to be looser, more fragmented and discontinuous than men's,[27] although it certainly cannot be said of Darwin's autobiography that it is "continuous," or of the last half of Mill's either. But on the strengths of these narratives, this view needs to be qualified: Martineau and Webb write with teleological clarity, moving through their intellectual and professional histories with some rigor toward the activities of profitable professional work among intellectual laborers.

Somerville, Martineau, and Webb had to fulfill the ideals of science as thoroughly as the men and, of course, with far more distractions. Somerville's achievements are perhaps the most remarkable of all because she learned her science against the wishes of her father, who she reports to have said to her mother, "Peg, we must put a stop to this, or we shall have Mary in a strait jacket one of these days. There was X., who went raving mad about the longitude!"[28] She continued to study as a mother herself in the interludes of domestic responsibility. When her first husband, who was entirely unsympathetic to her attempts to learn mathematics, died, she was, as she says, "much out of health . . . and chiefly occupied with my children, especially the one I was nursing." But, she goes on, "as I did not go into society, I rose early, and, having plenty of time, I resumed my mathematical studies" (p. 79). The work she did was prodigious, but since she had the "means," she continued at it:

> I was considered eccentric and foolish, and my conduct was highly disapproved of by many, especially by some members of my own family. . . . They expected me to entertain and keep a gay house for them, and in that they were disappointed. As I was quite independent, I did not care for their criticism. A great part of the day I was occupied with my children; in the evening I worked, played piquet with my father, or played on the piano, sometimes with violin accompaniment. (p. 80)

This is as close as Somerville seems to have come to rebellion in her personal life. Her daughter, however, remarks that "[i]t would be almost incredible were I to describe how much my mother contrived to do in the course of a day" (p. 5). Her rebellion, then, insofar as it might be called that, was in the firmness of her dedication to personal discipline and self-sacrifice (but to her work, not to her extended family). She never, of course, shirked her domestic responsibilities nor, in the public "Recollections," suggested that she wanted to. She learned to work furiously hard during hours of the day when she was not needed for domestic tasks. Though "quite independent" for a short time, she soon married again, but rather more cleverly and happily, one might infer, to a man who fully and unusually supported her work so that her domestic responsibilities and her pursuit of science were at least partly reconciled.

One notes in all these books the subdued tension between the commitment to the norms of women's domestic lives and the ambition to work effectively in a man's world. Martineau, with her ear horn and her celibate life, had something of the "eccentric" about her, but unlike the others she also had the burden of supporting herself, a burden that could help account for her extraordinary productivity as an intellectual worker. Webb, beautiful where Martineau was odd and rather unattractive, was nevertheless also an eccentric. She became increasingly ascetic (with some contradictions) in lifestyle, with only minimal attention to appearance, and with a sustained seriousness that many must have felt to be entirely humorless.[29] Her ascetic power gave her the strength to reject Chamberlain, one of the most powerful, charismatic, and attractive men in England, even though she seems to have loved him intensely. She eventually refocused her love in mutual work with a husband from outside her class.

Yet with all her strange ambition and self-restraint, Webb allowed herself to remain secondary to her husband, Sidney, as she had earlier postponed her commitment to him for the sake of her father, whom she nursed through his long decline. But she got from Sidney Webb the respect and attention she could not have got from Chamberlain, who would have required absolute obedience from her and who would almost certainly have resisted her self-denying vocation. But in her conversion from self-denying commitment of service within a capitalist economy to self-denying service for a socialist state, Webb remained antirevolutionary. Her Fabianism was, like the science from which she learned, gradualist. She stayed inside the world of Victorian reform, even in her commitment to Russian communism, as she stayed within the ideals of self-denying, objectivist science.

Webb's book is one of the most brilliantly and compactly written surveys of the main currents of Victorian thought and, to a lesser extent, politics, that we have; it is a public book. Virtually every element of its structure announces its aspirations to impersonality. It is framed as a battle between elements present in everyone, "the Ego that affirms and the Ego that denies,"[30] and is generalized immediately into that psychological division:

> Though for the purpose of describing my craft I quote pages from my MS diary, I have neither the desire nor the intention of writing an autobiography. Yet the very subject-matter of my science is society; its main instrument

is social intercourse; thus I can hardly leave out of the picture the experience I have gathered, not deliberately as a scientific worker, but casually as child, unmarried woman, wife and citizen. . . . Hence, if in describing my apprenticeship I tell too long and too egotistical a tale, the student can skip what appears to him irrelevant. (p. 18)

The self-description is controlled by a theory of development—as Galton's is controlled by a theory of heredity that immediately absorbs the individual into a wider context. The first long chapter, "Character and Circumstance," assumes that one must know the perspective from which a story is narrated; one must also know the social class and the family history of the protagonist. One cannot, on this theory, adequately judge Webb's treatment of her subject without understanding her "standpoint of observation" and her "inevitable bias" (p. 17). Only these problems require consideration of her self.

Here again, the strategies are familiar from the autobiographies considered in earlier chapters. The women, too, enact the idea that the responsibility of the autobiographer as of the scientist is to submit oneself to larger truths. Only by knowing nature can one know one's own powers or know what the genuinely moral act might be. Thus, they make no complaints about the astonishing amount of work they did; like Darwin, Trollope, and Mill they almost take it for granted. Similarly, Martineau and Webb accept the misery of their childhoods without recriminations (as Mill would never blame his father or Trollope his mother). At no point do they register resistance to the demands they and society put upon themselves, although Somerville frequently refers to society's misguided assumptions about the limitations of women.

The self-denial and repression of feeling that the narratives restrainedly describe are made to seem unironically consonant both with the culture's imagination of women and science's moral/epistemological ideal. Yet in fulfilling these, the writers demonstrate that as women they did not fit the cultural stereotypes. They self-evidently were not under the sway of desire (both Martineau and Webb gave up attractive lovers); they did the kind of work that demonstrates quite extraordinary powers of reason and abstract generalization. Few men could have shown more effectively that they were capable of estimating wisely without abandonment to feeling.

The narrative of this development gives to these autobiographies the texture of dying to know. Linda Peterson suggests that "in 1855, Martineau may well have been the first English autobiographer to substitute a scientific model of self-interpretation for the traditional pattern of Biblical hermeneutics."[31] Like Webb's and Somerville's, her story has a persistent upward movement, a sense of deep personal satisfaction, and a belief in the possibilities of a happy ending. For each, the entrance into the world of science marked the beginning of vocation and of meaning, and it marked as well the destruction of the preoccupations with self and subjectivity that persistently muddied the world.

In their narratives of scientific epistemology, the women dramatize more emphatically than the men the way humility is empowering. As Deirdre David puts it about Harriet Martineau, "The passive textual recording of an actively ameliorative world became a key to her adult intellectual life."[32] The "natural theological" imagination produced for these thinkers a kind of "realism" equivalent to that which had come to dominate the novel. The responsibility to tell what the world is, as precisely and accurately as possible, the vocation of Ruskin and George Eliot in their different ways, can be read (and has been consistently so read by twentieth-century critics) as capitulation to the dominant ideology. But the women showed themselves to be as hardheaded as the men, enthusiastically adopting the strategy of daring to take a good look at the worst. If poets and novelists like Hardy were willing to take the Darwinian vision and turn the moral world upside down, they did not surrender the pleasures of form and did not evade the implicit meaningfulness (even if a reversed one) of the world. These women intellectuals took the masculine risk and accepted the secondariness that realism requires of its practitioners on the faith that ultimately what was obscure would come clear, what was apparently evil would ultimately be recognizable as good.

But even that implicitly value-laden activity was not likely to be available to these women without either antisocial eccentricity or the great good luck of wealth and training. The acquiescence in the real that their work required led to the ironic reversal that made rebellion into knowledge become acquiescence in what is—in social norms. Women could enter this world and achieve this vision only by remaining "women," by fulfilling without complaint the demands laid upon them by families and society alike. Thus nobody, says Martha Somerville of her mother, "was more thoroughly and gracefully feminine than she was, both in manner and appearance. . . . no amount of scientific labour ever induced her to neglect her home duties."[33] But self-submission did not mean self-obliteration. In a passage echoing one of Darwin's charmingly self-deprecating affirmations of personal strengths, and yet with a distinctively gendered twist, Somerville writes:

> Timidity of character, probably owing to early education, had a great influence on my daily life; for I did not assume my place in society in my younger days; and in argument I was instantly silenced, although I often knew, and could have proved, that I was in the right. The only thing in which I was determined and inflexible was in the prosecution of my studies.

They were perpetually interrupted, but always resumed at the first opportunity. No analysis is so difficult as that of one's own mind, but I do not think I err much in saying that perseverance is a characteristic of mine.[34]

"Dogged is as does it." Somerville writes with a confidence (looking back from very old age) that belies her deferential relation to nature and other great scientists.[35] She became a distinguished scientist and mathematician by virtue not only of remarkable intelligence, but of astonishing powers of application in difficult circumstances, and by sufferance. While her influence was great, she always remained ancillary—a distinguished mind providing services for other minds that would have the opportunity not to be ancillary. But there is no evidence that she or any of the other accomplished scientific women outspokenly regretted their secondary positions. The knowledge was a strenuously acquired gift, a pleasure for its own sake, and a liberation. The problem was not to change the ideal but to make women privy to it.

Acquiescence in the ideals of objective science did not inevitably mean complicity with the social and political forces that were coming to dominate Victorian society. Reading backward, it is too easy to see science as culpably and obsequiously serving the new industrial capitalism. But just as realism, the literary mode now almost universally seen as effacing its own positioning and complicit with the world it describes, carried a weight of ironic critique of the represented world and became the weapon of nineteenth- and twentieth-century socialism, so science had its own dangerous and revolutionary implications in the nineteenth century. Science could not have been treated by these women or by any intelligent Victorian as unqualified servant of the new power. While its subversive activities can be seen as self-interested efforts at professionalization, it was nevertheless the guiding spirit, as Webb saw it, to a new free thought hostile to the dogmatic modes of established religions. In its revelations of the material base of all secular activity, it threatened to be thoroughly subversive of those restraints with which even the new economic order sought to control disturbance. We have seen that Webb in particular, but Martineau as well, sought scientific knowledge as a means to resist and transform the dominant political structures. Martineau might be said to have rebelled in the direction of a new dominant power, but even to achieve that power she broke the mold, and throughout her life she took up strong feminist positions, which seemed to her consistent with the new knowledge. Webb's gradual conversion to socialism grew in large part from the education in social observation she got from Charles Booth, who certainly would have rejected the political position she inferred from the use of his techniques.

Here again, it is a mistake to take a particular, abstract epistemological position as allied inevitably with a particular political position. If Martineau's adoption of Comte's positivism seemed to her compatible with political economy, it was also compatible with her feminist causes. Webb's commitment to the view that character is largely shaped by circumstance was compatible both with her socialism and with her signing an appeal against women's suffrage.

Like the Victorian sages who preached against the ideals of political economy but were often complicit with them, science was both complicit and part of the resistance. Webb and Martineau, like George Eliot and others among the intellectual avant-garde of the time, imagined science not as the problem but as the solution. But even those who worried about it as disease could use science as a possible antidote to some of the worst disasters of Victorian social organization. Dickens's celebration of science and technology in *Little Dorrit*, for example, is part of a critique of reigning economic practice, but it is also part of a fantasy of the fullest and most satisfying possible form of new capitalist activity, in which intellectual superiority and technological efficiency would be rewarded immediately, without obstructions created by self-interested and self-perpetuating institutional structures.

Yet if particular ideologies cannot be permanently associated with the particular ideas of these writers, clearly the pursuit of knowledge and the ideas they discovered once again entail the entanglement of epistemology and ethics—the Platonic separation of mind and body, the submission of self to the claims of the natural world, a liberating acquiescence in the necessity imposed by nature and reality. Despite Webb's ultimate skepticism about the moral efficacy of science, these writers all take the methods of science to be necessary to any genuine ethical activity. The difference is that the ideal, associated in the culture's imagination of women with deep feeling and powers of painful self-sacrifice, is associated in science with the power to resist feeling and thus to become fully rational and fully moral. But the women only act out (or write out) more explicitly the implications of the culture's commitments to the ideals of objectivity and impersonality.

Strikingly, while Somerville pursues knowledge without reference to social implications, and her "autobiography" might be assimilated to a largely domestic form,[36] both Martineau and Webb make their autobiographies public documents and expositions of theories of knowing in the interests of social reform. They are at least as unsentimental as the men in the way they figure knowledge and in their determination to move beyond the particularity often taken to be the mark of the feminine. But these narratives, as they follow the lives of three unusual women, in effect push to the surface elements that tend to be minimized in conventional narratives of scientific epistemology—as Mill found out at the price of a mental breakdown, abstract knowledge must be made vital by feeling and desire.

Thus "masculine" and "feminine" play out very similar stories. But playing out the narrative of scientific epistemology in somewhat different ways, with different affect, they illuminate each other in their difference. The male autobiographers find narrative strategies that minimize the claims of the self and yet end by implicitly affirming the self's power—their narratives in part depend on the readers' initial awareness of their authority, and the self-deprecations only increase that authority while at the same time they never budge on matters to which they have committed themselves. The "hermeneutic strategies" of all these autobiographies—male and female—as Peterson might put it, entailed the application of the methods of science to self-description. The autobiographies then participate in the great nineteenth-century project (spurred by positivism), the application of science to humans. For the men, that application proves to be consonant with the decorum of the gentleman: reticence about the self and its desires (with implicit argument that thus their scientific ideas were entirely disinterested). For the women, the vocation self-consciously stands in for the displaced religion and provides them a means to minimizing the self. There is about each of them, however, a healthy frankness about the value of their own work. They are willing to quote praise, modest without false self-deprecation. Mary Somerville quietly records almost every honor she receives, often in passing, usually without fanfare. The women tell their stories with the frankness of science itself, with an unembarrassed willingness to be domestic, ordinary, inadequate, but with an equal willingness to allow that they have done well and not be surprised by it.

Thus in other ways than those we have found in reading the literature of male autobiography, these biographies suggest that it is not easy to dismiss either dispassion or self-surrender as a condition of investigation. Evelyn Fox Keller suggests that the problem is not the willingness to attempt to put the self aside but what she calls "premature" anonymity. "Objectivist ideology," she says, in claiming disinterest and impersonality and "radically excluding the subject," "imposes a veil" on the day-to-day practices of science, which, like all other human activities, are impelled by emotional, social, and political commitments.[37] The problem is in the "veil," in the "premature" excluding of the subject. And the project of thinkers like Keller, Sandra Harding, Helen Longino, and Donna Haraway has been to imagine what Haraway calls "situated knowledge."[38] For each of these writers, a tension develops between surrender to epistemological relativism, which would leave even its claims for gender equality unsupported, and recognition that useful scientific activity requires something equivalent to objectivity, whose dangers they have powerfully exposed. In a sense, they seek a way to reinsert life into the quest for knowledge by coming to terms with the human quality of investigation, with the moral and emotional engagement of researchers, studying at historically particular times, from social and individual perspectives.

The moral impetus in the narratives becomes clearest when the women reject traditional modes of self-surrender for alternative ones. But for a self-made intellectual like Mary Somerville, emotional engagement and self-fulfillment were already in the abstract externality of mathematics and of scientific law. In considering how earlier women thinkers turned to science, it is necessary to account for the sort of exuberance we have seen in Harriet Martineau's praise of externality, objectivity, disinterest, and the idea of universality. In a society in which women were thought to be Sissy Jupes, too subject to emotions, too intellectually limited to see beyond the particular, too subjective to be rational, the scientific ideal of self-denial and objectivity looked very attractive, and was indeed very different from the domestic ideal of self-denial. Sissy makes her desire for her father a fact about her father. But for Martineau and Webb it is not desire that governs nature. Rationality and cool disinterested objectivity, the recognition that particulars derive from universal laws and can be understood best by seeing them in their complex interrelations with other particulars by means of those laws—these seemed to provide access to power, and to very special and exclusive pleasures.

There is, for example, an almost Gradgrindian pleasure in Somerville's comment on Buckland, who had tried to resist the new geological sense of time for the sake of biblical tradition, but who finally had to concede: "facts are such stubborn things," she says.[39] But Somerville's commitment to fact was always connected to her passion for "connection." Like Herschel and most nineteenth-century theorists of science, Somerville found deep satisfaction in release from the tyranny of particulars, from the concrete to which women were stereotypically bound.

All three of these autobiographies express belief that the natural world—which came to include the social world—could be understood in terms of fundamental principles, laws divine or merely natural that it was the function of the scientist to discover and apply. They believed that the world in all its separate, myriad, and marvelous manifestations was coherent, connected, "designed." (Somerville thought of the design as intentional, from the mind of God; so, in her more antidogmatic way, did Martineau.) They believed that these laws were discoverable through empirical study of the natural world and the application of abstract principle, largely mathematical, to the discovered facts. And they believed that the discoveries, empirical and theoretical, were only possible through disinterested openness to experience and the subjugation and denial of personal desires.

The irony of women's exclusion from the practice of science in the nineteenth century is intensified by these autobiographies, which fully work out the narrative that embodies gendered scientific epistemology. The stories they tell may give an unusual turn to this pervasive cultural phenomenon, yet each of them ultimately retells the story of self-effacement. In writing about themselves, each sought a way to release herself from the mere particularity of selfhood into a larger vision of the world. Like the men, these women were dying to know and, for better or worse, were pretty good at it.

— 7 —
The Test of Truth: *Our Mutual Friend*

Having moved from discursive to narrative prose by way of the strategies of self-representation in scientifically inclined intellectuals, I want now to take the leap to fiction, where direct concern with epistemological projects is further diminished. One of the attractions of the narrative of scientific epistemology is that it follows closely the lines of narratives that worry issues of ethics, religion, or aesthetics. It has the structure of the pilgrim's progress, which is fundamental to the structure of the Victorian novel and particularly the novel of development.[1] Another version of this narrative pattern, made notorious in Freud's use of it for other purposes, describes the Greek tragic hero who learns the truth at the point at which the story turns (*anagnorisis* occurs at the *peripeteia*)—Oedipus discovers that he has killed his own father and married his mother. Paring away the realistic details and the particularities of place and time that give Victorian novels much of their distinctive quality, one can find something of that pattern—often turned to "comic" purposes—operating still.

But this great oedipal story can also be read as an epistemological one. It is about questions of truth; the self-humiliation of the protagonists comes not as a *consequence* of the new knowledge but as a *condition for it*. And the problem of how to find things out, of uncovering what is hidden, is pervasive in Victorian fiction. That the detective novel is one of the generic inventions of the period is hardly surprising, since that form depends on the assumption that the conditions of ordinary life constantly disguise (even if in a Holmesian world they leave many traces of) an authentic reality, and since it produces a scientific protagonist who, like Mr. Bucket of *Bleak House,* is successful because there is about him something determinedly detached, almost inhuman.[2] Moreover, the classic Victorian narratives tell stories that rigorously punish efforts of characters to impose their desires on reality—images of the blinded and maimed Rochester or of the feverish Pip come immediately to mind. Knowledge is available either to Galtonian figures, like Tulkinghorn and Jaggers, who in effect deny their humanity and seek personal invisibility, or to those for whom detachment is combined with fellow feeling. In all cases, however, the best way to acquire knowledge is to be nowhere. For Victorian novels, the best investigators are the readers themselves, who often take overt or implicit advice from narrators: remaining outside the story, the readers are able to interpret the signs that engaged participants cannot.[3]

Virtually all the conventions of Victorian realism bend in the direction of the epistemological narrative; what makes the connection to epistemology more convincing, however, is the remarkable consistency with which "truth" is registered in Victorian narratives as the most fundamental of moral values. John Kucich points to John Morley's important essay, *On Compromise,* in which, in preparation for arguments about how to overcome the forces for compromise in contemporary society, Morley describes the current situation in England: "According to the current assumption of the writer and the preacher," Morley claims, "the one commanding law is that men should cling to truth and right, if the very heavens fall."[4] And then with a virtually religious rhetoric designed to sustain a rationalist position, he concludes, "Our day of small calculations and petty utilities must first pass away; our vision of the true expediencies must reach further and deeper; our resolution to search for the highest verities, to give up all and follow them, must first become part of ourselves."[5] This sounds rather like the George Eliot who regularly dramatizes the consequences of not making the ideal a habit, a "part of ourselves." Lydgate casting his vote against Farebrother for the hospital board is a perfect instance.

Although few novelists were as outspoken on the subject as George Eliot, the practice of realism itself, and critical demands for truthfulness, suggest how central to the Victorian novel was the enterprise of knowledge seeking and truth telling, how often plots turn on the power of protagonists to develop the proper temper and state of mind to allow realistic confrontation with the "object"—what one might see as acquisition of the proper "method." One can only achieve truth through objectivity; one can only be objective by virtue of the moral strength of self-restraint. The interdependence is visible from George Eliot's intervention in chapter 17 of *Adam Bede,* to Thackeray's self-conscious representation of his own artifices and their distance from the truth in *Vanity Fair,* to Trollope's determination never to trick the reader with surprises, to Charlotte Brontë's earnest insistence on making the narrative of *Shirley* as ordinary as Monday morning.[6]

The pervasiveness of such preoccupations encourages a reading of these novels as explorations of the limits of the possibilities of the narrative of scientific epistemology—as critique, or as perhaps unselfconscious reenactment. The formal constraints, the play of narrator and reader, the limits within which the protagonists are permitted to move, can be read as reflections of the values and implications of the epistemological ideal and indications of how that ideal constrains narrative. Just as the autobiographies suggest something of how it was possible to think of one's own temporally constrained life—at least publicly—within the limits of imagination of how access to truth might be possible, so Victorian fiction suggests the limits of the possibilities of self-representation. While the kinds of narratives the Victorians were writing tended to follow the form of the epistemological ideal I have been discussing, much of the interest of their work lies in the way they explore the limits and struggle to imagine alternatives. The fiction, to put it differently, is not ideologically, or even epistemologically, passive but richly and creatively involved in critique and rethinking.[7]

The "honesty" that Kucich finds central to Victorian self-conception is scientific as much as it is personal, and Victorian novels tend to require "saintlike" objectivity of their protagonists. Knowledge regularly comes, in the best-known works of Victorian fiction, at the point when the self is humbled or inactive (in Hardy, as an extreme instance, knowledge very often comes when the one who learns is totally unprepared and merely overhears—often by accident—discussions that bear on his or her own life). The ethical and the epistemological become indistinguishable.

In the chapters that follow I will, in the light of these epistemological issues, reexamine three familiar novels—Dickens's *Our Mutual Friend* (1864), George Eliot's *Daniel Deronda* (1876), and Hardy's *Jude the Obscure* (1896). Of course, there have been countless other more obvious (perhaps more sensible) ways to read them. But the perspective of the epistemology of dying to know should make clear how that epistemology penetrated the culture and helped shape it, and how narrative embodiment allows new perspectives on the epistemological ideal itself. Whatever else these narratives are, they are also explorations of the limits of the possibilities of self-surrender in the pursuit of moral and intellectual salvation. They are almost invariably thematically and formally preoccupied with questions of knowing—how, in *Our Mutual Friend*, to recognize intrinsic value and discover the truth about other people's feelings; how, in *Daniel Deronda*, to incorporate desire into the moral austerities of objectivity; how, in *Jude the Obscure*, simply to "know," or to defeat the pull of desire and of the body. All these novels implicitly question, more or less critically, the ideal of self-denial in pursuit of objectivity, as that ideal impinges on the lives of real people living in the material world.

Each of them is sensitive to the difficulties of truth—its disguises and elusiveness and dangers. In effect, they reenact, with the stakes higher than mere intellectual and scientific satisfaction, Descartes's determination not to trust any inherited knowledge or anything that simply *appears* to be the case. It might not be a dream, but it could easily be a trick. The novels frequently build their plots around the problems caused by the body and the passions in gaining access to the truth, except that, as novels, they can never dismiss the body as trivial or irrelevant. They are all, then, directly concerned with the passion to know—and with its relation to death.

Thus, when I talk about the "epistemology of death" in Dickens's novel, I mean to be quite literal about the word *epistemology*. Criticism is more habituated to thinking of Dickensian thematic and symbolic patterns in terms of myth or fairy tale; I am shifting the focus to philosophy, and this not in spite of the fact but largely *because* no novelist is further from abstract speculations than Dickens, who, in relation to criticism from high culture in his own time, existed as a rather vulgar if sometimes pleasant phenomenon. G. H. Lewes, for whom epistemology was a continuing preoccupation, concluded his review of Forster's *Life* of Dickens this way: "For the reader of cultivated taste there is little in his works beyond the stirring of their emotion." We enjoy his work "like children at a play," Lewes asserted.[8] Here is a critical test for the idea that philosophical issues, and the patterns and problems implicit in them, emerge in and constrain fiction that seems quite distant either from science or from epistemology. There is no disentangling from the complex fables of Dickens's imagination the mythic and philosophic, the social and the political. But *Our Mutual Friend* is, even more than most Victorian novels, deeply preoccupied with questions of "truth," and fits almost too neatly the conception of John Tyndall as I quoted him in the introduction to this book. All its social preoccupations, all its tensions between the personal and private, all its concern about the possibility of recuperation and resurrection, are connected to questions of method, knowledge, objectivity, truth, and, at the same time, questions of ethics and religion. The epistemology illuminates the narrative; the narrative puts the epistemology to the test.

Dickens was not at all ignorant about science, but rather an enthusiast for it, though, as Edgar Johnson long ago pointed out, you would find far more evidence of this in his periodicals than in his novels.[9] The mockery of science in *Pickwick Papers* and the ferocious assaults on the *merely* rational and analytic are paralleled by his enthusiasm for most forms of technological progress and by enormous respect for professional efficiency—not least in the detectives and lawyers, whose knowledge is their power. To put it crudely, Dickens wanted science, but with a heart, in touch with the lived experience of his characters.[10] His first interest was always people, but in the disciplining of their hearts, so important a motif in his work, his emphasis is not on suppressing reflection and analysis but on suppressing feeling. As

Kucich points out, these "repressions" have different consequences for male and female characters (a point clear from the women's' autobiographies considered in chapter 6)—the males able to "assert the secondary skills that repression brings them, whereas his women must hide or deny these skills."[11] This point has particular significance for *Our Mutual Friend*. Repression is essential to the ultimate values of the fiction. It is important to note, for example, that when David Copperfield complains of his undisciplined heart, he means among other things that Dora should have attended to the kinds of efficiency and rational competence necessary for the running of a working household. That is, Dickens sought "discipline," and, as for Tyndall, it is only within discipline that moral, intellectual, and spiritual success becomes possible.

Death hovers ominously over all the novels I will be considering (as it does over the narrative of scientific epistemology), but in no book more than in *Our Mutual Friend*, where it undergoes an extraordinary range of variations and almost obsessively asserts itself as it threatens, tempts, illuminates. So one can begin, simply, with death. Dickens, as all his biographers note, approached the writing of the novel with death all about him. Not only was his own physical condition deteriorating, but he survived an almost certainly traumatic train crash while he was writing the book, an accident that increased his awareness of danger and decline and required some real enactment of his motto, "never say die."[12] The novel's postscript gives a brief description of the accident: "I can never," Dickens wrote, "be much nearer parting company with my readers for ever, than I was then, until there shall be written against my life, the two words with which I have this day closed this book: THE END."[13] In addition, the friends and rivals who were his contemporaries were dying all about him—he seems to have been shaken by the death of Thackeray, although they were obviously not good friends. And his own son, Walter, had just died in India. If, as Patrick McCarthy has said, Dickens's letters at the time "reveal no morbidity,"[14] there can be little question that *Our Mutual Friend* is obsessed with death in ways quite different from those of even the most death-filled novels that preceded it. He had long since made a living out of death, through Little Nell's prolonged dying in *The Old Curiosity Shop* and Paul's not quite so prolonged death listening to what the waves were saying in *Dombey and Son*. Ruskin had counted off the dead in *Bleak House*[15] as a mark of its modernity and of its inferiority to Scott's novels. *Our Mutual Friend*, however, playing on death in new and abundant ways, takes it to the edges of epistemology and into the crises of possible action in an increasingly commercial society.

In Dickens, there had always been a tension that had made his conception of the good hero a very difficult one, between a vision of the contemporary capitalist world ferociously active and energetic and a moral ideal of noncompetition, love, and self-abnegation. The ideal of moral detachment incurs the danger of coldness, of the kind of inhumanity that Dickens dramatized in Tulkinghorn and Jaggers, and this possibility helps explain the problem of the protagonist's ambivalence toward engagement in the world of contemporary business. *Our Mutual Friend* pushes the moral resolution of detachment to the very point of death, as though the only way to escape contamination from the world of buying and selling were through death: the protagonist, John Harmon, is "dead" through most of the novel, and his more or less double, Eugene Wrayburn, originally designed to be killed at the end,[16] barely manages an enfeebled resurrection.

Thus Dickens's continuing difficulty in imagining active and socially engaged figures who could remain worthy heroes is exacerbated in *Our Mutual Friend*. Here, the crisis of modern capitalism, which had figured so importantly in *Little Dorrit*, takes the form of the impossibility of finding a vocation that is not compromised and corrupted. Throughout the novel, the values of the protagonists are at odds with the demands of commercial success. The scathing critique of society, noted by George Gissing and G. B. Shaw in particular, is often taken to be fundamentally compromised by Dickens, who invariably affirms the values of the very class that benefited most from that society.[17] Fascination Fledgeby may be exposed, but Harmon manages to exploit the dust heaps and protect Bella with the proceeds. Clearly, Dickens admires and is fascinated by energy, and yet he dreads its consequences. *Our Mutual Friend* plays fascinating variations on that theme, in the semiaristocratic and debasing indolence of Eugene Wrayburn, the self-improving and destructive energy of Bradley Headstone (a figure who embodies Dickens's ambivalences about disciplined efforts to rise above one's class and, thus, about the importance of self-restraint). In his later novels, he worried out this problem directly and forcefully by imagining protagonists like Pip, corrupted by their ambitions and successes. Distrusting the society in which, to live happily ever after, his protagonists needed to succeed, Dickens strained to imagine ways not to stain them with their own ambitions and with the very success the novel form required. In *Our Mutual Friend*, Harmon's status as "dead" is a perfect embodiment of Dickens's ambivalence toward contemporary social and economic organization, and at the same time a perfect embodiment of the epistemological strategy of self-annihilation to allow the truth to speak for itself.

Of the three novels to be considered, *Our Mutual Friend* seems the most intrinsically acquiescent in the dominant dying-to-know model because it so extravagantly pushes the limits of death itself, creating liminal spaces in which the paradoxes of the epistemological narrative can become dramatically present. While it is obviously not literally about epistemology, its focus being on social and economic issues, its plot sets up a central "experiment" in knowing: the impossible condition of knowing while dead; and it resolves itself in a sequence of self-surrenders that gradually reveal all and resurrect the dead. Even the basic Tyndallian paradox of the epistemological quest—requiring self-surrender *in order to* achieve personal and general betterment—is implicit in the novel's form and action. The protagonist of the novel is "dead." He is also the major experimenter, for being dead he is in a position to allow the subject of his experiment to manifest itself (or herself) without influencing it. Although Dickens was not self-consciously imagining the conditions of a scientific experiment, he makes the large moral and social issues of the book dependent on strategies of knowing, and these strategies are linked to self-effacement. The conjunction of the narrative of scientific epistemology and the crises of society about which Dickens wrote and worried is at the heart of *Our Mutual Friend*.

John Harmon (who disappears into George Radfoot, Julius Handford, and John Rokesmith) can only discover whether Bella Wilfer is worthy of being his wife if he is dead; this he can only accomplish if he can both see her in action and not be seen—that is, dead and not dead at the same time. He must watch her behavior like an anthropologist or zoologist, leaving her free to act as she would if he were in fact dead and she unobserved. Harmon, that is, attempts to set up the conditions of the perfect experiment, one that modern science has taught us is not quite possible[18] but that provides the model for virtually all scientific publication. Not surprisingly, the narrative delights in the paradox of Harmon's condition: "our mutual friend" turns out to be both protagonist and dead man. One of the great ironies of the experiment is that while the results seem good, Harmon wants to keep it going—longer than necessary. There is special pleasure in finding things out when one's own desires have no influence on a result that satisfies one's own desires.

Obviously, such a narrative is difficult to conduct within the limits of "realism," which is committed to creating a sense of the likely and possible. As a result, Dickens is driven to the most elaborate contrivances of plot, not only in John Harmon's transformations but in a whole series of more minor narratives in which characters waver between literal and figurative life and death. Here it is essential to Dickens's project that the impossible conditions of the epistemological project—seeing from nowhere—be made to seem possible.

Preoccupation with knowing is not confined to Harmon's story. The "Inspector" investigates with an almost perfectly affectless relation to his subjects—like Inspector Bucket in *Bleak House*. Despite Dickens's commitment to the "heart," he recurrently celebrates mere efficiency, even if it leans toward the affectlessness of the detective. Efficiency for the sake of money is usually associated with an inhuman coldness, even cruelty, but efficiency for the sake of truth and order—which always requires some kind of austere self-discipline, even in so loving a creature as Esther Summerson—is always highly valued in his novels.

Our Mutual Friend relentlessly returns to themes of learning and knowledge, as a whole range of characters acquire an education, more or less successfully: from Charlie (Lizzie Hexam's brother), to Floppy, to Bella, to Mr. Boffin, to Lizzie herself. The question of how the knowledge is to be acquired is critical in each case. Bradley Headstone, in some ways a Dickens surrogate, is the villain of the piece despite his rigorous self-discipline. Valued self-control seems to be the property primarily of the middle class, although lower-class women may also display it—most obviously Jenny Wren and Lizzie Hexam. But Bradley's passion bursts sporadically with the blood from his nose, and passion drives him, while education is only a means to something else. Various other villainous figures, like Wegg or Rogue Riderhood, seek evidence that will enhance their own positions (and along the way provide excellent examples of nonobjective knowledge, marred by the desire for short-term satisfactions and suggesting how the absence of objectivity is itself corrupt). In the pursuit of truth by other more honorable characters, there seems to be a fundamental division along the lines of gender. The men quite literally risk their lives, while the women learn to defer to the appropriate authority.

Another motif that relates directly to the dying-to-know narrative is the transformation of the human into an object for sale. The novel opens with the Hexams dredging the Thames for the valued residue of human life, and for bodies that will bring money in any of various legitimate and illegitimate ways. Mr. Venus buys and sells human parts. Bella Wilfer, of course, is an object for sale, as she is the object of the book's major experiment. The Lammles bicker like vultures over Georgianna Podsnap and try to do it again with Bella. The search for a replacement for Wegg's wooden leg runs throughout the novel. "You could easily buy all you see of Lady Tippins in Bond Street" (p. 164).

Our Mutual Friend is preoccupied with death in a new way in that it no longer lingers over the slow decline of innocents, building deep reservoirs of pity. There are only two deaths in it that reverberate with the old indulgence of sentiment for lost innocence and prodigal wisdom: Betsy Higden's, performed in the name of truth and the virtue of independence against the bureaucratic brutalities of social reform, and the orphan Johnny's (with his last breath he invokes Bella, the "boo'ful lady" and provides evidence, if anyone were to need it, that Bella is going to emerge beautifully from the experiment). But death is far more integral to the plot than these rather more familiar Dickensian virtuosities might suggest.

Everything depends on deaths, or near deaths. The liminal state between life and death that fills the novel's pages always suggests that the touch of death is refining, paring away the spots and defects of life. It confers value on the most wretched of figures and opens up the possibilities of true wisdom. This middle state is an exact narrative embodiment of the idea of "dying to know." Mysterious, the wise innocent, Jenny Wren, paradoxically child and parent, calls down from the roof of the house where she makes her dolls and tends to her drunken father, "Dolls," "Come up and be dead." When the nefarious Fascination Fledgeby asks Jenny what it feels like to be dead, she responds:

> "Oh, so tranquil!" cried the little creature, smiling. "Oh, so peaceful and so thankful! And you hear the people who are alive, crying, and working, and calling to one another down in the close dark streets, and you seem to pity them so! And such a chain has fallen from you, and such a strange good sorrowful happiness comes upon you!" (p. 334)

Encumbered by an inadequate body, Jenny recognizes death as the place where the chains of bodily restraint fall away. Only dead can one free oneself from the sorts of corruptions Jenny faces every day. Death becomes the ultimate standard of wisdom and goodness.

Of course, there is more than a touch of strange sentiment about Jenny's sense of death, but a Ruskinian chart of deaths and quasi-deaths in the book would suggest its relatively unsentimental significance and its great importance in establishing the framing values and norms of the narrative. Offstage, the novel begins with the death of the older John Harmon, whose estate is the subject of struggle throughout, and is built of the detritus of London—itself a form of death that can be transformed into wealth (or "illth," another form of death). But the novel in effect starts with the murder of the younger "John Harmon," though we learn that it is, rather, a lookalike sailor named George Radfoot who has been murdered. Thus Harmon dies, but is resurrected. Along the way, Gaffer Hexam is found drowned, entangled in the nets of his own parasitical boat; and Rogue Riderhood is drowned, only to return: better in death than in life, he is brought back, but "as he grows warm, the doctor and the four men [working to save him] cool" (p. 506). Rogue drowns a second time, accompanied in a death struggle by Bradley Headstone—figuring in their death embrace the price of uncontrollable greed and desire. Jenny's drunken father also dies as a kind of object: "'you had better send for something to cover it. All's over'" (p. 801).

Dolls goes out as an "it," no lingering, no middle state, no lamentations, even from Jenny. But the climax of one major stream of the book's narrative is the prolonged quasi-death of Eugene Wrayburn. Only in the condition of death can he marry Lizzie Hexam. The marriage itself takes place in that liminal state, and in the short scene that follows, Eugene says, "I have been thinking whether it is not the best thing I can do, to die" (p. 824).

The epistemological implications of the novel are most obvious in the experiment upon which the various other plots make continuing comment. The investigator, Harmon, transforms the metaphor of "dying" into something almost literal. He takes on the role of secretary to the Boffins, and the word reverberates with its roots in secrecy. As secretary, like a good experimenter, he can control the experiment, but he has to avoid biasing the results by making his "subject" aware that she is being tested. He can let Bella know his feelings for her, but only on the condition that she *not* know that he is Harmon. When she rejects him, Harmon buries himself yet more deeply: "John Harmon lay buried under a whole Alpine range; and still the Sexton Rokesmith accumulated mountains over him, lightening his labour with the dirge, 'Cover him, crush him, keep him down'" (p. 435).

In certain respects, of course, Harmon is hardly a disinterested investigator. But commentators on science, both before and after *Our Mutual Friend*, note two kinds of scientific observation, passive and active, although the distinction is not absolute. John Herschel points out that observation without intervention produces knowledge more slowly, but for both kinds of observation the responsibility is the same and the language is familiar—"to dismiss as idle prejudices . . . any preconceived notion of what might or might not be the order of nature in any proposed case."[19] The question, then, is not whether but how the scientist interferes. Claude Bernard, some years later, argues that the distinction between active and passive is not significant epistemologically. All scientists, he says, have the common and immediate object, in their investigations, "of establishing and noting facts and phenomena as rigorously as possible . . . the only difference is this,—as the fact which an experiment must verify does not present itself to him naturally, he must make it appear, i.e., induce it, for a special reason and with a definite object."[20] The testing of Bella Wilfer entails laboratory manipulations: Harmon's disguises, and Boffin's also.

But the narrative persists in the ideal, regarding Harmon as a "living dead man" (p. 430). Ironically, narrative here further exposes the contradictions in the enterprise of knowing, because in burying himself, Harmon in effect simply lies. Active manipulation translates into a form of lying so that built into experimental conditions is a kind of falsification that significantly compromises the ideal of truthfulness governing both the moral and the epistemological quest. Yet it is the perfect experimental condition: from the position of death, Harmon remains invisible. As he sees, he must deceive. His experimental life allows Harmon to see the real fidelity of the Boffins, certain that it is not an obsequious disguise for greed; from the position as living dead, he can learn "the revolting truth" that he would have purchased Bella, with her caring nothing for him, "as a Sultan buys a slave" (p. 429).

Catherine Gallagher reads *Our Mutual Friend* in terms of economic rather than epistemological theory and notes that in Ruskin and Dickens, the emphasis on death follows in certain economic theories (like that of Malthus, for example) "that claim to valorize life most fully." For those, she goes on, "are the tendencies stressing the commodity's need to have a life-giving potential that makes up for its life-draining origins."[21] Gallagher's emphasis on economics, indeed, "bio-economics," captures in another mode the process I am describing here. *Our Mutual Friend* depends on Dickens's feeling for the supreme value of life itself. It is evident in the way he treats murder from the start of his novel-writing career, particularly in the almost surreal horror of Sikes's clubbing of Nancy in *Oliver Twist* and in the powerful guilt that almost overwhelms the imaginations of both Sikes and Fagin—and of the novel itself.

The proximity and mutuality of life and death, both literally and symbolically in *Our Mutual Friend*, make it difficult often to distinguish the value of the two. Death, as Jenny Wren describes it, is a good place. But the value of death depends in part on the dominant values of life, and the moments of near death in *Our Mutual Friend* make this very clear. Gallagher has discussed the connection between value and life—"wealth"—in Ruskin, and opposed it to Ruskin's other term, "illth"—that which does not avail for life and therefore has no value. Value in Ruskin and Dickens is what avails for life. Thus, even when it is Riderhood dying, the possibility of death evokes sympathy. But Riderhood's life is worth more than he is, and, as I have noted before, as he grew "warm," the onlookers grew "cool." His daughter Pleasant found it "something new to see her father an object of sympathy and interest." She even entertains vaguely the idea "that if affairs could remain thus for a long time it would be a respectable change"(p. 505).

Yet more strikingly, when Lizzie Hexam runs back after hearing Headstone's assault on Wrayburn, she at first does not recognize Eugene. Her response, then, is not as a lover but as a truly disinterested figure, someone who values life regardless of her own needs and desires. In that melodramatic moment, Lizzie prays:

> Now, merciful Heaven be thanked for that old time [when she worked with her father dragging bodies from the Thames], and grant, O Blessed Lord,

that through thy wonderful workings it may turn to good at last! To whomsoever the drifting face belongs, be it man's or woman's, help my humble hands, Lord God, to raise it from death and restore it to some one to whom it must be dear! (p. 768)

That the selflessness and commitment to redeem work with death (where the value Gaffer Hexam produced was "illth") for the sake of life ("wealth") turns out to lead to Lizzie's own happiness is an exact embodiment of the pattern I have been describing. And yet there is a strange prelude to the scene of rescue, when Wrayburn kisses Lizzie goodbye. Lizzie, conscious of the questions of class and of Wrayburn's quite accurate evaluation of himself, will not quite say she loves him. But when they part, the sequence concludes, "He held her, almost as if she were sanctified to him by death, and kissed her, once, almost as he might have kissed the dead" (p. 764). The love itself is in a liminal state, pure *because* it is both alive and dead.

The ideal state is, then, neither life nor death but the liminal stage in between—from which Jenny calls, in which Lizzie works, where Riderhood lingers, and from which Wrayburn will ultimately marry Lizzie. The middle state matters so much exactly because of the supreme value of life itself, but that in turn derives from its tenuousness and from the way it is always threatened by the corruptions of corporeality. Death would be perfect if one could be alive in it.

It is important to note that the economic narrative, normally the focus in criticism of the novel, as it traces the transformation of death into life, of waste into vitality, is an epistemological transformation as well. The two are mutually dependent and virtually the same. Here again Tyndall's scientific method is reaffirmed: the striving self must abase itself in order to be vivified. Thus the novel plays out another, ironic aspect of the epistemological narrative, that is, the way in which it must emphasize self-humiliation, but in doing so opens the way to a kind of moralized self-aggrandizement. Some of the narratives in the novel explore alternative modes of self-humiliation—particularly in Bradley Headstone's case—almost as if in an effort to determine more precisely what constitutes an adequate and valid form of it.

Making her bio-economic argument about *Our Mutual Friend*, Gallagher accurately notes that in the book, "Apparent death becomes the only direct access to the essence and value of life. Apparent death is the condition of storytelling and regenerative change."[22] The paradox of being nowhere is the central formal and ethical paradox of the novel's narrative. For implicit in the paradox, as Dickens works it out, is that despite his deep passion for life and even for the pleasures of the body, in the end he, like the religious and epistemological *and* economic traditions embodied in his narrative, deeply distrusts the body. As Gallagher puts it, "John Harmon's plot . . . demonstrates that value, even the value of Life itself, is only discoverable from some vantage point outside of the body."[23] Nowhere.

Almost at the center of the novel, there is a remarkable passage that awkwardly, hence nicely, embodies this impossibility. The passage dramatizes John Harmon's excursion to "nowhere," and his attempt, then, to remain there in order to be able to conduct his experiment. In many respects it is a clumsy sequence, crying out for some more modernistic sense of narrative. The difficulties of exposition create a distinctly formal problem, but one that also throws light on the epistemological issues. The postscript in effect lays out the formal problem Dickens was facing:

> When I devised this story, I foresaw the likelihood that a class of readers and commentators would suppose that I was at great pains to conceal exactly what I was at great pains to suggest: namely that Mr John Harmon was not slain, and that Mr John Rokesmith was. . . . Its difficulty was much enhanced by the mode of publication; for it would be very unreasonable to expect that many readers, pursuing a story in portions from month to month through nineteen months, will, until they have it before them complete, perceive the relations of its finer threads to the whole pattern which is always before the eyes of the weaver at his loom. (p. 893)

By the time of book 2, chapter 13, then, the complication of John Harmon's disguises and of his various self-incarnations requires direct explanation. Thus, "A Solo and a Duet" is devoted to Harmon's ruminations about himself, and about what has actually happened, most of which the book has only allowed the reader to guess. Free indirect discourse—a method that allows the narrator to be inside and out at the same time ("nowhere," perhaps)— might have eased the awkwardness, but Harmon's thoughts are represented as a lecture to himself, a long stage soliloquy. It has the quality of the melodramatic imagination in its efforts to explain everything. Melodrama is the perfect medium for ideas, for in it, as Peter Brooks has argued, nothing is spared because nothing is left unsaid.[24] Laying it all out, Dickens faces the difficulties of the narrative of self-annihilation: how does one embody nonbeing?

What is most striking formally about the chapter is that Harmon is doing some of the work of the omniscient narrator and that the dramatic struggle in the passage might be described as Harmon's effort to achieve the condition of that narrator—the only position within the narrative that, unobserved itself, allows vision of everything. Only from an omniscient standpoint can the action of the narrative, in all of its complicated relations, be made comprehensible—and that standpoint should not, theoretically, be available to any character within the world of the novel. Yet it is precisely such a standpoint that the epistemological ideal requires. The omniscient narrator, as Audrey Jaffe points out, occupies a position rather like that liminal state I have been describing.

In her discussion of realism and of *Our Mutual Friend*, Elizabeth Ermarth claims that the lack of narrative connection between the stories of Wrayburn and Harmon, "despite the deep implicit similarities between them, dramatizes the fact that the saving mutual consciousness that sustains the human world exists, as it must exist, transcendently invisible, independent of immediate circumstances."[25] The problem for realistic narrative, and even for this passage, which hovers between novel and stage, between omniscience and limited perspective, is that once the efforts at full knowledge are dramatized, they must be shown to be impossible. The transcendent and invisible consciousness that might save the human world is, even for omniscient narrators, very difficult to come by, and always impossible for actors within the novels.

Watching Harmon and Dickens struggle to make the invisible visible is part of the fascination of the passage, which manages to summarize the main plot of the story up to its midpoint, while at the same time achieving an effectively stagy sort of drama that almost gleefully treasures its paradox: "I have no clue to the scene of my death," Harmon begins (p. 422). Aside from the paradox, it is immediately interesting that Harmon is represented as speaking aloud, although of course he is talking to himself. The narrative will not let him sink into the nothingness he almost occupies and leaves him in one of those liminal spaces in which *Our Mutual Friend* abounds. But the passage is carefully located in particular places, and much of its tension, so clearly "premodernist" in its method, is between the precise corporeal location of the "speaker" and his felt condition as "nowhere." So he notes how what he is feeling is a "sensation not experienced by many mortals,"

> to be looking into a churchyard on a wild windy night, and to feel that I no more hold a place among the living than these dead do, and even to know that I lie buried somewhere else, as they be buried here. Nothing uses me to it. A spirit that was once a man could hardly feel stranger or lonelier, going unrecognized among mankind, than I feel. (Ibid.)

It is a moment of almost Cartesian minimalism. Shaking off the conditions of the busy world he daily occupies, Harmon, alone, enjoins himself: "Don't evade it, John Harmon; don't evade it. Think it out!" (p. 423).

The struggle to know that follows is punctuated with moments when, again with almost Cartesian insistence on what is absolutely clear and distinct, Harmon stops: "Now, stop, and so far think it out, John Harmon. Is that so? That is exactly so" (ibid.). In the course of these passages, given the authority of Harmon's own conscious affirmations, Dickens unfolds to the reader not only what has happened but the state of mind Harmon experienced when the things were happening. The chapter becomes a summary at the same time that it provides the reader with new information that it would have been difficult to infer from the earlier parts of the narration. Now, however, this information has the authenticity of the first person, along with the external guarantee that the first person is, as it were, the third person, hence "dead." The knowledge is registered as clear and distinct.

The concern about method, which corresponds, by the way, to the kind of care John Rokesmith, as overseeing secretary, gives to Boffin's affairs, makes this nondramatic narration a kind of epistemological venture. The punctuations to assure himself, to clarify, to be certain, are accompanied by other sorts of intrusions in the narrative. At one point Harmon pauses: "But let me go on thinking the facts out, and avoid confusing them with my speculations" (p. 424). In quasi-scientific mode, he refuses to allow anything but the facts in. Clearly, he is not in this monologue a truly "disinterested" investigator, but that is the point of the insistence on facts. Method impedes "speculation" and Feuerbachian imposition of desire on matter. I do not make hypotheses, Newton said, and the tradition that hypotheses are not to be trusted (they are, after all, fictions) pervades nineteenth-century fiction itself. Bernard, however, turns away from this tradition, finding that hypotheses are virtually a condition of good experiment, and implying that the pure pursuit of truth by way of self-annihilation cannot be sustained for long.

It is in another liminal state that Harmon describes his decision to conduct his experiment on Bella. Between life and death he "proves" Bella. Appropriately, it was an earlier version of this condition that got Harmon into his current difficulty because he had decided to switch identities briefly with George Radfoot, and it was Radfoot who began the plot—and became the unsuspecting victim of it—leading to Harmon's "death." The death became a condition for conducting the experiment long term. "I have checked the calculation often," Harmon says to himself, "and it must have been two nights that I lay recovering in that public-house. Let me see. Yes. I am sure it was while I lay in that bed there, that the thought entered my head of turning the danger I had passed through, to the account of being for some time supposed to have disappeared mysteriously, and of proving Bella" (p. 427).

Of course, Harmon is not "dead" in the same sense that one might say that the narrator, seeing from "nowhere," is dead. Although Harmon has figured out his own past, he still cannot know the future. The retrospective, omniscient narrator sequence ends with another self-assurance, but also with the emergence of alternative possibilities that mark the instability of the future: "Now, is it all thought out? All to this time? Nothing omitted? No, nothing. But beyond this time? To think it out through the future is a harder though a much shorter task than to think it out through the past. John Harmon is dead. Should John Harmon come to life?" (p. 428)

The internal colloquy that follows, since it has to do with decisions that follow upon the clarification of the past and are yet to be made, no longer occupies the territory of the omniscient narrator. The choice to remain dead, however, is a tempting one because the condition of death allows Harmon to know the goodness of others without either imposing on them or clouding that knowledge by his real presence. In particular, it keeps him from "purchasing" Bella "as a Sultan buys a slave." Part of the problem with the novel, however, is that throughout the later part of the book Rokesmith, having "succeeded" in his experiment and having detachedly reconstructed the past, moves from his liminal position, despite keeping up the disguise and the experiment far longer than necessary, and acts with the confidence of omniscience. Rokesmith, having nothing more to learn, acts with absolute confidence. When the inspector, who is still in the dark, comes to "arrest" Rokesmith, Harmon responds with the absolute confidence of someone who knows more than anyone else. He is not surprised by the inspector's visit and he is undisturbed by the accusation. When the inspector warns him that "whatever you say, will be used against you," he calmly replies, "I don't think it will" (p. 832).

The transformation of an actor into someone who is rather like the weaver with all the threads in his hand changes the tone and significance of the narrative. Dickens needs this sort of transformation — one might say it is a natural result of the experiments Harmon and the book have been conducting throughout — to achieve that final resolution that will be consistent with the traditions of the happy ending and with the epistemological narrative ideal. It makes of the experimental and extravagant *Our Mutual Friend* an apparently uncritical exemplar of this epistemological tradition. The affirmation within the text of unchallengeable knowledge requires of the most-valued characters something like the self-humiliation that the experimenter has required before the facts. Harmon carries all the weight of authority that scientists would have attributed to nature, and yet the consequences, for those who defer to this authority, seem quite different in life than in science. In effect, while the men are, as Gallagher describes it, "knocked out, drowned, dried out, stored up" (p. 364), and eventually reanimated with new value and new wisdom, the women, like Bella and Lizzie, can only defer to their authority.

Yet despite what I would call this lapse into transcendence, the transformation of Rokesmith into a fully informed knower, virtually everything else outside the liminal states of quasi-death is clouded by interest, money, and corporeality. The novel's retreat, as it were, its need to ensure community and a general consensus, both moral and epistemological, through some "invisible consciousness," is impelled by the utter opacity of the material, human world. That world is marked by the fragmentary perspectives of eccentric individuals and fragmentations of self like that of Rokesmith/Harmon in his soliloquy. Ironically, Dickens's intense concentration on corporeality and the grotesque pushes the novel toward a radical critique of the conception of transcendence and of the moral triumphs of self-surrender. At the same time, given the corruption of the world of getting and spending, it dramatizes the importance of transcendence. In effect, the novel plays out more or less self-consciously the same ambivalence about the "view from nowhere" that marks the epistemological tradition of dying to know.

Even for the best of the people, like Harmon and the Boffins, disguise and deception is a condition of survival. Every plot and subplot is marked by deceptions, attempts to spy on the deceivers, attempts to spy on the spies. The pursuit of knowledge, then, as a means to break through the mysteries of self and sheer materiality (the greatest obstacles to the production of scientific knowledge and of social coherence) becomes the work both of narrator and of characters within the novel.

The network of spying begins in the central plot device, Harmon's fake death and his experiment: he is always "spying" on Bella. The inspector lurks through the novel, seeking the killer of Harmon, and he does not disappear until Harmon emerges from Rokesmith. The whole Boffin household turns into a spying nest. Boffin's disguise as a miser is designed as yet another test of Bella, and both Boffin and Rokesmith watch her closely. But this activity is complicated when Bella feels the need to watch Boffin, to see whether he is getting as bad as he seems, and "on one point connected with the watch she kept on Mr. Boffin she felt very inquisitive, and that was the question whether the secretary watched him too" (p. 533). Her difficulty, of course, is that she has determined not to have contact with Rokesmith after his first proposal to her. But, "it fell out insensibly . . . that her observation of Mr. Boffin involved a continual observation of Mr. Rokesmith" (ibid.). Rather more dramatically and comically, the book lingers over the brutal mutual observation of Wrayburn and Headstone, during which Wrayburn leads Headstone on a mocking wild goose chase, provoking yet wilder spasms of bitterness in the schoolmaster. This sequence succeeds to an even more serious one, in which a spying Headstone attempts to murder

Wrayburn. Everyone watches everyone watching everyone: Wegg watching Boffin, Boffin, through Venus, watching Wegg. Wrayburn, in his turn, manages to wrest information about Lizzie's whereabouts from Jenny's father.

Perhaps the best justification for Harmon's "experiment" is that deception is the condition of virtually all relationships, loving or not. The Veneerings and the Lammles, with names that suggest false surfaces, set the tone. The Lammle marriage is one of the most impressive and bitter achievements of the novel, and that, too, depends on the characters' concentration entirely on their own self-interest. Among those who attend the high social dinners that punctuate the novel only the bemused Mr. Twemlow, low in the social hierarchy, seems to have a sense of value at all commensurate with the values the novel attempts to affirm. Twemlow's hapless and confused condition actually reaffirms the value and the necessity of detachment, for he is constantly used as an object in others' machinations; outside of society (though in it), he can see beyond its limits. Fledgeby notoriously cloaks his petty viciousness in Riah's Jewishness, and imposes on Riah the burden of his own greediness. The Wilfers, all except the hapless father, are utterly false. But—and this is a complication that makes the question of the relation between virtue and knowledge more difficult than it would seem—the good guys, too, use disguises. There is a rather sour flavor even in Venus's spying on and deceiving the villainous Wegg. It is almost worse, however benevolent and cheery it is supposed to be, that Mr. Boffin himself, as part of the experiment on Bella, deceives her—and, worse, the readers. Rokesmith retains his disguise long after he needs to. Though the novel pretends to endorse the sort of downright openness that Abby Potterson represents, its plot endorses many deceptions. Although detachment and self-repression are the conditions of success in this crazed world, these qualities are regularly compromised by their participation in the same kinds of deceits and interested interventions that mark the behavior of, say, Fledgeby and Wegg.

Most of the deceptions in the novel are connected with that other major tendency in the book—buying and selling people and parts. If the material world is virtually unreadable through its opaque and thick wall of interest, it is partly because all people are merely objects. Bella complains from the start that she was left to John Harmon "like a dozen of spoons" (p. 81), and Georgiana becomes an object of sale for the Lammles. Gaffer and Riderhood notoriously make their livings off the corpses floating in the Thames, and Mr. Venus makes his from recycled body parts. Lizzie is herself virtually sold by her brother to Bradley. When the Boffins seek an orphan to support in place of their "lost" John, the market in orphans is appallingly exposed, so that "fluctuations of a wild and South-Sea nature were occasioned, by orphan-holders keeping back and then rushing into the market a dozen together. But, the uniform principle at the root of all these various operations was bargain and sale" (p. 244).

Thus, while the central experiment of *Our Mutual Friend* pushes to the limits the possibilities of "objectivity" and becomes an embodiment, however awkwardly constructed, of this epistemological tradition, the experiment is set within multiple narratives and, in the end, can only be sustained by not attending fully to what other parts of the narrative seem to embody. The ideal for all characters is something like Harmon's kind of self-annihilation (we have seen it already in Lizzie), but Dickens seems not to trust the ideal in lower-class figures and to shift its implications for women.

Bradley Headstone ought to be the key figure in asserting the value of self-repression in pursuit of knowledge. As is well known, Dickens is unforgiving of Bradley, who is contemptuously described in "his decent black coat and waistcoat, and decent white shirt, and decent formal black tie, and decent pantaloons of pepper and salt, with his decent silver watch in his pocket and its decent hair-guard round his neck." Decency, a Dickensian virtue, hisses off the page at this young man, whose respectability is marred by the signs he gives of being still a kind of "mechanic." Headstone's heroic activities of self-improvement by way of rigorous self-discipline, allowing him "a great store of teacher's knowledge" (p. 266), receive no commendation from the narrator. As surrogate for Dickens he is allowed no escape, and all his powers of discipline and self-repression are taken as signs of some lack. His "was the face belonging to a naturally slow or inattentive intellect that had toiled hard to get what it had won, and that had to hold it now that it was gotten" (p. 267). Yet the ideal of science, "organized common sense," implied a democracy of intellect with a moral hierarchy according to degrees of self-restraint.

Bradley, the most complex and interesting of the bad guys in the novel, displays all the virtues required for the acquisition of knowledge and yet is denied what Harmon and Wrayburn, both beaten close to death, are given. The "animal," "fiery (though smoldering), still visible in him" (ibid.), marks him as contaminated by corporeality. In the contest for Lizzie Hexam, the less worthy man, in part because he is beaten (and undoubtedly because he is to the manner born), wins. Headstone, within the formula of epistemological virtue, fails because he surrenders to passion and desire. John Kucich has argued that in Dickens, "both passion and repression are always condemned to exchange their self-transcending energies for self-interested ones. . . . Short of death, desire is necessarily gratified within selfhood."[26] Dickens will not allow Headstone that absolute resignation of selfhood required for epistemological and moral salvation—resignation such as that which is attributed to Wrayburn when he tells Lizzie that he wonders if it wouldn't be better if he were dead. But the law of the dying-to-know narrative is that true humility will be rewarded by elevation, so that ironically, almost deconstructively, all true resignation in the narrative must have about it a touch of "interest" in the reward. Headstone, then, the "decent" and rigorously disciplined and repressed young man, dramatizes the reality that repression always threatens to be a mere disguise for self-interest, from which no true knowledge can follow.

But in the stories of women new difficulties in the narrative of scientific epistemology begin to surface. As Gallagher points out, the women, never beaten like Wrayburn, never arrive at that point of illumination in which dying and knowing conflate and matter is transformed from illth to wealth, from death to life. As in the autobiographies, the self-abnegation of women never somehow became an equivalent of the self-abnegation required of the men pursuing science, so in *Our Mutual Friend*, the self-abnegation of women is never equivalent to Harmon's almost literal self-obliteration. This might be, at least in part, because the very "disinterest" of the activity of experimental observation is built on self-interest (as is the ideal of scientific epistemology), and women may *never* express "interest" of this kind. Harmon understands that in order to know, he will have to eliminate himself; but he only wants to eliminate himself so that he will be able to achieve a satisfactory marriage. Just as behind epistemology there is always an ethical impulse, so for the "scientist" Harmon there is a relatively self-interested drive, although the novel sympathetically treats it as benevolent and just. Women are not interesting or redeemable because, from the start, they *must* be disinterested and have no *Bildung* of humility to undergo. If they need redemption, they cannot find it. Bella's story is as far as Dickens will go in redeeming self-interested women, and she is redeemed from the start by her love of her father.

Nor are women allowed true scientific disinterest. Like the Lizzie Hexam who instinctively runs to the rescue of the anonymous dying man, they can only satisfy their desires by not pursuing them, and they must rather act on exclusively moral grounds—the saving of life, for Lizzie. Women were not to be trusted in the attempt to rise above the level of the local and the particular to the universal realm of truth, or in the novel, to a full understanding of the complications that even informed readers might have had trouble disentangling without Rokesmith. In the end, women must be dependent for such truths on higher authorities—like Rokesmith, or like Dickens. A purely moral transcendence is all that is allowed to Lizzie and Jenny, but the emotive intuition that allows for such transcendence is largely incompatible with the kind of authoritative and intellectual transcendence that is available to men, no matter how much—perhaps because of how much—they are beaten and drowned.

In one sense, the story of Bella Wilfer perfectly reenacts the alternate kind of narrative—the move to nescience, rather than science, through self-abnegation. Women are not experimenters but the object of the experiment. While Bella is transformed in the course of the narrative into the kind of self-abnegating woman who "merits" a good man, she is not allowed the kind of resurrection that is granted Wrayburn—nor, of course, is she beaten to within an inch of her life. Gallagher points out that the liminal state I have been describing here (she calls it "suspended animation") is "exclusively masculine" in the novel, although the condition is "naturally feminine."[27] The male becomes "the transcendent subject," while the women are left in all their corporeality and particularity, not allowed the "reanimation" that the men achieve, and therefore have no access to the transcendent truth. They succumb to "authority" at last, and thus fail the moral/intellectual test the Cartesian narrative imposes.

The two primary heroines embrace their utter secondariness. Lizzie has rejected an education in order to stay with her father. She accepts Wrayburn's offer because she loves him, but by the end she can only cross the class line when Wrayburn is in his liminal state. And as he recovers, when he asks her what he can do to repay her for bringing him back to life *and* marrying him, she can only say (quite touchingly), "Don't be ashamed of me . . . and you will have more than paid all." And, she enjoins, "live to see how hard I will try to improve myself, and never to discredit you" (p. 824), although down to Eugene's last suggestion that he would do better by her if he were dead, the book has dramatized her intellectual and moral superiority to Eugene.

But yet more painful (given that Dickens was taking something of a risk in crossing classes with Lizzie and Eugene, and emphasized it in the novel's final pages), is the fate of Bella. Her narrative does follow the "pilgrim's progress," but she is never alone in that progress. By the end, having married "Rokesmith," she acquiesces in ignorance and authority with an absoluteness that is "transcendent" only in its abjectness. Bella marries him without knowing his real name, and is undisturbed by the fact. When the police come for Rokesmith, she is unshaken. When he is under suspicion of murder, she can only say, "My beloved husband, how dare they!" (p. 829) Loving his experiment, Harmon buys her a new house without her knowing it.

The final twist to the epistemological plot here is that female "transcendence" is virtually descendental. Bella manifests her goodness through her ungrudging ignorance. That ignorance is only dispelled in a scene in which it is the Boffins who tell her to whom she is married and it is they, though even here by indirection, who explain the "experiment." Bella's appropriate response, after all the revelations are made, is to claim that she will never be able to "understand" how "John could love me when I so little deserved it" (p. 848). In the happy resolution, all pleasure stems from self-sacrifice and self-forgetting. In the end, to support epistemological virtues of the liminal deathlike state in Harmon and Wrayburn, the novel invokes two women whose vocation is precisely not to know.

Dickens's efforts to validate the move to transcendence of self that would affirm the epistemological ideal requires pyrotechnical extravagances of plot and characterization: only with such pyrotechnics can self-denial and self-humiliation be taken as conditions of success in the world the novel describes. The novel ends in virtual somersaults of self-denial, all affirming the deep distrust of personal interest that pushes some of the characters to the brink of death. The trial of men is to purge their way into the condition of women. As active and aggressive males they are inevitably contaminated both by desire and materiality (the "illth" that Gallagher invokes). Only when purged of selfish desire, disarmed by their liminality, do they gain access to the fullest knowledge. There is a long string of feverish men in Dickens's novels, Pip not the least of them, who have their selfishness burned away.

The trial of women is to rescue the men and submit to them. Women like Bella Wilfer, gesturing toward a greed that turns them immediately into objects for sale, are guided out of their mere embodied state by men who have themselves been disembodied. The men, in the vulnerability and passivity of their enforced liminality, become like women—but return like men. For the most part, then, the women need not go through the purgative experience that takes self-sacrifice toward death (and raises consciousness above the particulars at the same time). If they are to be heroines they occupy the material world without interest. They fall into the arms of the resurrected man, and they surrender the self. They belong, then, to a material world in which resurrections don't happen. Like Bella, they are left in the ignorance of particularities and in a selflessness that participates in (and as in Lizzie's case, can give the inspiration to) the transcendence acquired by the man. Bella Harmon is distinctly *not* dying to know, only to be good.

Yet neither the novel nor the epistemological story can do without women. The women's secondariness has a double effect on the experiment and the epistemological implications of *Our Mutual Friend*. Their move toward nescience rather than science is perhaps the most powerful evidence the book offers that its model of knowledge and resurrection does not work after all. The true heroines of self-denial demonstrate that absolute self-denial is incompatible with knowledge. In the end, the women, however marvelously Jenny and Lizzie participate in the resurrection of the protagonist who should have been dead, are dependent on men for the "elevation" that follows upon self-denial; female self-sacrifice has no epistemological virtue after all. And *Our Mutual Friend,* a powerful dramatization of the narrative of scientific epistemology, perhaps against its own intentions, dramatizes its inadequacy.

At the same time, the centrality of women to the epistemological narrative dramatizes how critical it is to the epistemological venture that the separateness of desiring and material selfhoods be transcended. If Dickens has not quite got it straight, setting up a sort of Cartesian dualism of his own (with which he is deeply uncomfortable), his novel labors to find ways to transcend the divisive and selfish energies that constitute the society and the personalities he embodies. To transcend requires *both* knowledge and the selflessness that allows acceptance of otherness. For the men to achieve acceptance requires violent beatings; for the women to acquire knowledge requires absolute submission. Something is wrong here, but the aspiration to an ideal of loving detachment, of objectivity energized by feeling, struggles, in *Our Mutual Friend,* to find embodiment.

— 8 —

Daniel Deronda: A New Epistemology

Unlike *Our Mutual Friend, Daniel Deronda* is explicitly about questions of knowledge. I have read Dickens's novel as though it represented, though in an extreme and precarious form, what scientific epistemology might look like if it were narrative, its ideal constrained by embodiment in particular characters within a particular society at a particular moment. There, questions of gender and class complicate the ideal; and there, too, the conjunction of moral motive and intellectual need fails at last, for the women sustain the moral qualities of the pursuit of knowledge, but only the men acquire knowledge, and control. Much more the intellectual than Dickens, George Eliot was, by the time of the writing of *Daniel Deronda*, self-consciously concerned with the implications of the limits of that narrative and with epistemology, and the problems suggested by Dickens's novel. In particular, she plays out a theme that had long preoccupied her: the question of the relation between knowledge and feeling, which in Dickens, as we have seen, produces a kind of gendered schizophrenia.

Daniel Deronda is an extremely dissatisfied book, and from the start, it pushes at the edges of the ideals of knowledge, ethics, narrative, and religion to which, at the same time, George Eliot wanted to cling. Notoriously, throughout her career, she took her often radical insights and experiences back into a conservative social political position. By the time of her last full novel, her disillusion with her own society led her to seek even more intensely for those ethical and epistemological ideals that might serve as stabilizing forces, both against the increasing complacency of a disenchanted contemporary society and against the corrosive possibilities of a growing relativism, even in her own restless mind.

So, while the ideal of "dying" lies at the heart of this novel also, it lies there with a difference, and within a narrative that struggles against it. Self-abnegation and the realist openness to the hard unaccommodating actual had gone together in her works. Moreover, her heroines—unlike those of *Our Mutual Friend*—cannot acquiesce in an uninformed deference, but they cannot assert their desires without being punished, either by society or by nature. In this respect George Eliot's treatment of the question is continuous with Dickens's. But *Daniel Deronda* is more ambitious, more restless about the possibility of large-scale heroism, more uneasy about dying, more self-conscious in its confrontation with the ideals of scientific epistemology. I want to talk about the novel here as though that confrontation were its fundamental subject.

The Cartesian narrative leads inevitability to the crisis of Western epistemology: remove God from the narrative sequence by which Descartes climbs from the cogito to an authoritative view of the truth, and solipsism reemerges. Western philosophy has struggled persistently with the problem of whether the mind is capable of getting outside of itself. How can we know anything about what isn't us when standing between us and the world is that enormous, overshadowing, often inchoate self, which filters all signals from the outside, obtrudes its desires on everything, limits the angles from which the outside can be perceived, and forces the very distinction between inside and outside? Within the framework of this threat, eliminating the embodied self is the task of epistemology if any science—and I extend the term here to mean any systematic natural knowledge—is to be possible. Obviously, any such effort entails ethical choices, not only because the self must be restrained, diminished, eliminated, but because only by breaking the constraints of the self can we make true contact with other people and not simply impose ourselves upon them.

But the task for George Eliot was particularly difficult because she already perceived that knowledge was dependent not on the fact out there, which Kant had demonstrated could not be known, but on the condition of the perceiver. Eliminating the self from a knowledge system in which the self constructs all knowledge might then be tantamount to eliminating all knowledge. George Eliot's sophistication in philosophical matters put her in touch with a "relativist" tradition, identified by Christopher Herbert, that was developing in the midst of the Victorian rage for objectivity. Herbert points out that the idea that knowledge is only of relations, and that it is mediated, and created entirely by consciousness, gained increasing power through the nineteenth century in England.[1] In her novels, George Eliot notoriously emphasizes the plurality of possible interpretation and in her philosophical reflections insists on two apparently incompatible elements that Herbert has shown to mark most relativist thought: the ultimate unity of all things and their virtually infinite multiplicity. As the narrator of *The Mill on the Floss* puts it, "In natural science, I have understood, there is nothing petty to the mind that has a large vision of relations, and to which every single object suggests a vast sum of conditions."[2]

The epistemological crisis, regardless of rather noisy and confident advocates on both sides, is still very much with us. The urgency of its implications, despite some views that epistemology is the least important of activities, is exacerbated by the ever-increasing crossing of boundaries that marks global society, and by the increased violence that marks the emergence from colonialism of cultures hitherto virtually unknown (or unreflected upon) in the West. Much contemporary social theory has focused on such events and has exposed the provinciality of Western thought about other cultures. The post–World War II assault on Enlightenment universalism and increased Western embarrassment at a long history of empire have produced, along the way, a salutary skepticism about attempts to know other cultures. But the ironies are great because, as we recognize increasingly the degree to which attempts to learn about other cultures had for many centuries been aspects of often brutal imperialism, the skepticism has produced a view that one *cannot* know them.

The other side of relativist (or quasi-relativist) doubt and the determination to rethink relations among cultures, abandon the notion of the "primitive," and welcome global diversity, has been a strong, even embarrassed admission that it is not possible to know "the other," that the mysteries of other cultures cannot be ethnographically discovered without imposing one's own culture on what one "discovers," that the very effort of discovery is corrupted by ideologies of conquest and empire. In the midst of an explosion of knowledge and international communication, certain aspects of intellectual work take the shape of Descartes's Great Doubt, without the saving qualification that while we wait to find out the truth, we must act as we always have acted, and without the sense that it is ultimately possible to find one's way to the foundation of the clear and distinct. Recent forms of skepticism are built on what is assumed to be sophisticated recognition that the ideal epistemological condition of *Selbstödung* is impossible in a world of contingency and that in its name a lot of pretty terrible things have been perpetrated.[3]

It is something like this crisis that George Eliot confronted when she wrote *Daniel Deronda*. In recent years, that novel has become the focus of attention in George Eliot studies after the many years of half-dismissal that followed on the criticism of Henry James and, seventy-five years later, of F. R. Leavis.[4] Unlike contemporary theorists and critics, George Eliot was unhappy in her skepticisms and *Daniel Deronda* has become especially interesting, not because it is a better novel than *Middlemarch* (virtually no nineteenth-century novel is) but because it is so extreme in its test of many of the attitudes that contemporary critics—political, feminist, poststructuralist, for example—find inimical. The "resolution" of the novel does not radically change George Eliot's attitude toward women, politics, or "reality," but it tests the limits. Just as *Our Mutual Friend* can be seen as an extreme experimental exploration of the possibilities of self-sacrifice and death, so *Daniel Deronda* can be seen as George Eliot's experimental reexploration of the possibilities of objectivity in the light of the cultural crises that she believed threatened to corrode society.

Disinterest and objectivity are always impossible and always necessary, as "aspiration," as Amanda Anderson suggests, rather than as achievement.[5] Dominick LaCapra makes a useful distinction between "objectivism" and "objectivity": objectivist epistemology implies the untenable "objectifying status of a transcendental spectator," but one can, he argues, "defend objectivity in a delimited sense that implies the need to counteract projective reprocessing or rewriting of the past and to listen attentively to the 'voices,' notably when they pose a genuine challenge by resisting one's desire to make them say what one wants them to say or to have them become vehicles for one's values and political agendas."[6]

The Victorians are often accused of dangerously scientizing a falsely "objective" anthropology. But "the late eighteenth- and nineteenth-century literature of pre-professional modern ethnography," as Christopher Herbert has shown, rather than "mystifying the moral and epistemological predicament of the European observer in primitive society . . . enacts the conflictual interplay of prejudice and ethnographic experience . . . with an anguished frankness."[7] Herbert shows that evangelical ethnographers were often caught in irreconcilable tensions between a determination to transform pagan cultures into Christian ones and the need to know those cultures with something like "objectivity." As the impetus toward creating the social *sciences* grew, the question of how to achieve the sort of objectivity required to get to know other cultures or even one's own became critical. Social activism, professionalization of science, epistemology, and ethics ran into one another, and clergy, philosophers, scientists, novelists, and poets worried the issue.

The practical urgency of the task of knowing others required strenuous intellectual and moral work. Herbert talks of how

> the cultural anthropologist's idea of the necessity of undergoing "an extremely personal traumatic kind of experience" as the prerequisite of shedding prejudice and thus attaining ethnographic truth (defined as entering into another conceptual world) reproduces closely the Evangelical salvation narrative in which an influx of awareness of sin is imagined to be the prerequisite of the shedding of egoistic selfhood and the spiritual new birth which follows.[8]

The parallel between this salvation narrative and the one I have been describing as the narrative of scientific epistemology is obvious.

The most famous Victorian writers, often now indicted for parading the bourgeois subject as a universal, were engaged in just the kind of struggle Herbert describes. When they found themselves skeptical in ways that anticipate our own hermeneutics of suspicion, they were provoked into anguished struggles to get it right. Thus again, epistemology was charged with the ethical as Victorians sought to overcome a splintering and selfish modernity given over, Matthew Arnold believed, to "doing as one likes." The moral failures condemned by everyone from Carlyle to George Eliot were the same as the epistemological ones: acquiescing in the obscuring desires of the self. It has been strongly argued that the Victorians' moralized resistance to self-assertion was another form of the bourgeois policing of personal identity.[9] Yet this claim itself is an assertion of the possibility of knowing not only the overt views of others but their hidden motives. To denounce the Victorians for their false ideology is to allow the possibility that they *could* actually know the other cultures that they are accused of repressing.

The Victorians are for us another culture, though they reside not in a different geography but in a different time. We know a lot about them, and to begin now denying what we know because there are obvious limits to any kind of knowing would be to undercut the very enterprise of critique that, for example, cultural studies now takes as their object. Moreover, as Herbert has suggested, Victorian culture harbored all the multiplicities and complexities that make talking about a *singular* culture problematic for us now. Finally—and here is where my own argument will focus—the Victorians form part of what Anderson calls the "genealogy" of our own relation to detachment, to the objectivity that manifests itself so often in the form of dying to know. George Eliot, more than most, explored the possibilities of detachment and tested the most forceful objections to it.

While the commitment to scientific knowing was very Victorian, so, too, was the romantic distrust of the potential heartlessness of detachment. Pure rationality cannot give access to the life of a past culture: after the appropriate Dryasdust saturation in the archives, one needs, as Carlyle insisted, an act of imaginative sympathy. "The journey into history," John Rosenberg claims, "is for Carlyle always a double journey, backward in time and downward into the self."[10] Self obscures, but it is also a condition for sympathetic knowing. Only those of George Eliot's protagonists who have strong egos are capable of the sympathy that produces real moral action, which is why Dorothea, in that opening scene as she looks at the jewels with her sister, is represented as vain about her manifest humility. But the way to know the other is through some universal uncontingent self, Schiller's "Person," for example, who would shortly become Arnold's "best self." As Carlyle affirms, access to what lies outside the self is at best fragmentary, but through the obstructions of "foolish noises," to perpetually hindered eyesight, "some real human figure is seen moving: very strange; whom we could hail if he would answer—and we look into a pair of eyes deep as our own, *imaging* our own, but all unconscious of us; to whom we, for the time, are become as spirits and invisible!"[11]

George Eliot's work belongs to this romantic extension of the Enlightenment enterprise. While attempting to resist the merely rationalizing and disenchanting methods of science (the image of her at work translating Straus' *Das Leben Jesu* with Thorwald's crucifix above her desk accurately captures this resistance), she put her faith in a passionate rationalism, what Arnold called "imaginative reason," and for both it was a means to true community. Nevertheless, her programmatic realism threatened to disenchant the world so completely that there would be no place for the passion, and the obliteration of self began to feel rather more like mere self-destruction than a means of access to the other.

If we can know an atom, a star, a geological formation, we can do so by overcoming or circumventing the obstacles of self. Lewes was impressed with the astronomers' use of a formula called "the personal equation," by which the inevitable physical limitations of any observer of the stars can be overcome by averaging the conclusions of many different observers.[12] We have seen the arguments of Daston and Galison about the ways nineteenth-century science refined instruments to overcome the personal limitations of scientists, and Theodore Porter describes how statistics, in a positivist tradition (in which George Eliot was much implicated), were designed to overcome both the fallibility of human experimenters and the irregularities of experimental data.[13] Scientific epistemology moves away from particulars and contingency toward broader, more inclusive, more "objective" laws.

Thus, the experiment of *Daniel Deronda* develops because George Eliot was determined to find narrative strategies that might allow for the possibility of knowing the other—the past, other cultures, the Jews of the novel—while at the same time honoring both the rational scientific project and the various kinds of opacity that her empiricist commitments seemed to entail. She needed a rigorous epistemology that did not demand utter self-annihilation but that allowed for passion, imaginative vision, and the satisfactions of desire. The stakes were high and the enterprise almost oxymoronic in its difficulty.

The structure of *Daniel Deronda* suggests that her disillusionment with contemporary society exacerbated what must have been her disillusionment with the epistemology that underlay her earlier work. Maggie Tulliver's story, for example, attempts to affirm "the onward tendency of human things" through the young who rise "above the generation before them" (p. 363), and gradually transform the culture; Dorothea Brooke's fate, echoing more feebly that hope, is responsible in part for "the growing good of the world." But this left neither with much to show in this life, and the ideal of such a future was mocked by current petty and painful realities—the great historical change of the Reform Act in *Middlemarch* reduces to a comment about "those times when reforms were begun with a young hopefulness of immediate good which has been much checked in our days."[14] The frustration of noble-minded heroines, rendered as morally ennobling, had become increasingly unsatisfactory. The culture's deep provincialism, parallel to what George Eliot saw as the overwhelming constraints of a limited and limiting self, seemed to her increasingly dangerous, both in personal relations and in social and international ones.

Daniel Deronda implies that one price of realism is a vision of a society emptied of meaning and a fate for its protagonist (in this case, heroine) utterly desolate. Achieving true knowledge of the other, conditional on a stern refusal to make any demands for the self, regularly entails failure or death. The narrative equivalent of the epistemological ideal of self-annihilation, novels like *The Mill on the Floss* or *Felix Holt* had suggested to George Eliot that something is wrong with the Carlylean ideal. One may not live adequately with a self-asserting self; but one cannot live at all without it. *Daniel Deronda* explores the possibility of selflessness derived from the energies of egoism. Egoistically annihilating egoism, the heroic figure might at last gain access to a world that is not self, learn from and do justice to that "other" who lies just beyond the walls of ego.

Whatever else it may be, *Daniel Deronda* is a novel about epistemology. It is, as Alexander Welsh has claimed, "a critique of knowledge."[15] From its first moment—in what might be called a prenarrative epigraph—it announces the epistemological difficulty. Having resisted from her earliest work the fantasies of mere fiction, George Eliot nevertheless begins by announcing that "[m]en can do nothing without the make-believe of a beginning."[16] Post-Lyellian science is the literal source of this truth that announces its own fictionality. The strongest realist truth is that its narratives are fictions. Truth and fiction are intricately coimplicated and any serious epistemology needs to take into account human limitation, which invariably is a subject of realist narrative. The epigraph announces not only the epistemological enterprise that the book will try to work out, but, by implication, the necessity for revisions of realist narrative form if an adequate epistemology—and therefore an adequate practice—is to be imagined.

It is interesting to juxtapose *Our Mutual Friend* with *Daniel Deronda* because their modes of treating questions of knowledge are linked to similar dominant motifs. In both novels, British society is utterly hollow, and their protagonists are afflicted with a neurotic incapacity to know themselves or what they should do. Wrayburn shares with Deronda this modern fate. His growth to moral knowledge leads him to marry outside his class; Deronda finds his identity and vocation through a relationship that entails a kind of mixing of "races." The female protagonists, Bella and Gwendolen, are educated from overwhelming egoism (though they are in other respects quite innocent) into a self-submission that validates their lives. Bella escapes with something because she is prevented from marrying a Grandcourt by the book's "experiment." Gwendolen ends with nothing but her new knowledge.

Both books, too, are importantly concerned with Jewishness. The way to knowledge is through another culture, although *Our Mutual Friend* isolates Jewishness in a single idealized figure. The juxtaposition of cultures operates in both, and certainly not to the favor of complacent Christianity. Both books aim at opening to another culture by representing Jewishness as misunderstood. Riah is no Mordecai, it is true, but his emblematic relation to Fledgeby suggests the epistemological difficulty of understanding and valuing another culture and also its importance. The striking parallels in novels almost antithetical to each other suggest how the narrative of scientific epistemology hooks into narrative forms and practical moral and social issues.

Dickens's solution to the linked crises, the gendered splitting of moral and epistemological project, does not work for George Eliot. Where Dickens squeezes his rather daring narrative into conventional comic form in which literal marriages resolve all, George Eliot refuses that form. Gwendolen does not marry, and though Daniel does, he moves immediately beyond the range of Victorian domestic realism, and the last words are from Milton's *Samson Agonistes*.

But it is in her engagement with the difficulties of truth-telling that George Eliot's novel is forced to break the molds of the traditional forms of Victorian domestic realism. Her sensitivity to the difficulty of getting it right—her awareness through metaphor of the "defective" or scratched mirror—emerges in a wide range of details and commentary in *Daniel Deronda*. Even in the most solemn and truthful moments interest might be the true driving force of truth claims, as when Deronda tries to persuade Hans not to exhibit his painting of Mirah in public.[17] The enterprise of mirroring, which had such metaphorical force for George Eliot in her earlier work, implies refusal to impose one's will and desires on the mirrored world; but in *Daniel Deronda* this refusal becomes bleak and potentially demoralizing. Telling the truth does not reward the heroine, as Bella and Lizzie are rewarded, with her heart's desire.

But even in *Middlemarch,* George Eliot works within comic form. Dorothea's dark night of the soul, which concludes with her awakening to "the largeness of the world and the manifold wakings of men to labour and endurance," was meant to be and is a moving climax to her story. It marks the turn that a chapter later will evoke from Rosamond Vincy the information that Ladislaw loves Dorothea, not Rosamond, and thus makes possible the more or less happy ending. But Dorothea is "elevated," in Tyndall's phrase, or gets her prize because she accepts the self-diminishment the vision implies: she comes to understand that "she was part of that involuntary, palpitating life, and could neither look on it from her luxurious shelter as a mere spectator, nor hide her eyes in selfish complaining."[18]

When an almost exactly similar moment befalls the less idealized heroine of *Daniel Deronda,* the bleakness of self-diminishment becomes clear. For Gwendolen it leads to no narrative prize. As Deronda reveals to her that he will be marrying Mirah, Gwendolen was "for the first time being dislodged from her supremacy in her own world, and getting a sense that her horizon was but a dipping onward of an existence with which her own was revolving" (p. 876). As for Dorothea, it is a moment of moral regeneration, but it also marks her final marginalization. Her perception of a world larger than herself from which she is in fact excluded is painful enough. But worse, the novel confirms that marginalization in its own form by turning away from her, as *Middlemarch* does not turn from Dorothea or as *Our Mutual Friend* does not turn from Bella. It focuses instead on another romance, or two romances, the wedding of Daniel with Mirah and the spiritual betrothal of Daniel and Mordecai. The last paragraphs are not about Gwendolen's "incalculably diffusive" good effects, but about world-historical enterprises. Gwendolen's pain is the end of her story.

The ideal of self-sacrifice was a troubled one for George Eliot. She had produced a series of women whose self-assertions could find outlet only in their capacity for self-sacrifice. They live out that ideal of true "resignation," as Comte puts it—"a permanent disposition to endure, steadily, and without hope of compensation, all inevitable evils." They can do this, Comte claims, because they have "a deep sense of the connection of all kinds of natural phenomena with invariable laws,"[19] laws, that is, that cannot be changed by human intervention. Yet *Romola,* a critical novel in George Eliot's development, is repeatedly about "the problem where the sacredness of obedience ended and where the sacredness of rebellion began,"[20] and this formulation explicitly recognizes that rebellion *is* potentially sacred. Both Dorothea and Gwendolen learn the Comtist lesson in their visions of a world spinning on without them, but even if we put aside the unlikeliness of Gwendolen's knowledge of invariable laws, she has nothing (but her mother) to sacrifice herself for. That Gwendolen's is not the novel's decisive story suggests that resignation without hope closed out too many possibilities. The "Jewish part" opens up, at last, those possibilities.

It is in the "Jewish part" that the results of George Eliot's epistemological experiments are represented, though it takes the entire book to create the experiment. The Jewish question is also an epistemological question, asked out of moral exigency. George Eliot seeks a new answer and breaks the generic mode of her earlier novels and of Dickens's because that mode had been made to correspond all too exactly to the "dying" narrative. For the ideal of self-annihilation that "redeems" Gwendolen supports a realist epistemology. But if the energies of self, desire, and imagination enter into the act of knowing, knowledge turns into action, and reality into a human construction. Deronda, then, turns into a Jew who will abandon the realist Gwendolen. Is reality out there like a hard determinate thing, or does it partake of the nature of human mind itself? The book's epistemological crisis is not abstract but has to do with nationhood, other cultures, and political action. And *Daniel Deronda*'s various plots replay a central issue: is it possible to know anything, but particularly other people and cultures, without imposing on them the distorting desires of the aspiring self, which science had tried to banish (yet without which there would have been no particular reason to understand the "other" in the first place)?

The "Jewish part" of the novel is no late-career aberration but a continuing concern, and it could not be disentangled from the question of epistemology. From George Eliot's point of view, readers' resistance to the "Jewish part"—which she had predicted—enacted the conditions the novel self-consciously set out to correct. As she wrote to Harriet Beecher Stowe,

> There is nothing I should care more to do, *if it were possible,* than to rouse the imagination of men and women to a vision of human claims in those races of their fellow-men who most differ from them in customs and beliefs.[21]

I am not claiming that the rejection of the "Jewish sections" by critics so distinguished as James and Leavis is based merely on unarticulated cultural prejudices—although Jane Irwin shows that the language of the rejection retains strong traces of anti-Semitism.[22] My own early resistance to the novel was comfortably Leavisian. However, there is much in the Jewish sections that is both "realistic" in the same sense that the Gwendolen sections are, and rigorously thought through. The epistemology at which George Eliot arrives in her struggle to find a way to know *sympathetically* another culture (and one whose history is marked by the antipathy of her own) is no pipe dream. The problems in constructing such an epistemology were enormous, and the easy notion of "failure" says nothing about what matters in the effort and the achievement of the novel. George Eliot's belief that it would be rejected was not merely artistic defensiveness but an appropriate alertness to the profundity of cultural prejudice. The problem of where intellectual coherence ends and cultural and social prejudice begins is one of the problems worked out in this epistemological venture.

It would be a mistake, given the now prevalent recognition of George Eliot's ideological conservatism, to overlook the degree to which the experiment was a deliberate risk. George Eliot risked the kind of controversy that marks our own elaborately multicultural enterprises and that now dominates the criticism emerging from cultural and postcolonial studies. She wrote to Stowe that the "impious and stupid" attitudes of Christians not only toward Jews but toward "oriental peoples with whom we come into contact."[23] impelled her to write sympathetically of the Jews. Grandcourt in dinner conversation talks of the Jamaican negro as "a beastly sort of Baptist Caliban," while Deronda "said he had always felt a little with Caliban, who naturally had his own point of view and could sing a good song" (p. 376). This polite table talk suggests George Eliot's embittered relationship to the contemporary society that sanctioned great ignorance and unselfconscious brutality.

The alternative to Grandcourtian exploitation is not the ideology of the "white man's burden." The alternative, on George Eliot's view, is to understand the other, and this is only possible if one works at it. "Only learned Rabbis," wrote Lewes about George Eliot, "are so profoundly versed in Jewish history and literature as she is."[24] Knowledge precedes just action, she, like Matthew Arnold, seems to believe, but the energy generating knowledge is moral, and it is determined by feeling. The epistemological question in *Daniel Deronda* is also a question of personal feeling. If George Eliot failed to rouse the imaginations of her readers to understand the Jews, it would, even on her own accounting, have been the consequence of an absence of desire to know the truth. Fiction itself requires the rigors of intense intellectual work and scholarship. In her preparations for the novel George Eliot enacted the very work of trying to know another culture that is the novel's subject. Her labors were no assertion of power or easy ecumenism but of resistance to the ignorance and condescension that marked English relations to Jews.

The difficulties on the way to this knowledge fill the book. Virtually every scene focuses on failures of understanding, obstructions put up by desire, custom, habit, prejudice, and ignorance to recognizing the complex reality of others—all of Bacon's caves are crowded. As against the omniscient stance of the traditional nineteenth-century writer and the kind of quasi-omniscience to which Rokesmith aspires, the primary narrative mode of *Daniel Deronda* is psychological exploration through free indirect discourse. This allows for constant juxtaposition of alternating and diverging perceptions and implies, in the minuteness of the rendering of the characters' minds, both that knowledge is almost always tenuous and dependent on the particular angles of perception and feeling of the characters *and* that access to the not-self, to the reality of others, might be possible. Minimally, what is required is a constant alertness to one's own inevitable partialities, to the intrusions of interest, prejudice, intellectual laziness and habit. The brilliant manipulations of the time sequence in the first fifteen chapters—replaced later in the novel by juxtapositions of the two major narrative lines—slowly open a way of access both to Grandcourt and to Gwendolen and to the significance of the events and questions of the first pages.

The voice of the narrator certainly has, for the most part, the quality of omniscience. At times it seems the voice of the view from nowhere,[25] but it registers the diversity of thoughts, the limitations of feeling and perspective, in ways that make the novel's struggle to imagine a mind capable of transcending the limits of empiricism and the limitations of perspective intensely dramatic and problematic—problematic enough, the history of George Eliot criticism makes plain—so that only a few readers have in the end been willing to accept either Mordecai or the mode of knowledge he represents. But where the epistemological crisis in *Our Mutual Friend* produced the awkwardness of the critical soliloquy in which Harmon hovers between the condition of omniscient narration and of befuddled actor on the scene, George Eliot refuses the strategies of melodrama. The difficulties push her to other sorts of awkwardness, but the narrator is fully conscious of the problems (and the awkwardness). The narration repeatedly dramatizes the struggle and the intellectual and moral responsibility to know the other, and that is also the burden George Eliot wanted her reader to bear: to know the other, to believe in the possibility, and thus to know and understand the Jews, Daniel Deronda, George Eliot, and her text.

Empiricism always threatens to pull back from reality to the sensation of reality to a solipsistic self. Like Lewes, strongly committed to empiricism, George Eliot had repeatedly resisted its solipsistic pull, usually by tracing the enlightenment of a major character but always through the narrator's increasingly complicated rhetoric. Is it possible really to know another object, another person, another culture, another text? The question implies a belief that there is a discoverable reality rather than one shaped by our consciousness of it.

Insisting on the possibility and necessity of sympathetically knowing an alien culture, the novel attempts to teach its readers how to read it. It, too, is an alien text. The opening epigraph begins the process of forcing the reader to consider the conditions of representation and the problem of the relation between fictions—"lies," as Carlyle would have it—and truth. It first deceives the reader into thinking Gwendolen the heroine and into wondering whatever happened to the titular hero. Gwendolen bears the ordinary egoistic consciousness that is disrupted at every moment by events, and the reader is led, like Gwendolen, into the surprises that transform the novel into an exploration of utterly alien territory. A strange thing talking about strange things, the novel attempts to induce the reader to know and value the strange by breaking through the constraints of an obscuring self such as that which determines Gwendolen's fate early in the novel. Her emergence from the depths of selfhood figures the progress of the reader into an understanding of how to read the "Jewish part."

Gwendolen's story is as critical in addressing the problem of knowing the other as Daniel's search for Mirah's brother and for his own mother. The two stories depend on each other in the epistemological experiment. The discovery that Deronda had a life altogether apart from her gave Gwendolen "a shock which went deeper than personal jealousy" (p. 876). The otherness of Deronda's life is for Gwendolen and the reader something to be discovered, beyond the limitations of the other selves who have busily interpreted and misinterpreted him throughout the novel. Deronda, though known inwardly by the narrative, is Gwendolen's other, and ultimately the reader's. Opening all "facts" to interpretation, the narrator never minimizes the difficulty of discovery. But her aim *is* discovery. As Suzy Anger argues, George Eliot "enacts interpretive conflicts not to demonstrate indeterminacy, but instead to reveal the conditions for correct interpretation."[26] The problem, then, is how best to make such discoveries, and its urgency and seriousness is conveyed to a skeptical reader by the way the story demonstrates that our moral actions depend on what is there to be discovered. The connection of knowledge with action is manifested in the fact that Deronda's discovery of his Jewishness determines his decision to work to implement Mordecai's vision.

The narrative of Daniel's growth and change is a critical part of the epistemological (and generic) experiment. His problem has a distinctively modern flavor, new for a male protagonist in George Eliot's work, in that it reverses the realist disenchantment narrative by *beginning* in disenchantment. Franco Moretti claims that in *Daniel Deronda* "the hero is so mature from the very start as to dissociate himself suspiciously from anything connected with youthful restlessness."[27] Whereas the quest narrative of the bildungsroman leads a misguided youth through disillusioning adventures to maturity, from country to city, Deronda moves maturely inside a romance narrative that takes him through a series of self-abnegating adventures out of the metropolis entirely, to the Near East. Utterly selfless, Daniel begins not by imposing his desires on the world, but without any clearly defined desires at all, with a defect of will, an inadequate selfhood. Thus, while Daniel's mind is rendered in minute detail, its strangeness creates a distance that, for many readers, has never been overcome.

Each of his adventures places him in a secondary position. His story seems gendered female, down to his loverlike relation to Mordecai, to whom he is more firmly wedded than he is to Mirah. When Daniel and Mordecai meet, they do so "with as intense a consciousness as if they had been two undeclared lovers" (p. 552). Gwendolen is the star of the opening chapters, Deronda her secret conscience; Hans is made the star at Cambridge through Deronda's sacrifices for him; Mirah, rescued and protected by Deronda, becomes the singing star of high-society London. Ironically, then, Deronda's story requires the reverse of the moral chastening of, say, an Adam Bede or an Arthur Donnithorne or even a Maggie Tulliver, who is far more rebellious and adventurous than Daniel. Unlike figures like Pip or Pendennis, but like the protagonists of the female bildungsroman, Deronda must learn not to deny himself but to assert himself; he must learn how to desire.

The first major epistemological crux in the novel comes in the description of Deronda's mind as he floats down the Thames on the day he rescues Mirah. For the most part, the passage is in free indirect discourse, as is much of Gwendolen's experience. But where in the Gwendolen passages that remarkably rich and flexible technique allows for juxtapositions and ironies, in the Deronda passages George Eliot tilts back toward Dickens's technique in *Our Mutual Friend:* his thinking is sometimes rendered inside quotation marks, almost as interior speech. The narrative immersion works like a strong endorsement, rather than a dramatic representation, of the movements of his consciousness. Psychological realism is joined to lyric intensity in a way that has disappointed critics for one hundred years, but that is certainly central to the novel's project. In this passage, just that distinction between intellectual and passional work is deliberately blurred, and we are told that

> It was his habit to indulge himself in that solemn passivity which easily comes with the lengthening shadows and mellowing light, when thinking

and desiring melt together imperceptibly, and what in other hours may have seemed argument takes the quality of passionate vision. (p. 229)

The moment deliberately anticipates Mordecai's experience, chapters later, of standing on the bridge and almost evoking Deronda from the water, a moment that stands as an endorsement of this blend of thought and desire. Here there are none of the usual ironies about the way cause and effect work without regard to desire. The suggestion is that the visionary is going to be validated, going to have priority over the merely "thought."

In the sequence that follows, Deronda undergoes that very act of self-obliteration that would be taken as the ideal condition of scientific objectivity. Knowledge becomes possible because the self is precisely, consciously decentered:

> He was forgetting everything else in a half-speculative, half-involuntary identification of himself with the objects he was looking at, thinking how far it might be possible habitually to shift his centre till his own personality would be no less outside than the landscape. (p. 229)

What is missing from this ideal intellectual prostration, this visionary embodiment of objectivity, is the impetus to act at all. It can be read as another version of the crisis described in Mill's *Autobiography* in which he realizes that he has no motive force because he has never actually considered his own desires. He does not know if he has them. In a painful passage so insistently analytic that it almost becomes a parody of analysis, Mill reflects on the difference between knowledge and feeling:

> [T]o know that a feeling would make me happy if I had it, did not give me the feeling. My education, I thought, had failed to create these feelings in sufficient strength to resist the dissolving influence of analysis, while the whole course of my intellectual cultivation had made precocious and premature analysis the inveterate habit of my mind. I was thus, as I said to myself, left stranded at the commencement of my voyage, with a well-equipped ship and a rudder, but no sail.[28]

The personal crisis is an epistemological crisis as well. Of course, Mill is very unlike the romantic Deronda, who apparently doesn't lack for feeling, although like Mill he lacks desire. Floating down the Thames he seems an embodiment of Mill's metaphor of aimlessness.

Where Mill is rescued by Marmontel and Wordsworth, George Eliot allows Deronda's selflessness and openness to the world to produce a different kind of rescue for him. It is at this moment of utter desirelessness and decentering that she rewards her protagonist with a reason to act—Mirah. Finding her provides not only another of Deronda's knightly selfless rescues of damsels in distress but the opening to true vocation, and that opening is understood to be a moment of self-assertion, at least assertion of personal desire. Mill's crisis begins as a crisis of vocation; and vocation, rather than romance, is the true object of the Victorian *Bildung*.[29] The implication of the narrative is that the act of decentering self is the preliminary moral and intellectual act, allowing new knowledge and new possibilities to enter. Mirah brings with her not only her own life, which would make the typical subject of romance, but the vocation of her brother, which, at last, fully opens for Deronda his energies of desire and will. As he tells Gwendolen during their last interview, "what makes life dreary is the want of motive" (p. 839). So in this first sympathetic exploration of Deronda's epistemological condition the narrative makes it clear that what is wanting is a strong selfhood, strong desire, without which one can only float chancily downstream.

For both Mill and Deronda, the first step to epistemological recuperation from the austere rationalities of objectivist thinking is the efflorescence of feeling and desire. As early as an 1855 essay on Milton, George Eliot had argued against that absolute ideal of impersonality that reigned, at least theoretically, in the realms both of knowledge and of ethics. She claimed that "there is much unreasonable prejudice against this blending of personal interest with a general protest." For, she argued, "if we waited for the impulse of abstract benevolence or justice, we fear that most reforms would be postponed to the Greek Kalends." "Personal interest," she concluded, "may lead to exaggeration . . . but *in* itself it is assuredly not a ground for silence but for speech, until we have reached that stage in which the work of this world will be all done vicariously, everybody acting for some one else, and nobody for himself."[30] The argument is surprising, given George Eliot's refusal so often in her narratives to leave ground for "interest"; yet it fairly represents an important aspect of her thought, an oddly almost anti-intellectual strain: it is never rationality that determines action. The distrust of rationality that is a recurrent aspect of George Eliot's passion for knowledge (eerily played out in "The Lifted Veil") emerges more centrally in *Daniel Deronda* than in any of the earlier novels.

Deronda, as actor rather than as pursuer of knowledge, would never have made a successful governor of a colony, and his *Bildung* requires not learning to accommodate a harsh reality but learning to assert himself against it. Deronda's education is enacted inside a society whose desiccated culture is shown to derive in large part from its complacent, provincial ignorance of other cultures and its own past. Like Grandcourt, who becomes its evil representative, it is indifferent to anything but its own phlegmatic will: "he had no imagination of anything . . . but what affected the gratification of his own will; but on this point he had the sensibility which seems like divination" (p. 616). Grandcourt's oddly sensitive and yet largely dormant will complicates the problem and suggests the dangers in allowing the quest for truth to be governed by "will." And yet, an adequate epistemology, even in the face of Grandcourt, must according to George Eliot resist the curve of Cartesian objectivity into self-annihilation, but build toward an active will (a will that, ironically, has the power to restrain the self).

Of course, George Eliot is not "disinterested." Her new enterprise of knowing is not merely committed to the reality of the other but directed at serving her own culture. Implicitly, the novel suggests that the deadly, spiritless provinciality of British society needs to be expelled through a new animation, a reenchantment, and that it can find new spirit only by understanding its own history and the history of its religious inheritance. Sir Hugo's stables, remodeled from an old chapel, where Deronda takes off his hat and Grandcourt sneers, is a case in point. Deronda in the end will take the reader back to biblical origins. The critique that blames George Eliot for leaving Gwendolen stranded in the midst of this small social drama, bereft of everything but the revivified and lacerating conscience Deronda has evoked from her, misses the point of the book's larger issues. The reader does not have to like it (not liking it would be part of George Eliot's point), but within the epistemological scope of the novel, Gwendolen's fate registers the consequences of absence of knowledge of wider relations, the consequence of ignorance of the other. Hers is a "small social drama almost as little penetrated by a feeling of wider relations as if it had been a puppet-show" (p. 186).

The dangers of the kind of epistemological argument the book seems to be making are great, and it would be a mistake to minimize them. I am not so much recommending what I take to be George Eliot's solution to the larger epistemological problems as showing the urgency of some revision of the strict epistemological Pilgrim's Progress toward self-annihilation—a revision that sustains the possibility of some kind of humane but authoritative knowing. The enterprise may be in the end undoable, but recognition of the costs of the scientific model and the dangers of the alternative is valuable in its own right. Nevertheless, George Eliot's imagination of "scientific knowing" is not incompatible with that of many textually oriented contemporary students of anthropology, most notably James Clifford. Clifford and George Marcus, arguing that current anthropology has become aware of the inescapable presence of self in ethnographical "representations" of other cultures, claim that "acute political and epistemological self-consciousness need not lead to ethnographic self-absorption, or to the conclusion that it is impossible to know anything about other people."[31] In a similar vein, Gillian Beer has argued that rejection of all Victorian anthropologists as racists merely "flatters ourselves." "Moreover," she adds, "such blanket rejection makes it impossible for us to hear a range of available meanings."[32] It is not difficult to discover *Daniel Deronda*'s complicity in the nationalist provincialism it rejects and satirizes; but the novel, like most literature, gets interesting exactly where it resists complicity.

The danger that in dying to know other cultures we in effect assert our power over them is real. Indeed, as John McClure has argued,[33] the romance quest of Westerners into other cultures for the redeeming sacred that has been banished from the disenchanted post-Weberian modern world reenacts in more subtle ways the strategies of imperial domination and exploitation. Ignoring the particularities of non-Western cultures, the terrible politics of starvation and repression and exploitation, new postcolonial romancers like Joan Didion can find spiritual solace in a world they hope is not yet quite subject to the Weberian rationalizing of modern Western society. Certainly, George Eliot seeks in Jewish culture just that sort of reenchantment, whose absence from her own culture produces at the end of her career her only novel in which the tone of embittered satire is dominant for long stretches.

Alexander Welsh argues that in *Daniel Deronda* the liberal theory of intellectual progress is rejected for a theory of knowledge that is, as he calls it, "ideological." The liberal theory is consistent with George Eliot's earlier work, imagining a freedom parallel to the scientific commitment to disinterest. George Eliot's critique of English culture is most clear in her insistence on the need to fuse knowledge with feeling. No knowledge can be effective or even matter, unless the knower desires, wills, and feels intensely. As Welsh puts it, "the fusion of knowledge with feeling, something George Eliot has always advised in moral relations, has [in *Daniel Deronda*] now become programmatic and an alternative to the commonplace and threatening exchange of information."[34] In other words, the fusion carries over from ethics to epistemology. The stories of Mordecai and his Zionist vision and of Deronda and his discovery of his parenthood provide the ideological and personal energy to make knowledge possible.

But what matters here is not the particular ideology of Zionism, nor certainly that ideology as it is understood in our own time. Zionism, in the real history of our times, has been singled out by its enemies as an obvious example of the way the Western moral imagination imposes on other cultures. Its quest for spiritual redemption has created serious conflicts with indigenous populations. There may even be a large irony in the fact that George Eliot, seeking to move beyond narrow, insensitive provincialism by reanimating knowledge, seems not to have thought of the possible displacement of yet another culture. The rejection of the "Jewish part" now is as likely to be on the grounds of its insensitivity to other cultures as on the formal Jamesian (and mildly anti-Semitic) grounds. It is not, then, that George Eliot is sentimental and preachy but that her sentiments do not spread out widely enough and that her preachiness fails to take into account the future she could not foresee.

My interest here, however, is in the nature of the struggle required to reinvent a way of knowing that transcends the limits of the rationalizing of knowledge that ends in self-annihilation. George Eliot's treatment of Zionism is sympathetic but not blind; it offers itself to her as an option by which the provincialism and blindness of her own culture might be transcended. Jane Irwin's arguments against the charges of false ideology are worth quoting:

> The "modern reader" has no need to look for irony in George Eliot's inability to foresee "the accident that [the] scheme for a national Jewish home in Palestine [would] become a desperately difficult political issue." George Eliot was well aware that the history of relations between Jews and non-Jews bears a heavy burden of irony. She had no need of second-sight to foresee that their relations would continue to be disputed ground.[35]

The critical point for the novel is not our contemporary politics of Zionism but its efforts to transform the stupidity, inhumanity, and insensitivity of British society toward the Jews by forcing it to transcend its provinciality through a new knowledge.

Mordecai's romantic nationalism can be dangerously blind to other cultures in just the ways George Eliot's new epistemology would seem designed to avoid. Amanda Anderson elaborates the argument that Deronda's relationship to romantic nationalism is equivocal. The novel explores and values the cosmopolitanism normally associated with the Jew (Catherine affectionately praises Klesmer's "cosmopolitanism" to Mr. Bult, for example): its emphasis on rootlessness is precisely antinationalist, as it is exactly Deronda's condition at the moment he meets Mirah. Anderson notes that Deronda becomes a "Zionist" as Mordecai would have him, but as a questioning and not thoroughly converted Jew. He tells Kalonymos, "I will call myself a Jew. . . . But I will not say that I shall profess to believe exactly as my fathers have believed" (p. 792). He insists on preserving the valued culture of the Christianity he grew up with, and he goes to Palestine not to establish the Jewish state but to inquire into its possibility. His relation to nationalism is "reflective and dialogical"[36] rather than visionary and essentialist. Zionism melds with the epistemological project of the novel: when Mordecai asks that his thought be "born again" in Deronda, Daniel replies, "You must not ask me to promise that" (p. 820). Rather, Daniel "must be convinced."[37] He explains to Gwendolen that he is going to attempt "to become better acquainted with the condition of my race in various countries there" (p. 875). His admiration and affection for Mordecai do not make him a Zionist but make Zionism a possibility to be pursued only *after* a voyage of discovery.

Reading, then, with the kind of openness the strategies of the novel ask for in order to discover the "available meanings" requires an enormous leap of the moral imagination—just the roused imagination that George Eliot described to Stowe. For this reason the Jewish sections open with the crass and banal Cohen family rather than with the messianic Mordecai. The novel assumes a resistant audience, a Bultlike, Arrowpointlike culture that is vaguely but consistently anti-Semitic. The qualities that lead to the dead individualism of a disenchanted contemporary England are qualities that also close out the potentialities for spiritual revitalization that the novel locates in art, in Deronda, and in Judaism. Encountering Judaism requires not a mystified appropriation of a people still unknown and disdained or romanticized, but risk-taking efforts at knowledge.

The novel registers an almost uncanny alertness to the dangers of such efforts to reenchant a disenchanted world and often anticipates the kinds of critique of George Eliot's enterprise that modern readers have produced. She recognizes the threat of fanaticism, of hidden motives and greeds, of the inevitable limitations of particular human consciousnesses masked in disinterest. She tries to distinguish between the practical realities of contemporary Jewish life—threatened itself by commercial rationalization—and the visionary ideal of Mordecai. (In fact, it is not difficult to detect in the narrator's almost prissy, white-glove consideration of the Cohen family and of Jewish commercial life on the East End something of the sorts of prejudice the book explicitly tries to resist.) Most particularly, she concentrates on how Mordecai's intellectual enterprise might be invalidated by his passion.

George Eliot's alertness to possible objections to her developing positions is paralleled by—likely generated by—her almost preternatural sensitivity to particulars and to the inevitability of multiple interpretations. If the novel begins with a series of questions about Gwendolen that flash through

Deronda's mind—questions that require the entire novel to be answered—it also introduces even the rather melodramatically conceived Mirah as a cluster of possibilities of interpretation. "There is no guarding against interpretation," Deronda knows and the narrator asserts (p. 323). For one example, Deronda interprets Sir Hugo's silence about his paternity and the fact that Sir Hugo imagined him as having a career as a singer as evidence that he is a bastard. The relation between Deronda and Gwendolen, which other characters assume is erotic, provides a more complicated example. The novel-loving reader will also assume that there will be a romantic conclusion between the two, and it is not certain that Deronda himself feels no romantic attachment. The later entanglements between Deronda and Gwendolen are important because Gwendolen, however much she transforms him into a spiritual counselor, does indeed feel, as Grandcourt has suspected, something other than spiritual attraction to Deronda. Attempting to keep Deronda above the conventions of the romantic plot, George Eliot does not ignore its power and attractions.

Under the scrutiny of the multiple perspectives and interests, facts seem at times almost to disappear. And the difficulty of "all at once describing another human being," the puzzlement that follows the *discovery* of another human being, the minuteness and ephemeral nature of causes, makes knowledge of virtually anything enormously difficult. Gillian Beer emphasizes George Eliot's attempt in *Daniel Deronda* to dislocate the sorts of sequential and perhaps determinist narrative explanations that mark the earlier novels.[38] But while it is true that the novel points toward an undetermined future and that its manipulation of chronology and juxtaposition of apparently incompatible plots overturn the narrative modes of her earlier novels, these efforts seem directed at a reimagination of epistemology, not at a rejection of the possibility of knowing.

Yet, while many of George Eliot's novels seem to imply that "interpretations are illimitable," her narrators speak sagely and authoritatively. Rosemarie Bodenheimer claims that George Eliot's final action after *Daniel Deronda* was an "explicit abandonment of the all-wise narrative voice,"[39] because, she claims, in *Daniel Deronda*, George Eliot had deconstructed sympathy. In Daniel's mentoring of Gwendolen, the voice is tested, and its limits exposed: Daniel cannot sustain fully sympathetic action, whatever he might say and feel, because he has a life with conflicting commitments and desires. Gwendolen sees him as a potential lover, but Daniel *must* abandon her for another woman and another vocation. The sequence puts to the test the narrator's own authority and all-embracing sympathy by giving Deronda parallel narrative authority with Gwendolen and then dramatizing the impossibility of self-sacrifice unqualified by the compromising conditions of a particular life under particular conditions. Ambivalences and uncertainties emanate from a text so attentive to the ways the desiring self encroaches on wisdom and sympathy.

But such strenuousness and doublings back create the condition for a new epistemology. In another epigraph (to chapter 16), George Eliot talks of how "men, like planets, have both a visible and an invisible history. The astronomer threads the darkness with strict deduction, accounting so for every visible arc in the wanderer's orbit," and the "narrator of human actions" would have "to thread the hidden pathways of feeling and thought which lead up to every moment of action" (p. 202). Narrator and scientist, as with Lydgate in *Middlemarch*, who "wanted to pierce the obscurity of those minute processes which prepare human misery and joy,"[40] must discover and disentangle the multiple threads of connection; a broader and more creative epistemology can grow only from such laborious work at the disturbing contingencies that make the fact difficult to pin down.

George Eliot's conviction that mere rational and systematic thought was dehumanizing determines her rethinking of epistemology. Her notebooks, essays, and novels insist on limitation and the inadequacy of systems and formulae: "There is no general doctrine which is not capable of eating out our morality if un-checked by the deep-seated habit of direct fellow-feeling with individual fellow-men."[41] But it is only in *Daniel Deronda* that she risks challenging the ideal of objectivity with her deep-seated belief that knowledge is always implicated in and sustained by feeling. In a passage titled "Feeling is a sort of Knowledge," she wrote in her notebooks that

> what seems eminently wanted is a closer comparison between the knowledge which we call rational & the experience which we call emotional. The sequences which are forced upon us by perception, which establish fundamental associations, & are classed as knowledge are accompanied in varying degrees by satisfaction, & denial or suffering, to the organism in proportion as the established sequences are affirmed or disturbed.[42]

That "closer comparison" is the focus of the drama between Mordecai and Daniel, and the epistemological crisis of the book.

Given the way Mordecai's wishful enterprise of turning Daniel into a Jew is validated by the narrator and by Deronda, it is striking that the book elsewhere continues to endorse George Eliot's old realist and objectivist vision. "The truth is sometimes different from the habitual lazy combinations begotten by our wishes" (p. 286), she notes about the Arrowpoints' assumptions concerning their daughter's marital prospects. And in yet another epistemological epigraph, beginning with the common view that "knowledge is power," George Eliot insists again on the necessity to follow particulars, to have precise knowledge. It is dangerous intellectually, morally, and politically to have what she calls "false conceit of means whereby sequences may be compelled" (p. 268).

In the very part of the story in which it seems as though sequences *are* being compelled, George Eliot does not ignore the point of the epigraph. In her treatment of Mordecai, it is evident that she wants to emphasize the way his mind lives outside the rules of causality, system, sequence, self-restraint. The difference between the moral and intellectual pressure she puts on the realist narrative of Gwendolen and the passional endorsement she gives to Mordecai's prophetic wishes is not played down. Rather, it and the reasons for it make the very substance of the novel. In her careful rendering of Mordecai's visionary condition, she tries to demonstrate her continued responsibility to the rigors of objectivity. The "Jewish parts" attempt to show that disinterested commitment to the ideals of objectivity is compatible with the visionary epistemology that seems to defy such commitment.

Few readers have agreed with this approach, if only because to imagine it George Eliot had to break the rhythms, structure, and narrative strategies of the realist fictions that had marked her whole career and that occupy at least half of the novel itself. After all, had she not, in her essay on "The Future of German Philosophy," praised Otto Gruppe for his argument that philosophy must "renounce metaphysics: it must renounce the ambitious attempt to form a theory of the universe, to know things in their causes and first principles."[43] But the aspect of her thought that gets played out most fully in the Mordecai sequences is an idea she had spelled out in her essay on the evangelical preacher, Dr. Cumming:

> that highest moral habit, the constant preference of truth both theoretically and practically, preeminently demands the cooperation of the intellect with the impulses; as is indicated by the fact that it is only found in anything like completeness in the highest class of minds.[44]

Mordecai is to be understood as someone with one of those highest classes of minds. But as part of its attempt to open to other cultures, the novel is careful to conduct the reader into that mind by way of a mediator, Deronda, whose receptiveness to otherness had already been well dramatized. It is exactly the ordinary Englishman who could not take Mordecai seriously, and Deronda's specialness (and, indeed, his un-English nature, which is marked well before he meets his mother) is essential to any possibility of narrative success. In this sense, Deronda functions something like the "Editor" of *Sartor Resartus*, in mediating Teufelsdröckh's life and work for a wider public. In the face of apparent fanaticism or delusion, Deronda is described as having a nature "too large, too ready to conceive regions beyond his own experience, to rest once in the easy explanation, 'madness,' whenever a consciousness showed some fullness and conviction where his own was blank" (p. 551). Moreover, the text is consistent in comparing Mordecai to a scientific investigator so that, when he sees Deronda rowing toward him, "His exultation was not widely different from that of the experimenter, bending over the first stirrings of change that correspond to what in the fervour of concentrated prevision his thought had foreshadowed" (p. 550).

"Visions are creators," Mordecai says, and the book just lets him get away with it.

"Concentrated prevision" is the scientist's passionately attentive imagination of possibility, a condition somewhat different from the official notion that the work of the scientist was predominantly to *observe* with a "passionately attentive imagination." But note how carefully George Eliot keeps the idea of "attention" at the center, and how subtly the notion of "passionate" attention is intruded. That passion, I would argue, has always been there—in Descartes's "method," in John Herschel's theory, in Darwin's practice.

Apart from the "concentrated prevision," Mordecai is shown to be practical and sensitive to the particulars of everyday life. Attempting to create a figure with almost transcendent intellectual powers, George Eliot does not neglect her persistent belief that attention to *all* particulars is a condition of true knowledge. He wins the affection of the little Cohen boy and he is sensitive to his responsibility to the Cohen family. He is shrewd and humane in dealing with his father: "He had a mind energetically moving with the larger march of human destinies, but not the less full of conscience and tender heart for the footsteps that tread near and need a leaning place" (p. 605). So the judgment to be expected of English society, even of the good but morally rather dense Sir Hugo, who would have thought of him as a "consumptive Jew, possessed by . . . fanaticism" (p. 568), is juxtaposed with the practical realities of his life and Deronda's receptiveness.

Despite the obvious idealizing strain in George Eliot's representation of Mordecai, the problems with the "Jewish section" of the book are not created by failure to think through the issues. They are caused by the sheer difficulty of the enterprise, of working from within a rationalist-empiricist model of knowledge to achieve a model that does not violate its rigors and particularity but that transcends its dehumanizing—its literarily deadly—limits. In moving from the interpretations of the Jew rendered by those with common English eyes, George Eliot tries to suggest to the reader that the common reading has alternatives that history has justified. Greatness, the narrator reminds the reader, might be and has been confused with madness. Once again the comparison is with scientists:

> Reduce the grandest type of man hitherto known to an abstract statement of his qualities and efforts, and he appears in dangerous company: say that, like Copernicus and Galileo, he was immovably convinced in the face of hissing incredulity; but so is the contriver of perpetual motion. We cannot fairly try the spirits by this sort of test. If we want to avoid giving the dose of hemlock or the sentence of banishment in the wrong case, nothing will do but a capacity to understand the subject-matter on which the immovable man is convinced, and fellowship with human travail, both near and afar, to hinder us from scanning any deep experience lightly. (p. 569)

The passage is both a set of instructions to the reader and a reminder that the range of available meanings can open up only after a gesture of fellowship makes possible real understanding. Like Deronda, the reader is being prepared to believe in the possibility that Mordecai's "wish-begotten belief" in his Jewish birth might in fact be a preparation for a real "discovery." "The world would be a poor place," says Mrs. Meyrick, "if there were nothing but common sense in it" (p. 629).

At last, through free indirect discourse again, George Eliot approaches the key passage in determining the possibility of taking seriously an epistemology that would seem to depend upon something in addition to the self-abnegating work of rational objectivity. Deronda wrestles with Mordecai's need for him as a disciple and finds "a more plausible reason for putting discipleship out of the question":

> the strain of visionary excitement in Mordecai, which turned his wishes into overmastering impressions, and made him read outward facts as fulfillment. Was such a temper of mind likely to accompany that wise estimate of consequences which is the only safeguard from fatal error, even to ennobling motive? But it remained to be seen whether that rare conjunction existed or not in Mordecai: perhaps his might be one of the natures where a wise estimate of consequences is fused in the fires of that passionate belief which determines the consequences it believes in. The inspirations of the world have come in that way too: even strictly-measuring science could hardly have got on without that forecasting ardour which feels the agitations of discovery

> beforehand, and has a faith in its preconception that surmounts many failures of experiment. And in relation to human motives and actions, passionate belief has a fuller efficacy. (p. 572)

It is not, then, as though George Eliot failed to consider the dangers of this way of thinking or to recognize, with modern readers, that desire and "ardour" threaten not only responsible science but any forms of responsible thinking. This is not mere visionary hope. Ironically, the analysis here rather accurately describes many episodes in the history of science in which scientists pursued ideas with "concentrated prevision" beyond their apparent disproof. Taking an example immediately relevant to George Eliot and to the arguments of this book, I would point to Darwin, who worked out his theory without ever finding the smoking gun. The idea for evolution that hit him decisively in 1838 required twenty years for publication and many more for refutations. Against an extraordinary array of arguments, against the failure to find the evidence necessary in the fossil record, against the perhaps decisive fact that species evolution had never been seen, Darwin persisted in his theory and as a result transformed biological science, and much else.

Moreover, almost all scientific work is conducted against the evidence until the evidence can be created to affirm the hypothesis. Hypothesis, that is, fiction, often driven by wish, is the motive force of scientific experiment. Michael Ghiselin has argued that Darwin's "entire scientific accomplishment must be attributed not to the collection of facts, but to the development of theory."[45] Scientists do not, despite the prevalence of the theory, lay themselves innocently open to "vast accumulations of fact," on the Baconian method. They read themselves and their fictions into the facts and, if they can, make their hypotheses true. In his fine study of the fate of the Baconian ideal in the nineteenth century, Jonathan Smith demonstrates not only that Darwin did not, as he claimed, work on true Baconian principles, but that "real science was more interesting, more imaginative, and more difficult than Bacon made it out to be."[46] On this accounting, Deronda's displacement of self into the objects he observed—itself a kind of dying—while morally impressive, would have been insufficient (at least in this novel) for the acquisition of knowledge or for moral action. That sort of selflessness may be a preliminary condition, but in itself it stops far too short of the requirements for imagination and passion that George Eliot insists are the mark of the exceptional, the "rare," thinker.

Constrained, myself, by the epistemological conventions that have dominated the West since Bacon and Descartes, and convinced by the very rigor that George Eliot exercises in the Gwendolen sections of *Daniel Deronda*, I, too, find it difficult to acquiesce in her "new" epistemology even as I recognize in it elements that are important to any effort to humanize our relation to knowledge, escape from the tradition of "dying," and allow for the realities of human desire that drive every action. All George Eliot has done here is attempt to bring together what anybody with any common sense knows is unofficially there anyway—the moral energy and the true passion that go into serious philosophical and scientific work all the time. Beyond that, she has described a process that corresponds quite well to what many great scientists have done but that doesn't quite make the official epistemological record—use their intuitions, seek to prove what they want to be true in the first place, but also wait until they are confirmed in other ways until they make their truth claims. As with good science, for Mordecai the vision he has is confirmed by the empirical fact: Deronda really *is* Jewish. And as for the Zionist vision, the utopia of a new nation, even Deronda, as we have shown, remains skeptical at the end. The novel doesn't say Mordecai is right; it simply allows the possibility. That the particulars of this narrative have not satisfied every reader's sense of the credible is beside the point of the more general concern here: the new epistemology is not nearly so fantastic as its presence in the Jewish half of the novel has made many readers believe through the years.

The greatest problem with this humanized epistemology, an epistemology that would not allow, for example, Galton's cool detached relation to violence and brutality in the quest for a more general truth, is that the burden of proof shifts from the evidence of the investigator to a question of who he is. Here again, George Eliot anticipates a point made earlier in the present text when I quoted Steven Shapin as arguing, "Every culture must put in place some solution to the problem of whom to trust and on what bases."[47] How would we know, before the fact, whether we have a Galileo or a charlatan? Can we trust Mordecai? Can we trust Galileo? What is to prevent someone with messianic claims (and one who might well actually believe the authenticity of his claims) from assuming the authority that the book confers on Mordecai? The narrative of *Daniel Deronda* attempts to dramatize how we know. In fact, by emphasizing the intensity of Mordecai's studies, the seriousness of his thought, the persistence with which he argues the issues, his attention to details, George Eliot dramatizes a character who has all the qualities necessary to the scientist, even as conventionally imagined.

Ironically, the strenuousness with which she worked to establish Mordecai's credentials pushes her and him back to the point from which my whole argument has taken its departure. That is, the book ends with death. Mordecai, like the true bearers of the old scientific objectivist ideal, dies, just as Moses died in sight of the promised land. The moral urgency that underlay the more traditional dying-to-know narrative, the nineteenth-century objectivity that Lorraine Daston and Peter Galison have described, reappears in Mordecai's story, for George Eliot requires an utterly self-sacrificing intensity of intellectual labor if she is to allow Mordecai the unusual and dangerous authority he (almost) acquires. The epistemological and religious patterns merge again in a scene of death, in a liminal moment (parallel to those liminal moments in *Our Mutual Friend*) from which new life and knowledge might emerge.

The difficulties and complications of this strenuous effort to imagine a more vital, creative, and personally satisfying way to know put an enormous strain both on the characters and on the narrative itself. How, in a serious narrative, to embody a character whose visionary powers, combined with "realistic" human qualities, lead him to an authoritative vision of the truth. Against the vulgarities that such a narrative must dramatize, the ignorant anti-Semitism, the commercial crudeness, the pervasive cynicism, such an ideal figure must seem "unrealistic," but cannot move without encountering the roughness of the ordinary. The struggle for the vision of the promised land of true knowledge remains, in *Daniel Deronda,* so difficult and strenuous that it, too, requires dying to know it.

Yet Daniel learns and survives and embarks on a project of potentially (if ambiguously) world-historical significance. No earlier George Eliot protagonist could have been imagined in such a way. The Savonarola of *Romola* is burnt and Romola achieves only a new revised domesticity. The possibility emerges from the fact that Daniel is himself *not* a visionary, like Mordecai. He bears, as mediator, some of that empirical skepticism that keeps him from surrendering entirely to the Zionist vision and his own Jewishness. Deronda again lives liminally, somewhere between the two narratives and the two ways of knowing.

Deronda had at last found his "motive," a motive that rescues him from the dreariness of life that Gwendolen endures, and that spurs both investigation and action. In this respect, the new epistemology is quite convincing, for the spur to investigation is, as Mill himself saw, motive. In turn, Mordecai's hypothesis about Deronda was only possible because of the driving motive of his visionary work. The epistemology is satisfying as well because, embodied by the enterprise of the Jewish half of the novel, it forces a recognition of an "alien" culture and of alternatives to the virtually solipsistic self-enclosure and self-satisfaction of contemporary England. Mordecai is thus a doubly useful sort of figure for George Eliot, not only because he is a visionary whose nature can recall biblical struggles, but because he is so opaque to his society and to the reader. George Eliot never sacrificed her belief in the possibility of knowing, of reaching beyond the limitations of self to facts, and texts, and cultures alien and remote from her own. That need seems more intensely important today. But she understood that any knowledge system unqualified by the complexities of human feeling and the particularities of each separate instance could never attain sympathetically to that obscure otherness of different consciousnesses, different ways of being, different cultures. She insisted on "the inevitable makeshift of our human thinking" (p. 572), but at the same time she recognized and dramatized in *Daniel Deronda* the deadly cost of remaining provincial, of not trying to know.

It is a position iterated, if rather differently, by James Clifford, who, against those inevitable obstacles of self and cultural difference and disguised ambitions for power, asserts the importance of the ethnological enterprise. "It is more than ever crucial," he claims, "for different peoples to form complex concrete images of one another, as well as the relationships of knowledge and power that connect them; but no sovereign scientific method or ethical stance can guarantee the truth of such images."[48] *Daniel Deronda*, and the epistemological theory it narratively unfolds, may leave the reader in doubt: the tension between desire and disinterest—between feeling and thought—remains after this extraordinary attempt, but it is an attempt that contains within it much that requires assent. If the Baconian model of scientific objectivity and receptivity to the facts entailed, finally, a condition of "dying to know," George Eliot sought in her last great novel a way of thinking that could lead at last to the promised land in which instead of dying to know, we live.

— 9 —
The Cartesian Hardy: I Think, Therefore I'm Doomed

Daniel Deronda revises the Cartesian narrative of scientific epistemology in order to avoid the deadly consequences of knowing. George Eliot is rigorous in her confrontation of the cost of knowledge but unwilling to let go of the view that the world "means." In that novel, she takes the tradition of comic realism that marks the main stream of Victorian fiction and turns it into something like tragic realism, with a conclusion that invokes the painful meaningfulness of *Samson Agonistes*—"nothing but well and fair, And what may quiet us in a death so noble." The narrative ends in death, but death still has the nobility of the quest that marks all narratives of scientific epistemology. The world that will not yield to human desire yields, at least a little, under the power of human nobility—and desire.

The common complaint about *Daniel Deronda* is that in it George Eliot cheats, for she allows herself to imagine a material reality that is ultimately consonant with the desires of her key investigators—Mordecai and Daniel. It is not so much Mordecai who turns Daniel into a Jew as George Eliot herself. But by the end of the nineteenth century, material reality—the world to which scientific epistemology opened itself selflessly—was coming to seem to many absolutely inimical, not only to human desire but to meaning itself. Notoriously, Thomas Hardy was one of those who, as they engaged with the narrative of scientific epistemology, opened themselves to the inhuman coldness of material reality. In his novels, no visionaries like Mordecai can ever be satisfied by what reality finally offers them, and the human nobility of their deaths is compromised—or perhaps intensified—by the cosmic meaninglessness their deaths expose. What each protagonist thinks the world might be turns out to be exclusively his (or her) creation: one need think only of Angel Clare, who, after Tess's confession, realizes that he has invented her and that the material reality is completely different from his imagination: "You were one person," he says to her—"now you are another."[1] Hardy's novels, like George Eliot's, explore the way consciousness can be seen to shape reality, but at the same time, as we have heard Swithin tell Lady Constantine, they describe a world in which "nothing is made for man."

The turn of science by the end of the century, as it entered popular discourse demanding "free thought" and urging that nobility lay in facing the facts as they were, was increasingly unfriendly to the human. It was not only that Darwinian evolution had decentered the human (with implications less morally satisfying than the decentering that Gwendolen feels), but that the theory of the conservation of energy and entropy led William Thompson (ironically in part to confute Darwin's theory) to predict the death of the sun,[2] and evolution in society was increasingly understood as degenerative.

Hardy's work, then, pushes the questions that *Daniel Deronda* does not allow itself to ask. What happens in a world where moral order does not lie behind the veil of matter, if the unknowable thing in itself is incompatible with the human? What happens if the novelist refuses to bend the discovered reality to the needs of his protagonists? What happens if a narrative unremittingly pursues the idea that truth does not give way to desire, and that, as narratives might be expected to imply, every intellectual pursuit is physically embodied and every act of knowing is contingent? The result—it seems obvious when posed in this way—is something like the extravagant bleakness of Thomas Hardy's *Jude the Obscure,* which I will be reading not as an attempt to revise the narrative of scientific epistemology, but as an attempt to push it to its logical and merciless conclusion. One literally dies in order to know.

But this leaves us with a perhaps too familiar Hardy, sardonically happy in his extravagant naturalistic pessimism, and it fails to capture the uncanny power of his novels. This power, I will be arguing, is connected to another aspect of the narrative of scientific epistemology. George Eliot and Hardy, both committed empiricists, were almost hypersensitively aware of the way consciousness shapes reality. It is not only Angel Clare who "creates" the world he moves in, only to be disenchanted, but the protagonist of *The Well-Beloved,* Hardy's last prose fiction, who as artist creates three generations of ideal women. Jocelyn Pierston's loyalty is to a Platonic idea, not to the embodied women in whom he, from time to time, finds the ideal.[3] And it is not only Angel and Jocelyn, but Hardy himself. While this turn to questions of art and the ideal may seem a radical swerve for Hardy, it only brings to the surface an underlying preoccupation of virtually all the novels. The opaque material world is incompatible with the human. In empiricist theory it is not the world but the sensation "of it," as it excites subjectivity, that is the known reality. My argument is that the redemption that George Eliot finds in imagining the power of consciousness to shape a moral universe, Hardy finds by surrendering to the hard, unaccommodating actual and committing himself to the possibilities of the ideal—that is, of art. The ideal is all there is for human consciousness. Instead of focusing on the catastrophic collision between ideal and material, Pierston the artist virtually leaves the material behind—or rather, simply exploits the material as more or less satisfactory embodiment of the ideal. So in effect, Hardy's novels revisit the new epistemology of *Daniel Deronda*, focusing on the creative power of the imagination, although with no sense that such power can transform the material world. The material world, in its meaningless obduracy, becomes a cluster of possible significances. Meaning resides in the human and human creation.

"The letter killeth," reads the epigraph to *Jude the Obscure*, and its narrative traces the way adherence to the letter of the law does indeed kill. But with all the possible meanings of the epigraph as they have been explored by Dennis Taylor,[4] I will be treating that killing letter as the merciless narrative of self-sacrifice that is the narrative of scientific epistemology. The letter of the epistemological law is unremittingly deadly, an echo of the laws of ethics and of the religion Hardy clearly despised—and could not do without. In preparation for the writing of *Jude* he reread the Bible.

But Jude the visionary occupies a world without visionaries. *Jude the Obscure* seems to stick to the letter of the realist program, so closely, indeed, that its artfulness becomes visible. From start to end, the protagonist, Jude, seeks knowledge; from start to end, the narrative tests out the uses of self-restraint, the possibility of returning to self-annihilation in pursuit of reality; from start to end, Jude is burdened by the corporeality that narrative requires, and that burden leaves him short of the knowledge he sought, until he quite literally dies.

Thus, *Jude the Obscure* might be taken as a final test among fictions of the dying-to-know model, an antiheroic inversion of it. If Dickens takes his characters to the edge of death, Hardy has no hesitation in pushing them over, with no lingering in a liminal state. The book never misses either the sordidness or the irrelevance (to any of the official moral and epistemological goals it has invoked) of that liminal state but plays out the "dying" metaphor to its literal and nescient conclusion. The bravado of the heroic willingness to take a good look at the worst reveals the worst, and the heroic remains only in the persistence with which Jude clings to an ideal utterly unrealizable. The book's original title was *The Simpletons*. Bertrand Russell's "A Free Man's Worship" was just around the corner, but Hardy was having none of that sort of whistling in the dark. There was no defense against the disaster of scientific epistemology and the materialist explanation of nature that it seemed to entail. None, perhaps, but the activity of imagining it, and the power of imagination to make symmetries and meanings of the catastrophe of matter. The self-abnegation that theoretically makes full knowledge possible is pushed into the realities of the body and demonstrated impossible. The virtues of self-restraint are dramatized as the vices of self-destruction. Knowledge that extends beyond animal cunning into Feuerbachian self-consciousness has nothing to do with life.

Extravagant as it might seem, then, I will be reading *Jude the Obscure* as a kind of inverse *Discourse on Method*, something of a philosophical novel, in which the philosophy is embedded in the narrative particulars (emerging occasionally in narrative commentary). Although Hardy always insisted that his novels were "impressions," not "arguments," the very form of narrative, as Hardy knew, put ideas to the test, turned them into ideals, even made them dangerous. Like Descartes's *Discourse* and *Daniel Deronda*, *Jude* is a quest novel. The Cartesian narrative seeks the truth by moving from the body through consciousness and back at last to the body, now redeemed from error, virtually resurrected. *Jude* broods morosely, with calculated symmetry, on this narrative tradition and on the fundamental issues of Western epistemology with which it is allied.

As one might have predicted from Hardy's handling of questions of knowledge in *Two on a Tower*, this late novel reaffirms the appalling consequences for the self of disinterested pursuit of knowledge; it dramatizes the bad fit between thought and "reality"; and it moves beyond epistemology and questions of truth as it revises or undercuts the literary and philosophical traditions to which it belongs. *Jude* is a novel of crisis, built in part on the collapse and the almost irrational persistence of Cartesian rationalism, a kind of literalness that, the novel suggests, really kills.

Within the quest-narrative tradition, the story of Jude's fall from the ideal into the entanglements of the flesh is obviously not new. But the novel should not be read as a characteristic example of the *Bildung* tradition.[5] David Copperfield and Arthur Pendennis, representative *Bildung* protagonists, succumb for some time to distortions of objectivity out of idealizing ignorance and youthful romanticism. But Jude is hit by a pig's pizzle, and sexual relations with Arabella begin quickly. In place of the comic, moving, and misguided passions of Penn for the Fotheringay or David for Dora, Jude's hitherto untapped sexual energies are aroused by Arabella's seductions, although he was already aware "in the secret centre of his brain, that Arabella was not worth a great deal as a specimen of womankind" (p. 102). But having made love to her, he feels bound by the "custom of the rural districts." Unlike his bildungsroman predecessors, Jude is immediately aware that he is "enchanted," and he "sometimes said laconically" that "the idea of her was the thing of most consequence, not Arabella herself" (ibid.). The *idea* of the physical woman is Jude's reality, as the *idea* of Tess is Angel's.

J. H. Buckley argues that *Jude* is the first English bildungsroman whose form is adapted to "the true and proper ends of tragedy."[6] The bleak, inevitably destructive direction of Jude's life, originally conceived as a short story that would end in the protagonist's suicide, is not at all characteristic of the bildungsroman. As Franco Moretti has argued, that form, as adapted by the English, invariably moved toward a conservative comic ending and a compromised accommodation of protagonist to society, in effect of spirit to body. In this respect, the bildungsroman follows the curve of Descartes's revolutionary but conservative program, his dismissal of all the assumptions of Western thought accompanied by a pragmatic acceptance, in life and action, of just those assumptions. It does not, however, in its more optimistic phases, follow out to the full the implications of the self-negation and the mind/body split that were inherent in the Cartesian progress. As Descartes comes through his meditations, he achieves a perfect Victorian happy ending, for he finds that commonsense assumptions about the material world are confirmed by what is "clear and distinct."

The Victorian bildungsroman—and Hardy's late version of it, in particular—makes an excellent narrative vehicle for testing this Cartesian system of negotiating the world and the mind, for like Descartes's thought it was driven by social transformations that made inherited cultural and intellectual conventions inadequate. It implies alternative ways of maneuvering through the new conditions as it traces the development of young (usually) men reared in the rural past into new urban societies. It, too, requires a new skepticism about the past and a new way to act and think, while, again like Descartes's *Discourse,* it dramatizes the risk of letting go of governing conventions. Ultimately, it makes its accommodations. As Descartes rebuilt the world from a radical (if disembodied) subjectivity, the bildungsroman builds upon a romantic subjectivity that must define itself against traditions no longer vital. But within the context of new bourgeois society, it is distinctly not revolutionary: like Descartes's *Meditations,* it tends—certainly in its English variations—to end in accommodation to the embodied world, to the economic transformations and dislocations that come with new social instabilities.[7]

Disenchantment, the displacement of desire by reason, is usually in the English bildungsroman the way back into a comfortable, ordered, and thoroughly compromised life. But in *Jude* disenchantment leads either, as in Jude's case, to further disenchantment and catastrophe, or in Sue's to a radical self-denial that is far more disastrous than the disenchantment. The simple conditions of daily life, the physical fact of being, one's place in class and gender are determining. Born poor, one cannot retreat into comfort from the ideal. In *Jude,* there is no escape from the constraints of the body except real physical death. Unlike Mordecai, who passes his spirit on, Jude simply dies. And with this absolute defeat Hardy embodies, at its extreme, the ideal of dying to know. To stay in contact with something equivalent to the ideal of knowing that grows from the act of self-negation, Hardy must stand outside the narrative, safely sheltered from the constraints of sexuality, money, and class. Redemption, if there is to be any, comes only in knowing that subjectivity has no access to the ultimate truth, that the ideal can only be *real*ized in the constructions of the human mind, and then in making art out of that knowledge. The instability and uncertainty of the conditions of a physical life require some other, nonepistemological authority: although *Jude the Obscure* only engages directly with art in its treatment of Jude's work as stoneworker or Sue's in making cakes, the book itself, in its relentless progress toward death, shapes the opacity and ugliness of the material and social world into almost perfectly formed patterns of irony, contrast, and symmetry. Hardy finds the consolation in the activity of describing a world without consolation—in the mind's power of responding to and creating "impressions" that are the only source of empiricist knowledge.

Hardy's cogito is no rationally stripped-down ego but, in its sensations, is implicated in the folds of desire and material need. It lives not in intention and conscious direction but in desire, which is notoriously unstable and perhaps even inversely related to truth. John Kucich notes that "we can find on the surface of [Hardy's] work a systematic diagnosis of honesty's inadequacy as an ethical ideal."[8] I would add that, as in Jude's transformation through the middle of the novel, it is a diagnosis of rationality's inadequacy as an epistemological ideal. Whereas the dominant position of Victorian scientists was that truth-gathering entailed moral discipline, Hardy read the natural world as having nothing to do with the human and rational, and that value lay in the powers of human imagination. As Darwin had shown Hardy that life itself might be random and beautiful and controlled by a powerful, mindless, impersonal force,[9] so characters like Jude and Sue move in apparent randomness through a symmetrically disastrous narrative that gives the appearance, like Darwin's pointless world, of being teleological. Hardy turns the metaphorical death to the flesh of the traditions of spiritual and intellectual redemption into a quite literal death.[10] The irrationality of the material world is offset only by the shape of the narrative itself, rather as the deadly natural selection makes a pattern out of the random variations upon which it works. But as a detached voice above the narratives, observing the randomness and the patterns, the narrator participates in yet another version of the ideal of self-annihilation—a new style of narrative omniscience and the view from nowhere.

Jude's story spins on a more absolute division and combat between body and mind than Descartes himself might have imagined. No other Hardy novel is so insistently preoccupied with the difference and the tension between mind and body. Jude's pilgrimage to Christminster begins and is sustained by extraordinary feats of self-denial, almost of disembodiment. As he absorbs himself in the studies, "[h]e had become entirely lost to his bodily situation during this mental leap" (p. 63). Within the first pages of the novel, he loses contact with the body at least three times: when he allows the crows to eat in Farmer Southam's fields (which leads, in turn, to *corporal* punishment), when he tries to observe the distant Christminster, and when he is struck by "a piece of flesh, the characteristic part of a barrow-pig" (p. 80). Descartes could have done no better, or no worse.

The mind/body dualism is necessary for Descartes if he is to find a way to overcome the confusions and uncertainties that always come with embodiment; the limitation of perspective, the failure or incompleteness of the senses, the distortions imposed by desire or by the mistakes of the past. Only by imagining the body and mind as separate can Descartes imagine his progress toward unqualified truth. Hardy exploits dualism for entirely opposite effects, however. Body may be "mistaken," but mind is more so. The body never leaves the mind alone, and at its best and most austere, the mind can only make up stories. The abstraction of the young Jude in his compassion for the birds is suddenly and immediately relocated in the material world as the farmer's paddle descends on him. Consciousness—in this case Jude's sympathy with the birds—is almost always represented as a kind of cosmic mistake. That, at least, is not how natural selection works.

The novel dramatizes with bitter force the degree to which the culture imagines itself above the demands of the body. The spiritual (so it seemed to the young Jude) university of Christminster is grounded in the stones of its buildings. Those gleaming spires of the town are decaying material constructions that require an "underworld" of manual laborers to sustain them—an underworld that the narrator identifies as the "real" Christminster. The class snobbery that denigrates physical work makes it impossible for the don who rejects Jude, T. Tetuphenay (not worth much more than a ha'penny, on which his authority is ultimately based), to value the pure epistemological energy that marks Jude's self-training.

Hardy himself was a man who did not like to be touched, and this may account for his intense exploration of the mind/body dualism. So, for example, in 1892, provoked by hurting a tooth at breakfast, he writes, "I look in the glass. Am conscious of the humiliating sorriness of my earthly tabernacle, and of the sad fact that the best of parents could do no better for me. . . . Why should a man's mind have been thrown into such close, sad, sensational, inexplicable relations with such a precarious object as his own body."[11] The language of "earthly tabernacle" registers both the traditional religious background that Hardy understood so well, and his ironic sense that his body might have spiritual significance: once again, the tradition of epistemological skepticism resonates with religious language. While it implies discomfort with the material world, it offers or implies no vision of ultimate conciliation of mind and body. When the body dies, the spirit dies also. For Descartes, the seat of the soul is the pineal gland. For late-Victorian thinkers, the soul could, in theory, have no "embodiment," could as nonmatter have no effect on the material body; and nobody seemed to be able to describe sensibly the relation between mind and body that Hardy felt so gloomily in his tooth. Seeking always some material explanation, naturalists could talk of mind, utterly immaterial, as a kind of secretion of the body, and its work is to make conscious its own anomalousness. In *Jude* every moment of forgetfulness of flesh leads to its punishment, every indulgence of flesh (for Jude and Sue, at least) leads to guilt and recriminations.

But while separate, body and spirit are also always linked. Even in the brightest of Hardy's early stories, no spiritual aspiration or erotic passion can ever be satisfied inside the condition that narratives allow. The body, the arena of the contingent, can make no spaces for the ideal, although as Elaine Scarry has beautifully demonstrated, the material always bears traces of its past, and in so doing carries the marks of the spirit.[12] So Grace Melbury's footprints in *The Woodlanders* become icons for her father's worship, and in *Under the Greenwood Tree,* Fancy Day's boot, revealing as it does through its "tell-tale leather" its owner's character, evokes Dick Dewy's "delicate feeling."[13] The ideal, sprung from the bodies with which the battlefield of nature and contingency is strewn, can only shadow forth what does not exist in nature. Every gesture of incorporation or embodiment, recoupling spirit with the natural body that conceived it, leads to an ejection. Those mindless fossils that stare into the eyes of the perilously clinging intellectual, Knight, of *A Pair of Blue Eyes,* might be taken as the ultimate condition of consciousness—the reduction of the ideal that Knight seeks to the state of petrified marks of prehistoric crustaceans. The mind plants the spirit in material objects, like Grace's footsteps, and then after idealizing the object, destroys it with a brutal rigor. The story of consciousness imagining the reality of what itself has planted in nature (as Feuerbach saw Christianity planting God in nature) is the story of dying to know, over and over again.

The inevitable interconnection and incompatibility of mind and body pushes the novel toward a schematic bleakness beyond realism, while at the same time the persistent registration of their mutual dependence anchors it in the realist tradition of the mixed nature of things. At the time he meets Arabella, Jude is ruminating on his intellectual achievements and dreaming of the knowledge he will acquire in Christminster. He has set aside the next day, Sunday, for rereading Griesbach's text of the New Testament, but Arabella interrupts: "She was a complete and substantial female animal—no more, no less" (p. 81). Here is the first of the often almost crude symmetries, the juxtaposition of Arabella's pure animality with Jude's spiritual vocation. Invoking the body in the tradition of realist narrative, Hardy makes it work within an outrageously contrived scheme of balances and ironies. The emphasis is unmistakable: Jude "had singled [Arabella] out . . . as a woman is singled out in such cases, for no reasoned purpose of further acquaintance, but in commonplace obedience to conjunctive orders from headquarters, unconsciously received by unfortunate men when the last intention of their lives is to be occupied with the feminine" (ibid.).

This notorious crisis in Jude, the body belying the quest for knowledge (and a place at Christminster), is a characteristic instance of Hardy's transformation of the materials of realism into patterned artifice. The novel's angry defiance of the Cartesian quest narrative (on which, nevertheless, its assumptions are so obviously built) is suggested by the overtly schematic structure. It is one of a large, careful set of parallels, juxtapositions, crossings, and bitter ironies that mark the narrative as a whole.

Among these formal sets of contrasts, Jude and Sue trace opposite intellectual paths: she begins enlightened, with a fundamental disbelief in Christianity, and Jude moves to that disbelief as Sue returns to the constraints of piety. The crossings within the Sue story are anticipated by the gross juxtapositions between that story and Arabella's, marked also by geography. As against the Arabella passages, thick with flesh and carnality, in which Jude cannot begin to hope to sustain his intellectual interests and his pursuit of the ideal, there follow the Sue Bridehead passages. If Arabella is all flesh (and it should be noted that her flesh is a constant provocation to lies, as in her famously unreal dimples and hair), Sue is a "disembodied spirit" (p. 309). If at Marygreen Jude aspires to Oxford—"the tree of knowledge grows there," he says (p. 25)—and is weighed down by carnality and sexuality, in Oxford he becomes a shade among shades, an invisible man: there he is "impressed with the isolation of his own personality, as with a self-spectre, the sensations being that of one who walked but could not make himself seen or heard. He drew his breath pensively, and, seeming thus almost his own ghost, gave his thoughts to the other ghostly presences with which the nooks were haunted" (p. 126).[14] The conditions of Marygreen recur and "the mean bread and cheese question" impels him "to smother high thinking . . . and seek for . . . manual work" (p. 130). His own embodiment, his "rough working jacket and dusty trousers," keep him from trying to talk to his "ideal" cousin, Sue (p. 136). But ironically, what impels Jude to Sue is, finally, not the ideal but desire. Even the ethereal Sue is embodied, gets pregnant, has breasts that the narrator chooses to note and that Jude admires ("the small, tight, apple-like convexities of her bodice, so different from Arabella's amplitudes" [p. 245]). And if Arabella half destroys Jude by exploiting that desire, Sue half destroys him by not indulging it. Body and spirit are constantly at odds, always in play together, but never reconciled, either by the characters or by the narrator himself—all except Arabella are uncomfortable in their bodies.

The body also bears the burden of class—is marked, that is, by entirely human categories. Recognizing that the real creative work in Christminster goes on among the workers, not among the scholars he wants to join, Jude is bitter at not being able to enter the buildings he repairs. But one of the grossest ironies comes when, at the pub, he finds himself drunkenly trying to establish his credentials among the students, who see him only as a working-class drunkard. That he does in fact know more than they, but that he demonstrates it while drunk, only emphasizes more forcefully that what is at stake when the epistemological quest is turned into narrative is not, after all, knowledge, but social power. The novel will not allow Jude to escape physical contingency, as Deronda so easily can, as John Harmon finds a way to do. Rising to the Credo in Latin in a pub is no triumph of knowledge, only humiliation and defeat.

Jude's story, as he succumbs to Arabella's seduction or is rejected by Christminster, in effect dramatizes the death of the Enlightenment dream of disinterested knowledge. Privilege gave Descartes the conditions under which he could think through to the cogito without disturbance. But for Jude and Sue, all thought, all work, is conditioned by material context or personal history. At one point when the community is harassing them from their ecclesiastical work because they are not married, Sue exclaims, "I wish we could both follow an occupation in which personal circumstances don't count" (p. 376). On the one hand, there is no connection between abstract intellectual merit and the way the world works and, on the other, what one thinks and does is inevitably connected with one's personal history, with the body.

Sue's right-minded openness and hostility to convention are undercut by what might be thought of as a visceral incapacity to live out her ideas. In the fates of Sue and Jude, Hardy tests his at least quasi-materialism against the radical subjectivity that allows his protagonists to value and imagine the ideal. Each is passionately committed to the truth and each is diverted by matter. The incompatibility between mind and matter produces a Nietzschean sense that all ideas of order are fictions imposed by the phenomenon of human consciousness on a matter entirely indifferent to it. While Hardy dramatizes this development insistently, he never treats it with Nietzschean exuberance, but always as a loss. A world in which mind and body are incompatible but yoked is deeply uncomfortable, one in which the realist's truth, through which worldly difficulties might be overcome, looks more like fiction than science.

From the depths of empiricist subjectivity, the narrator cannot affirm confidently some clear and distinct truth that might in the end overcome false conventions, the way Descartes had done four hundred years before. The flaw in human experience is not the result of some historical aberration but of the permanent incoherence between consciousness and the matter that generates it. Although *Jude the Obscure* seems ready in its dark conclusion to abandon hope for meaning and order in a world so incoherent, Hardy's allegiances remain ambivalent. The young Jude, driven from the farmer's field and understanding that the world does not rhyme, faces directly the vision we have seen take shape for the abandoned Gwendolen Harleth, and falls into an awareness of himself, "not at a point in [his world's] circumference," but at the "centre." Yet the centre is not the egoism that Gwendolen escapes but a terrible recognition of an unintelligible world. "All around you there seemed to be something glaring, garish, rattling, and the noises and glares hit upon the little cell called your life, and shook it, and scorched it" (p. 18). He is the Cartesian cogito revised, stripped down again, embattled, and incapable of knowing the not-self except as appalling sensations.

Yet Jude dwells in the ideal even as all his aspirations are thwarted and his body collapses, and his narrative symmetrically fulfills the unhappy forebodings of its beginning as he continues to love and pursue the conventions after they have destroyed him. He dies with the words of Job on his lips in the midst of the sacred city he had sought from the first chapter. Jude's quest for knowledge, an escape from the "glaring, garish, rattling" something, always encounters that something, and it once again dramatizes the mortal price of the old fictions of self-sacrifice. Yet he cannot help finding at least some of them still beautiful even though he knows them to be false.

Hardy's preoccupation with detachment, with not being touched, is worked out with an obsessive, even cruel insistence in the story of the relations between Phillotson and Sue. This is the book's most obvious instance of the importance to Hardy of the mind/body dualism that both drives and repels him. The ethereal Sue, the reader knows, is always reluctant about sexual contact: she held her first lover away until he died; she only yields to Jude out of fear of losing him to Arabella. But with Phillotson the revulsion from the physical is absolute. Instinctively, she even jumps out the window when Phillotson wanders by mistake into her room after they are married.

The climax of this last relation dramatizes most forcefully the book's preoccupation with the mind/body dualism. Sue reads the suicide of Father

Time as "nemesis," as though moral value inheres in the external world ("nemesis" is an important element in George Eliot's imagination of the world). She sees the murder of her children by the younger Jude as a transcendental commentary on her having had sexual relations out of wedlock, and as a consequence she gives up Jude, whom she loves, and returns to Phillotson, whose physical presence makes her ill. In its grotesque moral (and narrative) symmetry, that event marks another of the moments of extreme patterning. The extravagance of Father Time's self-elimination horrifies Sue so much that it drives her to her own horrendous, but bitterly convincing, leap to self-annihilation. She balances the novel's pattern of marriage-divorce-return by reassuming her wifely relation with Phillotson, as Arabella will remarry Jude. Although the move would seem to focus on sheer physicality, its import is the reverse. Sue is attempting to deny her body absolutely, since the experience of it with Phillotson is so awful for her. The act signifies both the sinfulness of the body and the necessity of shucking it off if one is to adhere to the "ideal" demands of spiritual exercise. Moreover, death by this allegorical intrusion from a child who has inherited the deeply sad wisdom of the modern, is marked as the only way she can imagine to extricate herself from the awful sufferings imposed by the material world, driven as it is by desire and the body.[15] The ultimate irony is that Sue escapes the horrors of the material world by surrendering physically to a man who appalls her. Here, if anywhere, the narrative of scientific epistemology, seeking the rarefied air of the spirit, is discredited.

It is relevant that the suicide seems to have been provoked by the "truth." Sue had, as she says, "talked to the child as one should only talk to people of mature age. I said the world was against us, that it was better to be out of life than in it at this price; and he took it literally." When Jude asks why she had talked that way to him, she says, "I wanted to be truthful" (p. 412). "Pleasant untruths" would have been far superior. The episode of Father Time works allegorically precisely against that Enlightenment and New Testament ideal of truth—"ye shall know the truth and the truth shall make ye free."

The realist aspects of the novel complicate the patterned contrasts and ironies. This is particularly so in its treatment of Phillotson. There are no rational reasons for Sue's horror of physical intimacy with Phillotson. The sheer, morally neutral, unintellectual force of her revulsion is part of what the book wants to emphasize. But Phillotson is a man of decent generosity, willing to expose himself to slander. He has behaved, indeed, nobly, and in his rational and self-sacrificing consideration of the situation, he allows Sue to leave him for Jude at the cost of what was left of his career after his failure to get into Oxford. His is another action provoked by justice and reason, those great Enlightenment ideals, which is punished by the profoundly non-rational, hard, unaccommodating realities within which all ostensibly disinterested acts actually take place.

But when Sue leaves Jude for Phillotson in another of the ironic narrative contrasts of movements of the couples, as in a kind of quadrille, she has become, as Jude would have it, superstitious. Secular epistemology lapses into religious narrative. When she leaves Jude, she announces the doctrine of disinterest and self-sacrifice:

> Don't think me hard because I have acted on conviction. Your generous devotion to me is unparalleled, Jude! Your worldly failure, if you have failed, is to your credit rather than to your blame. Remember that the best and greatest among mankind are those who do themselves no worldly good. Every successful man is more or less a selfish man. The devoted fail. . . . "Charity seeketh not her own." (p. 438)

In another form, Sue repeats the traditional narrative: selflessness gives access to the truth and is the highest moral good. And there, also ironically, Jude agrees with her (ibid.).

"Self-abnegation is the higher road," Sue says, as she recovers from the illness following the death of her children. And later: "I cannot humiliate myself too much. I should like to prick myself all over with pins and bleed the badness out of me" (p. 420). With a Nietzschean hostility to the ascetic, the narrative describes a Sue who cannot forgive her body—"This pretty body of mine has been the ruin of me already" (p. 474). She then seeks that ultimate spiritual purification which in its most benign secular form is disinterest but in its most extravagant religious form is self-flagellation. So in the patterns of symmetry here is one more: Jude begins his career with a self-sacrificing commitment to knowledge; Sue ends her career with a self-sacrificing commitment to moral and religious law. In both cases, bitterly, the book registers the cost of self-denial. "My good heavens," Jude exclaims to Sue, "how we are changing places" (p. 422). For his part, he seeks release from the schematics of dogma, but also from any systematic thinking, since that thinking seems to have nothing to do with whatever elusive reality is out there.

> I had a neat stock of fixed opinions, but they dropped away one by one: and the further I get the less sure I am. I doubt if I have anything more in my present rule of life than following inclinations which do me and nobody else any harm, and actually give pleasure to those I love best. . . . I perceive there is something wrong somewhere in our social formulas. (p. 399)

The touching vagueness of Jude's new sense of things restates the mind/body tension as Hardy (indirectly) registers his deep distrust of systems and formulas. The "inclinations" that do no harm are visceral, outside of the intellectual "formulas" that structure society. The world lies outside any conceptual scheme or representation of it. The mind, as it were, should keep its hands off the body, and then, inside its unformulaic imagination, value and meaning may indeed reside.

When the mind seizes the body, as it does in Sue's case (contrasting, of course, with the way the body seizes Jude's mind with Arabella), catastrophe follows. The last image of Sue in her physical surrender to Phillotson is one of the most horrendous in English literature. It is counterposed to the deeply unpleasant final joining of Arabella with Jude when she gets Jude drunk to win him back. Sue's final capitulation to Phillotson comes as he claims her in the name of human institutions that impose themselves on the reality of her revulsion. As she comes to him in his bedroom, Phillotson asks,

> "You wish to come in here?"
> "Yes."
> "You still bear in mind what it means?"
> "Yes. It is my duty!"
> Placing the candlestick on the chest of drawers he led her through the doorway, and lifting her bodily, kissed her. A quick look of aversion passed over her face, but clenching her teeth she uttered no cry. (p. 479)

Nowhere is the split between mind and body more clearly affirmed, and nowhere are the dreadful consequences of their necessary collocation more awfully dramatized. Sue rejects her body by surrendering it to Phillotson's sexual embrace. This terrible surrender of self to an idea re-evokes the self-surrender that drives the Cartesian epistemological quest narrative and in dramatizing it exposes its deadly implications. The crucial difference between this version and earlier ones is the inescapable presence of a body that by its very existence mocks the idea of spirit, and yet registers the power of its presence.

The extraordinary sequence of ironic contrasts continues to the final pages. Jude literally dies as Sue in effect and metaphorically dies in Phillotson's lovemaking. The embodied world of marrying and giving in marriage, the world of desire, becomes deadly: "Weddings be funerals," Mrs. Edlin says (p. 398). As against the physical embrace of Sue's newly remarried husband, Jude finds himself abandoned by his newly remarried wife. The effect of both embrace and withdrawal is the same. Jude's death, however, recalls for an attentive reader the opening scene in which the young Jude looks down into the waters of that deep well that comes to symbolize his ideal object, the waters of truth which throughout the novel have evaded or betrayed him. That scene had left Jude staring into the bottom of an ancient well at "a shining disk of quivering water at a distance of a hundred feet down" (p. 49). "It is proverbial," says Alexander Welsh, "that truth will be found at the bottom of a well."[16] That image of unapproachable shimmering distance predicts the glimmering seductive distance of Christminster; the tear Jude dropped into the well embodies his ultimate relation both to Christminster and to the "truth." And at the end, as he lies dying, he murmurs, "A little water, please" (p. 485). He calls out for both Arabella and Sue, and with his parched lips he begins the incantation from Job: "Let the day perish wherein I was born" (ibid.). Jude as protagonist never, even in the moment of death, approaches the wisdom of the usual *Bildung* protagonist. In life as in death, his body largely determines his fate. There is no liminal state, and no particular virtue in self-restraint. The world Hardy creates clearly rewards those who can unselfconsciously accept the terms that their bodies offer, and Arabella—gifted, it is true, with an animal shrewdness that makes her in some ways the intellectual equal of anyone else in the novel—exits fresh with life while Jude dies. The novel implies no alternative to the traditional epistemological tradition of self-sacrifice; but it imagines, dreadfully, its consequences.

The intensity of Hardy's refusal to bend the discovered reality to the need of his protagonists corresponds to the intensity of his pursuit of the ideal. As Descartes sought an unequivocally true knowledge and pursued it from the depths of consciousness to the heights of his God, so Hardy, throughout his novelistic career, was preoccupied with the ideal. No writer was more powerfully attracted to it or more bitterly aware of the impossibility of its narrative embodiment. Perhaps the richest image of the persistence and value of the ideal in face of the recognition of its entanglement with death occurs on the last page of the novel. Arabella and Mrs. Edlin look down at Jude's corpse while through the windows they hear a speech being made during the ceremonies of "Remembrance Week." As the words enter the room of the deathbed "there seemed to be a smile of some sort upon the marble features of Jude." Here, Hardy-style, is the "elevation" that John Tyndall describes as following upon the sacrifices necessary for knowledge. Yet here again Hardy dramatizes the attractiveness of those who pursue the ideal. Jude returns to Christminster, although he has abandoned all hope of his intellectual ambitions, and the smile of his corpse is eerily appropriate. But one must be dead to smile.

Pursuit of ideal truth in effect kills Jude, but it is also what makes him the sympathetic hero of his novel. The novel is far more subversive in its refusal to commit itself to the redemptive powers of truthfulness than it is in the frankness of its treatment of sexuality. Nevertheless, Hardy did insist on his own truthfulness. In reenacting the quest narrative, the novel carries out the self-effacing tradition of pursuit of the truth in at least two ways: first, by rendering the action in that narrative voice we have noted, a voice that implies distance and disinterest, detachment from the world being described; and second, by making pursuit of an ideal truth the leading motif and by making Jude take a good look at the worst, even at the price of his life. But it also short-circuits the bildungsroman's normal move to enlightenment and redemption, as Jude ends in what Patricia Galivan calls "helpless proneness."[17] Enlightenment leaves him powerless rather than empowered, and disenchantment in *Jude* leads not to Quixotic or realistic compromise but to destruction. Although the novel suggests that Jude dies believing that his ideas were "fifty years too soon" (p. 482), it allows no space for even such restrained optimism as this implies. The novel's reality, formally shaped by the strenuousness not of its fidelity to the empirically verifiable but of its relentless and symmetrical variations on bad fortune, is of a world in which the body subverts the mind, in which mind is a mode of human self-torture, and in which the traditional notion of triumphant spirit is belied and mocked. What remains of the truth subscribed to by the world begins to seem like a Nietzschean set of conventions by which the fictions of the spirit are imposed brutally and self-maimingly upon the material particulars of the living.[18] *Jude the Obscure* describes a world, as the preface to the first edition puts it, in which there is "a deadly war waged between flesh and spirit" (p. 39).

Famously, after *Jude the Obscure*, Hardy turned from the novel to poetry. And *Jude* itself seems one of the least self-consciously poetic of Hardy's great fictions. It is as though he had decided to separate out his instinct for poetry, which makes his most agonizingly painful fictions wondrously beautiful, and at last expose the brutality of the world, of nature unredeemed by art, of rationality unredeemed by meaning. And yet, as I have been trying to show throughout this analysis, *Jude* is a strangely beautiful book and one in which the artifice of narrative construction is brought radically forward as a way of dramatizing the irrationality of the commitment to rationality. The novel leaves virtually no space at all for a satisfactory dying-to-know narrative because it demands recognition of the irrationality of schemes of order and meaning and their disastrous consequences amidst the opacity of the social and material world.

Clearly, *Jude* is not a book about art and yet, I have been arguing, it is the art that survives its nihilism—*"Let the day perish in which I was born, and the night in which it was said, There is a man child conceived."* So Jude mumbles, dying (p. 403). These Jobian laments do in fact elevate the death of this obscure workingman, but only in confirming the enormity of the darkness that the novel describes, the absoluteness and meaninglessness of the death. The biblical language, which Jude had long since denied and that has left Sue "defiled," as Jude thinks at the end, resonates with a power that belies its participation in the myths that have absolutely failed to come to terms with the embodiment of all ideals.

I want to argue that *Jude the Obscure* participates in key developments in late-century England in which the philosophy of empiricism, the dualism of mind and body, and the radical bleakness of the news that science was disseminating joined together in a new celebration of art. Not, of course, that *Jude* is celebratory, but it is, as Hardy evasively wrote in his preface to the book, "simply an endeavour to give shape and coherence to a series of seemings, or personal impressions" (pp. 3–4). Of course, Hardy was busy evading accusations that his novel was doctrinaire, another one of the many antimarriage novels appearing at the time. And yet the notion of "seemings" or "impressions" recurs frequently throughout his writing about his writing. Art is perhaps the most honest response to the empirical: a setting in order of a set of sensations, not an engagement with the material reality beyond.

Against the ultimate descent into disorder that the incoherence of mind and matter seems to entail, Hardy poses not some better truth but the art of shaping its story. The truth becomes secondary to art. In effect, like Nietzsche contemporaneously, Hardy severs the connection between truth and morality. The pursuit of truth for its own sake depends on a prior faith in the moral value of truth—exactly the value that one finds underlying the Cartesian epistemological venture, and the one that underlies the nineteenth-century moralizing of epistemology. At a historical moment when questions are being raised about exactly that value, from the critique of Nietzsche to Oscar Wilde's praise of lying, epistemology could slip over into aesthetics, which have a similar narrative to recount. John Kucich argues that Hardy affirms "aesthetic consciousness as the single area in which honesty can survive—and then only by candidly denying, within art, the moral possibility of truth."[19]

Hardy's epistemology becomes a mixture of Feuerbachian elements that one can detect in aestheticism and in its apparent antithesis, positivism. The knowable resides not in the world out there but in the shaping imagination. T. H. Huxley makes a similar move in his "Prolegomena" to *Evolution and Ethics,* which asserts the necessity of developing an ethical process wholly inimical to nature. In both cases, the natural world is hostile to human creativity. Written into this context, *Jude* is much more persistently "interior," much less strenuously (and lyrically) engaged in registering the quality and texture of the natural world, than the earlier novels. What counts in the end is the very consciousness that matter persistently defeats; and meaning is by definition human.

Hardy stands ambivalently between science, for which he remained an enthusiast, and art, to which he gave the fullest priorities. His notebooks are full of quotations from essays about science, and the novels clearly reflect that reading.[20] At the start of his little essay on "The Science of Fiction," Hardy applauds the "exercise of the Daedalian faculty for selection and cunning manipulation."[21] While the essay itself implies the accessibility of truth "to science," Hardy argues for an artistic kind of truth that no mere powers of systematic observation could achieve. In the end, he celebrates "a power of observation informed by a living art" (1:138). The living art is the priority, for it is the art that makes what is observed humanly real. The stars are governed by immutable laws, and they fill the vast universe, whose magnitude science allows us to understand. But Hardy's novels are virtually always about mere human love, the accidents and trivia that make love-narratives. Mere human love—insofar as it aspires ideally to a condition beyond the erotic—is not out there with the stars; it only complicates their laws. Implicitly, by the end of Hardy's novel-writing career, value has shifted from the truth that kills to the fictions that revivify by making unreal order out of the randomness of the material world and, through human feeling, imposing a value in which it is impossible to believe but about which one continues to care.

Novels themselves, as Hardy's insistent antinaturalism implied, are forced into self-division: aspiring to truthfulness, they cannot be taken as representations of the "real world" but are rather ideal constructions of a world that, when detected at all, seems without ideals or ideas. At one point, in his notes, Hardy underlines a key phrase from a discussion of Friedrich Albert Lange's work, *The History of Materialism and the Criticism of its Present Importance:* "all happiness lies in self-deception." He emphasizes that "fancies" that do not answer to things in themselves "can afford him a felicity which nothing can replace" (1:140). The "felicity," an odd word for *Jude the Obscure*, is the felicity of the pleasure of creation of wonderfully contrived objects, the felicity exactly of what the world, without mind, cannot supply.

The material world, so unideal, is inextricably enmeshed in the ideal, which it generates. But whereas Descartes's Enlightenment vision gives primacy to rationality and the idea and then takes them as capable of representing adequately the irrational material world, Hardy—following von Hartmann and Schopenhauer—finds that consciousness and feeling are simply incompatible with their progenitor.

In one of his best-known comments on art, Hardy writes,

> Art is a changing of the actual proportions and order of things, so as to bring out more forcibly than might otherwise be done that feature in them which appeals most strongly to the idiosyncrasy of the artist. . . . Art is a disproportioning—(I. e., distorting, throwing out of proportion)—of realities, to show more clearly the features that matter in those realities, which, if merely copied or reported inventorially, might possibly be observed, but would more probably be overlooked.[22]

The curious inconsistency and yet formal symmetry of *Jude* evolves from this mixture of epistemology, aestheticism, and morality. Hardy had not quite given up on reality. If anything, he was convinced that science was giving him a pretty accurate reading of the natural world. But art defies the conditions science seemed to be reporting. The aesthetic—which historically, of course, was an attempt to respond to the body's feelings—arises in Hardy as a resistance to reality, a resistance to the body; and in effect it marks the recurring Cartesian split between mind and body once more, though in a different, almost an antiepistemological way. For Hardy's artful good look at the worst partly belies his insight into the way the body's interference makes knowing the truth an ultimate impossibility. The Hardyan narrator aspires to that condition of disembodied release from desire (as a means, at least in part, of protection from the consequences of desire) that he will not allow to any of his characters. He seems almost pleased to announce, rather ponderously at the start of one chapter, "The purpose of a chronicler of moods and deeds does not require him to express his personal views upon the grave controversy above given" (p. 357).

The detachment of Hardy's narrative stance and his determination to juxtapose human meaning to nature's meaninglessness gives another twist to the epistemological venture that he ends in Jude's death. Jude's is the death of the epistemological ideal; the clear and distinct are not to be found except in the constructions of imagination. Knowledge is not discovered in the hard reality to which the body must defer, but constructed out of the consciousness of embodiment that narrative entails. If Jude dies, the narrator, anxiously withdrawn from the world that allows no disembodiment, escapes into the act of creation itself.

In *Jude*, as I have already suggested, Hardy's normally austere, Latinate, and remote narrator is more self-effacing than usual. He refers to himself as "the chronicler of these lives" (p. 484), and the narrative moves predominantly through dialogue, without the natural description that turns so many of Hardy's novels into long poems. Thus, while Jude's story demonstrates the discontinuity between body and mind and the unreliability of mind, the narrator's voice establishes a clear vision of *that* truth: it is true that the truth is not available; it is true that whatever knowledge is available tells us nothing good. Thus, in a voice saturated with its resistances to desire, *Jude the Obscure* dramatizes the material impossibility and brutal self-destructiveness of that radical self-denial that marks Descartes's progress through the *Meditations* and that was for the nineteenth century the indispensable condition of scientific knowledge. It challenges the ideal rejection of the material, for narrative fundamentally implicates its characters in the act of knowing. And yet, in the end, the artist occupies the space of the sought-after truth, becoming the figure of epistemological and moral eminence, dispelling the fantasies of order and the enchantments imposed on reality by desire disguised as consciousness, yet becoming so only through his own "seemings," ordered perhaps more rigorously than those he dispels. Once again, the vision is possible only by disembodying the visionary, by dying to desire.

10

Daring to Know: Karl Pearson and the Romance of Science

In the last chapters, I want to focus on the literary and critical work of Karl Pearson despite the fact that he is relatively little known in the worlds of literature and cultural theory. I do so because his career and work can be read as a complicated summing up of the problems of the dying-to-know narrative with which this book has been concerned. Best known for his contributions to the science of statistics and his work in eugenics, he was also a man of letters, and author of a novel and a huge passion play in blank verse. The relations among all these works allow a peculiarly valuable insight into the fate of the *narrative* of scientific epistemology and the full significance of epistemological passion for self-denial. Pearson's work can be read in part as a variation on and a response to Hardy's way of handling the crises that empiricism and the epistemological narrative produced at the end of the nineteenth century. He understood the increasing distrust of the objectivist stance, the cost of "objectivity," the pain produced by revelation of a world utterly incompatible with the human, the need to erect a new kind of knowledge, the attractions of the aesthetic through which it might be possible to impose arbitrary meaning on an unintelligible world.

Pearson's *Grammar of Science* (1892) is a superb updating of the epistemology of dying to know, and it will be the focus of these chapters. But the rich and surprising arguments of that book become more interesting when they are seen in the context of the early work that led to them. This work—literary, historical, and philological essays, a novel, a drama, social commentary—makes clear that there is indeed a narrative to this epistemology, and that the ethics of knowing powerfully influence the theory of knowing. Pearson's epistemological journey places his work in the center of the great romantic and Victorian engagements with social, religious, and ethical crises, and *The Grammar of Science* can be seen as in dialogue not only with Hardy but with Walter Pater. The book richly conceptualizes the major paradox of the narrative of scientific epistemology: the impossibility of the dying-to-know model and its indispensability as an aspiration or approximation. Finally, Pearson's work reaches forward from nineteenth-century crises to early modernist positivism and modern deconstruction and cultural theory.

But it is important from the start not to overlook the most obvious point about Pearson's book: it is the triumphal voice of a new kind of positivism. The Cartesian strategy by which doubt turns into authoritative knowledge had, through the voices of propagandists for science, developed into a dominant scientific positivism (which was triggering brilliant reactions from a wide range of important thinkers, like A. J. Balfour, Samuel Butler, W. H. Mallock, and Bernard Shaw). Paring away metaphysical assumptions and traditional "superstition," positivism and scientific naturalism had written the Cartesian epistemological romance into history and, with predictable regularity, told the story of the decline of knowledge through the Dark and Middle Ages, where religious authority crushed it, and then of its Renaissance, during which period a new secularized freedom of thought allowed for the development of modern science. The great propagandists for this story in England were often unwilling to call themselves positivists—witness T. H. Huxley's constant denigration of Comte and Herbert Spencer's huffy refusal of any connection—but the rejection of authority this story entailed was a critical part of positivist mythology. On the strength of the preliminary denial of "authority," of the courageous sacrifice of dearly held prejudices and deep-seated desires, science staked its late-century claim.

Positivist triumphalism, however, is not as distant from Hardy's dark reading of the world as it might at first seem. Although positivism is popularly conceived as a dogmatic and fact-oriented philosophy, it shares with more overtly radical styles of destabilizing philosophies and theories of our own time a deep-rooted skepticism about quite literally everything—except, perhaps, its own authority. Positivism, though a theory of knowledge built on the Lockean view that all knowledge is derived from experience, and on an absolute commitment to empiricism, develops in part out of the post-Kantian recognition that "reality" is not knowable and that claims about it are metaphysical nonsense. The Comtean version of this epistemology is immediately concerned with "laws" as well, for laws allow for a generalizing of the significance of facts. So Comte says,

> In the final, the positive, state, the mind has given over the vain search after absolute notions, the original destination of the universe, and the causes of phenomena, and applies itself to the study of their laws—that is, their invariable relations of succession and resemblance. What is now understood when we speak of an explanation of facts is simply the establishment of a connection between single phenomena and some general facts, the number of which continually diminishes with the progress of science.[1]

So simple a formulation opens the way to a wide range of modern epistemological developments—from Pearson's kind of positivism to deconstructionist theory to contemporary reductionism.[2]

Although Comte's "positivism" was already deeply theoretical and historical, it, too, began with a rejection of past knowledges—"theological" and "metaphysical"—for the knowledge that could be confirmed by the senses. Although it was "positive" in all senses, indeed positively utopian, it is constructed not as a theory of strong belief but of strong disbelief. Tyndall, W. K. Clifford, and Huxley, the most famous of the "scientific naturalists," had established an iconoclastic style; their rhetoric represented them (and science) as heroically, boldly looking truth squarely in the face at whatever cost; they were unafraid to claim that all animals, including humans, are automata; and they championed the idea that humans were products of evolution and natural selection.[3] The question was not whether science would reveal a world comfortable for human self-esteem, but whether it would tell the truth. And the kinds of claims truth tellers might make had less to do with the real constitution of the world out there—a subject for metaphysics—than with the nature of experience and phenomena.

Reliance on the empirical, however, has always been problematic, inevitably underdetermined and subject to critiques like Hume's. Pearson and the scientists had to consider whether it was adequate to the acquisition of knowledge. Among other things a careful student of Kant's, Pearson valued Kant most for his insistence that there can be no access to the *Ding an sich;* and despite Kant's complex reintroduction of the metaphysical, Pearson and the naturalists ruled the "metaphysical" out of scientific court.[4] But then if the empirical is inadequate and metaphysics are nonsense, where is one to find true knowledge?

In a way, Pearson's career was shaped by that question, and *The Grammar of Science* was his long, considered answer. It is surely one of the most interesting books of the nineteenth century, and Pearson himself was one of the most famous and influential scientists of his time, a man who thought he had found a way to unify all the sciences—the Comtean project in a new, less obviously religious model—who developed formulae and methods that remain fundamental to statistics as a discipline and to the social sciences, who promoted the eugenics movement and in effect was the leader of its research arm. Something of his stature can be inferred from the way *The Encyclopedia of Social Science* describes his career: "Few men in all the history of science have stimulated so many people to cultivate and to enlarge the fields they themselves had planted. He provided scientists with the conception of a general methodology underlying all science, one of the great contributions to modern thought."[5]

For those who think of positivism as a simple (or complex) materialist empiricism, or of recent constructivist thought as a sudden liberation from things like positivism, or an inheritance from Nietzsche, Pearson's *Grammar of Science* will come as a shock. For those who think of positivism as an epistemological ally of a very conservative ideology, Pearson's socialism and feminism will be surprising (though not, perhaps, his commitment to eugenics). Despite its positivism, it offers a relativist and constructivist understanding of science. It defends science not by insisting that it has a direct and authoritative line to reality but by arguing for constructivism.

Although his turn to eugenics may confirm the views of many contemporary theorists that the ways of the Enlightenment and the commitment to rationalizing modes of understanding and analysis lead inevitably to oppression and disguised plays for social and political power, Pearson's thought seems at many points startlingly like that of the most extreme critics of the Enlightenment and of positivism. The weapons developed to criticize Western rationalism, ideas such as constructivism, the instability of the self, the role that "embodiment" and perspective play in epistemology, the impossibility of getting beyond language to "reality," are central to his positivist thinking and, *mon semblable, mon frère*, to those who attack positivism as well.

Pearson revises the narrative of scientific epistemology by thrusting mind back into nature. Most other late nineteenth-century representations of the indifference of nature tended to dramatize sympathetically the human consequences of that indifference; and the scientific, or perhaps "scientistic," tradition of self-repressing objectivity often encouraged a high moral disdain for anything that suggests the presence of the mind in nature. One characteristic element of the dying-to-know narrative, intensified in the nineteenth century, is that this moralizing of disinterest entailed great pride in the willingness to hear and deliver bad news to human pride. Current reductivist and sociobiological accounts sometimes come couched in a rhetoric of moral and intellectual contempt for those who cannot or will not accept rigorously mechanistic accounts of human behavior, and who worry the ethical consequences of such accounts.[6] Implicitly they argue (and this position is already latent in Pearson) that immorality is in the refusal to face the worst.

It does not feel like an accident that the history of modern science marks a progressive decentering of the human. The story, moving from Galileo to Darwin to Freud to modern discoveries in cosmology and biology, is a familiar one. Darwin figures centrally in *The Grammar of Science*, and "natural selection" provides an almost perfect example of a scientific principle that dehumanizes the world. Pearson, however, in a characteristic move of late nineteenth-century thinkers, in trying to wrestle the idea into meaning, assimilated it to a philosophical tradition that branched from empiricism to idealism and allowed him a fundamentally optimistic understanding of the relation of science to knowledge, and of the nature of that knowledge. Pearson's revision of the dying-to-know narrative, while it, too, requires a totally selfless look at the worst the world can offer, has a very different affect from that of the less philosophical thinkers who preceded him.

In the long run, for Darwin, the very inhuman indifference of natural selection became a consolation to him, since he could not accept a divine being capable of inflicting the pain he saw all around him and that he experienced, particularly in the death of his daughter.[7] The process by which speciation develops might be taken itself as the work of an ideal scientific investigator: since it has no place and embodiment, its view is really from "nowhere"; it observes with mindless and affectless rigor, and it has no interest in the results.[8] The individual, Darwin's source of transformative variation, does not matter to natural selection, but all individuals die and in dying carry out its work. The individual is lost in a great set of species-sized crowds, measurable by what Pearson and Galton would call biometrics. Daniel Dennett is certainly justified in suggesting, on the basis of Darwin's description of the evolutionary process in *The Origin of Species*, that the world is governed not by intelligence, not by design, not by intention, but by an algorithm.[9]

I invoke Dennett here because, despite a decided difference in rhetorical strategies and in the ways they choose finally to resolve what problems they see with the Darwinian model, Dennett certainly echoes the fiercely antimetaphysical aspects of Pearson's theory. Insofar as Dennett means by *algorithm* a nonmetaphysical description of sequences rather than a causal explanation of natural processes, he is following Pearson's way of thinking. Pearson would have been careful not to imply that his "algorithms" explained or even referred directly to "reality," for reality belongs in the world of "metaphysics." The only claim about reality that Pearson could comfortably make is that he could not know it. Beginning with a declaration of *ignoramus*, as he did, he would likely have said that regarding the world as though an algorithm governed it would be productive. As scientific observer, Pearson imagined himself a relentless gatherer of facts, but—and here Pearson would seem to deviate from the pattern Dennett follows—facts are not elements in the world; they are the sensations available to human consciousness.

Pearson's epistemology minimizes, or circumvents, both the humanity of the investigator (whose function, it might be said, is precisely to create algorithms that allow adequate interpretation of organic and social life) and the humanity of the humans under investigation. That is, just as natural selection works regardless of individual fates, so Pearson, as he built his statistical methods, sought ways to absorb the individual into a smooth and decodable curve while refusing the expected emotional response to the particularities that go into the construction of the curve. Behind such amoralism in Pearson there is a moral intensity that manifests itself in the power of the investigator to resist the temptation to look at the world as though it were designed to be compatible with human consciousness. Only by facing the worst will the best be possible. Only by facing the limitations imposed by the phenomenal world can one be free.

Pearson is, then, an almost perfect spokesman for the ideals of scientific objectivity. His work has special resonance for the dying-to-know story because his primary project is measurement of humans; he aims to develop methods that will make the social sciences as valid as the physical sciences and to show that the real knowledge produced by traditional science is as governed by statistical probability as the social sciences. On the one hand, his entire enterprise is built on the Enlightenment ideal that rationality is a condition of moral action, and therefore that science is a condition of good citizenship. On the other, rationality leads to a fundamental anti-individualism with complicated consequences. The investigator has to deny his unique selfhood. At the same time, in the interests of scientific generalization, the investigator must be ready to measure himself as he would measure his subjects, since his perspective and limitations can influence what is "seen." Moreover, the subjects themselves get reduced from their status as individual humans whose needs of spirit and body require attention, to elements in a larger community whose nature may be charted in the curves and planes of statistics.

Here, as elsewhere, Pearson follows Darwin and bolsters what might have been taken as unsubstantiated moralizing with scientific evidence. Tennyson's anticipation of natural selection, in the famous passage from *In Memoriam* in which he laments nature's care for "the type" but indifference to the individual, provides the kind of emotive response to the process of decentering the human that Darwin had to anticipate and that was developed yet more fully by Hardy. Darwin does not do a very persuasive job of defending against such responses, and his feeble resistance to the dark implications is well known: "When we reflect on this struggle, we may console ourselves with the full belief, that the war of nature is not incessant, that no fear is felt, that death is generally prompt, and the vigorous, the healthy, and the happy survive and multiply."[10] There may be a winning charm in Darwin's obvious attempt to make his readers feel better, but the act of decentering and reduction is continuous. His later move, in *The Descent of Man*, to naturalize virtually all the spiritual qualities of humanity, particularly the ethical and the aesthetic, spelled out what was implicit in the *Origin of Species*, that human morality derives from the social behavior of animals: "Any animal whatever, endowed with well-marked social instincts, would inevitably acquire a moral sense or conscience as soon as its intellectual powers had become as well developed, as in man."[11] The notion of the human as a uniquely and spiritually endowed being would seem to die in Darwin's science, and Pearson's science pushes this development to a logical and scientific extreme.

Before moving to fuller explication of Pearson's constructivist and idealist theories, I believe it important to understand the path he took in coming to terms with all the bad news the world was yielding to science. Pearson, like Hardy, needed to find a way out of the darkness, and in *The Grammar of Science* he opens that way by making the epistemological a more overtly moral enterprise while cleansing it of its metaphysical obscurities. Pearson's "canons of legitimate inference," as he calls them, illuminate the whole history of the epistemological romance as they blend the moral and the intellectual at every step. Here, for example, is the first canon: "Where it is impossible to apply man's reason, that is to criticize and investigate at all, there it is not only unprofitable, but anti-social to believe."[12] Against the delicate and difficult arguments of an earlier "Grammar," Newman's *Grammar of Assent*, as Theodore Porter has pointed out,[13] Pearson invokes the higher morality of truth-seeking. Newman moves toward probability in order to be able to affirm with confidence what cannot be absolutely given rationally and empirically.[14] Probability is the key to statistics as well, and working with probability, with roots in the skepticism of Hume, Pearson refuses anything not given rationally and empirically. Probability, that is, virtually substitutes for transcendental truth.

The fourth canon elaborates on the moral urgency of correct epistemological procedures: "While it is reasonable in the minor actions of life, where rapidity of decision is important, to infer on slight evidence and believe on small balances of probability, it is opposed to the interests of society to take as a permanent standard of conduct a belief based on inadequate testimony" (p. 73). Here again, the moral urgency is extended from the individual to the community. As individuals, perhaps, it doesn't matter except to ourselves—and in Newman's case to our personal salvation—if we guess at probability; but to accept such standards as "permanent" is antisocial. A false epistemology is damaging not only to the knower, but to society. With a deep Enlightenment commitment to Reason as the means to social and moral betterment, Pearson argues that a true epistemology—scientific epistemology—is a condition of good citizenship.

Thus, the critical question—what constitutes adequate testimony? what constitutes adequate evidence for belief?—is both social and epistemological. "Is there any other sphere," Pearson asks, "outside that of ideal citizenship, in which there is habitual use of this method of classifying facts and forming judgments upon them?" The answer, of course, is yes, and the field is science itself, where in an injunction that echoes the demand of objectivist epistemology from Descartes on, Pearson says, "The scientific man has above all things to strive at self-elimination in his judgments to provide an argument which is as true for each individual mind as for his own" (pp. 7–8). Thus even an epistemologist who denies the traditional form of self-effacing objectivity before a world that he does not interpret but only registers returns to the pattern of self-effacement that objectivity had required. An anti-antianthropomorphist, Pearson still insists on the need to diminish the role of the desiring self in the activity of knowing.

It is striking how, in a text coming more than three hundred years after Bacon and Descartes, and in a context in which the very tradition of epistemology is at risk, the idea of self-elimination recurs so baldly. But in such a new context, Pearson is aware of the critiques implicit in George Eliot and Hardy; surely he knows his *Middlemarch* and the fate of Lydgate, and he knows the inescapable corporeality that was to make the substance of Jude's failure. Moreover, he anticipates the more direct critiques of the Enlightenment quest for the view from nowhere with which in our times we are much acquainted. In his proposal of an ideal scientific method against the reality of human behavior, Pearson acknowledges the practical impossibility of that total self-elimination:

> It by no means follows that, because a man has won a name for himself in the field of natural science, his judgments on such problems as Socialism, Home Rule, or Biblical Theology will necessarily be sound. They will be sound or not according as he has carried his scientific method into these fields. . . . It is the scientific habit of mind as an essential for good citizenship and not the scientist as a sound politician that I wish to emphasize. (pp. 9–10)

Scientists, like everyone else, are vulnerable to desire, passion, bad intentions, and dishonesty. The question is whether one can separate the person who suffers from the thinker who creates, the scientist from the science.

That kind of separation is obviously the work of the ideal of scientific objectivity, but it is evident that his theory is built on a full and unhappy recognition of the limitations of the human (the subject of much of his nonscientific writing), hence his anti-individualism. Pearson's, like all positivism, is built on a sense of limitation and ignorance. His method is meant to overcome limitations by taking them into account and reducing the claims that the scientist should be willing to make in pursuit of knowledge. The empiricist self-elimination Pearson calls for extends so far that even the passion for knowing, for genuinely "knowing reality," is to be sacrificed in the urgency of getting at whatever little truths might finally be accessible. In the pursuit of truth he is determined to eliminate the static caused by selfhood, individuality, and the perceiving self.

The new science is as rigorously self-effacing as the old. It retells the story of dying to know, and Pearson's intellectual history is a pilgrimage through some of the familiar places on the way. First, Pearson prides himself on his ignorance. The Cartesian story begins in ignorance and triumphs in knowledge. Pearson both uses and transforms the story, for his version of it begins in knowledge but progresses through ignorance (to return again to a new kind of knowledge, more limited, more reliable). The epigraph to his journal, *Biometrika*, was "Ignoramus, in hoc signo laboremus." Let us work *because* we are ignorant, is the implication. In that light, or perhaps darkness, *The Grammar of Science* gives to the epistemological romance another turn. Affirming the evolutionary progress of humanity and of science itself as the most potent indicator and creator of human evolution, Pearson praises the great scientists who have reshaped the world because they were able to chase superstition from it: "To our fathers belongs not only the honour of solving these problems but the credit of having borne the brunt of that long and weary battle by which science freed itself from the tyranny of tradition. Their task was the difficult one of daring to know" (pp. 473–74).

Those older heroes have now been replaced: "we enter," he says, upon the heritage of the earlier scientific heroes, and "no longer fear tradition, no longer find that to know requires courage." But the epistemological pilgrimage is not close to being over, for there is another kind of heroism required. That heroism begins by recognizing that the heroes of science have left us still in muddles of metaphysics and in a world still pervaded by superstition, so that current scientists turn out to be *more* heroic than their heroic predecessors. "We must accomplish a task more difficult to many minds than daring to know," Pearson proclaims. "We dare to be ignorant" (p. 474).

The preliminary epistemological and moral act—perhaps parallel to the humiliation of self, or *Selbstödung*—is a Cartesian declaration of one's own limitations. Having worked through Descartes and Kant, through Galileo and Darwin, one must begin yet *again*. The triumphs of past progressive narratives turn out to have been in some way defeats, for while they have been effective in overcoming much "superstition," they have not in the end moved beyond "ignorance." So, now at a higher level of self-conscious sophistication, Pearson replays the strategy of self-denial on the way to objective truth. For him, as for Carlyle before him, the idea of "renunciation," or *Entsagen*, compellingly affirmed in German philosophy and in Goethe, was central to his work of renunciation. Pearson writes as though compelled to tell the dying-to-know story once more, with immediately diminished expectations—probability replaces the clear and distinct—but with a rhetorical stance implying even greater bravery, and with an almost imperial assertion of power: renunciation and hard work would do no less than unify all knowledge and produce a better citizenry.

The juxtaposition of the icy positivist with a figure like Carlyle is not arbitrary. The romantic and moral energies that drove Carlyle were extremely important also in shaping Pearson's ideals and ideas. While "romanticism" is understood in some of its aspects as the cultural explosion of individual desire and aspiration, it imagined the object as a *princesse lointaine*, unreachable as the thing in itself, and it contained or bred within itself an equally strong revulsion from the merely individual. In these respects it might be thought of as an unacknowledged ancestor of positivism. It is certainly the primary source of Pearson's commitment to positivism. In its worst late manifestations, romantic revulsion from the merely individual manifested itself in the celebration of the *Volk* and the absorption of the individual within the sharply defined and brutally exclusive larger community, but romanticism has its links with a more democratic program as well. If Byronic heroism is one aspect of the romantic, Wordsworth's quest for the language spoken by common people is another. The particular comes to reflect the general and may even be thought of as its fullest expression. The tensions between the newly discovered "infinite" possibilities of individual desire, which can never be satisfied, and the deep need to sacrifice the self to something larger, are explicit in Carlyle. Thus, if celebration of the romantic Byronic hero can be taken as characteristic, so, too, can socialism. "Close thy Byron," exclaims Carlyle's Teufelsdröckh, "Open thy Goethe." Carlyle and Mill flirted with Saint-Simonianism, out of which Comte himself emerged.

Pearson, with what we might call a Byronic passion for the truth, brought science and romanticism together. For if on the one hand he was a staggeringly ambitious man, whose energies made him a true polymath—philologist, philosopher, social critic, scientist, novelist, poet—on the other, the need for self-restraint spoke to his temperament. He was reported to be a man who treated every "emotional pleasure as a weakness," and reminded Olive Schreiner of "a lump of ice."[15]

Such restraint, Pearson believed, was a condition of knowledge and social order. In *The Grammar of Science*, he invokes natural selection to justify his view that knowledge and self-discipline are essential to society. And in him we find what might seem an odd conjunction of social Darwinism with a rejection of capitalism, whose injustices he decried. Not only scientific discipline but its social counterpart, socialism, was "a morality."[16] The first duty of man, says Pearson, is to "the group of humans to which he belongs."[17] Whereas Christianity emphasized personal salvation, Pearson emphasized mutual dependence and the group. He found in the life of the community the objective that social Darwinist apologists for capitalism saw in the individual.[18] At its lowest level, natural selection encourages a ferocious battle of each against each; but there is another social level, one that affirms socialism, because individuals require societies in order to survive and societies require "educating, training, and organizing the powers of all their individual members"; and beyond this there is a need for "solidarity of humanity in the struggle with its environment," which "is no less a feature than Individualism or Socialism of the law of evolution. We may perhaps term it *Humanism*."[19]

Pearson's social, literary, and philosophical essays prepare the cultural context for the version of the dying-to-know narrative that appears in *The Grammar of Science*. At every point, questions of knowledge are questions of ethics, for—again consistent with the dying-to-know tradition—Pearson saw knowledge as imperative because it was a condition for rational action in the world. He talks about "the almost sacred nature of doubt," and he advocates "a rational enthusiasm and a rational basis of morals."[20] In his essay called "The Market-Place and the Study," he argues that "the freethinker's religion is the pursuit of truth and his sole guide the reason, so his morality consists in the application of that truth to the practical side of life." He celebrates "the rational guidance of the social impulse" as infinitely superior to the "emotional appeals of a dogmatic faith" (p. 113).

If a profession of ignorance and a commitment to the practical importance of reason are two crucial elements in Pearson's story and that of the narrative of scientific epistemology, its most pronounced element is "Renunciation." His essay of that title argues that it is essential to human development. Tracing traditions of thought about renunciation from Buddha, to Meister Eckhart, to Maimonides, Spinoza, and Kant, he finds in them all, except Eckhart and Kant, a commitment to rationality, a recognition that the pains of this world are better managed with broader knowledge and understanding. Christianity, however, sees rationality as a danger and, insisting on a total emotional and indeed mystical rejection of the phenomenal world, inflicts great pain on those who subscribe to it.

But Pearson is explicit about the problems created by "renunciation." He resists the self-surrender that would mean the end of all sensuous pleasure and desire. By working through distinctions that preserve the idea of renunciation but self-consciously attempt to eliminate its most self-killing implications, Pearson illustrates another important argument of this book— that while the epistemological ideal can be deadly, it has sustained itself because it has always seemed also to offer otherwise unattainable possibilities for human improvement.

Pearson did not think of renunciation as destructive of the self but as the fullest expression of it. He begins his exploration of the subject with a question: "By some rational process on the one hand, or some transcendental rebirth on the other, cannot man render himself indifferent to the ever-changing phases of phenomenal slavery, and withdraw himself from the world in which fate has placed him? The means to this great end may be fitly termed, *Renunciation*" (p. 67). He admires Buddha's handling of the problem, for it is certainly Pearson's as well: "In order that the individual may free himself from this condition of slavery he must renounce his desires, his delusions; the only means to this end is the extermination of ignorance and predisposition" (p. 68). "Make your numerator zero."

Ironically, but predictably, the renunciant begins, as Buddha does, by refusing to believe anything for which a "rational origin" cannot be discovered. But the Buddha, in insisting that "knowledge alone is the key to the higher path, the one thing worth pursuing in life" (p. 74), rejects "sensuality" (an indulgence that inevitably multiplies pain unless it is disciplined by maintaining good health and keeping the mind in full energy), individuality (man is "an ever changing combination of material properties, which when believed in as a soul filled self is the source of endless misery"), and ritual (formal worship that hinders the growth of self-culture). "So long," says Pearson, "as . . . a man has no time for the development of his intellectual nature, he cannot be moral" (pp. 115–16). Renunciation in Pearson becomes an almost mystical expansion of self, even as the self as an entity virtually disappears (in part, perhaps, *because* it disappears).

Renunciation is connected also with religion, an issue that would seem settled for so outspoken a freethinker and scientist as Pearson. But in his essays on Maimonides and Spinoza renunciation connects with religion in complex ways. These philosophers were attractive to Pearson because they offered a religious vision forcefully undogmatic and rational. Both philosophers imagined a God who is *im*personal and who can be seen as "an intellectual cause or law running through all phenomena" (p. 129). Here, religion and epistemology become indistinguishable. Maimonides' arguments prepare the way: "You can never see matter without quality, nor quality without matter, and it is only the understanding of man which abstractedly parts the existing body and knows that it is composed of matter and quality" (p. 130). For Pearson, then, God becomes equivalent to knowledge, the ground and the object of all human intellectual activity. Maimonides, Pearson argues, sees the highest good as "the intellectual love of God" (p. 131), and for both Maimonides and Spinoza, God is "the knower, the known, and knowledge itself" (p. 135). "The intellect of God is *all*" (p. 136).

In the identification of God with intellect, Pearson lays out an idea that ultimately supports his modern empiricist sense that the outside world is a construction of human intelligence. For Maimonides, Pearson points out, "all existing things, from the first degree of intelligence to the smallest insect which may be found in the centre of the earth, exist by the power of God's truth" (p. 134), which is virtually to say that the natural world is a "construction" of God's intelligence. On Pearson's interpretation, Maimonides is arguing that

> [b]ehind the succession of material phenomena is a succession of ideas following logically the one on the other. This thought-logic is the only *form* wherein the mind can co-ordinate phenomena because it is itself a thinking entity, and so subject to the logic of thought. The "pure thought" which has a logic of its own inner necessity is thus the cause, and an intellectual one, of all phenomena. That system, which identifies this "pure thought" with the Godhead, may be fitly termed an intellectual pantheism or a pantheistic idealism. (ibid.)

The blending of religion and epistemology leaves Pearson with a religion, but a religion specifically intellectual, the religion of the mind.

At last, Pearson's early speculations on these issues lead directly to his science. Knowledge of God, not so surprisingly, is "associated with renunciation of all worldly passions, all temporal strivings and fleshly appetites; it is the replacing of the obscure by clear ideas, the seeing things under the aspect of eternity, *i.e.* in their relation to God" (p. 132). If one takes God here as a metaphor—and in Pearson it is hard to do anything else—this becomes the ideal description of scientific objectivity.

Pearson's theories of science are sustained by religious and ethical speculations of this kind: he felt no incompatibility between the idea that science is morally interested and the idea that it is valid—the only valid form of knowledge. The selflessness he sought in social and political behavior, he sought also in scientific behavior. Although even in his own time, critics were not certain that for Pearson the truth took precedence over all other ethical priorities, the morality he affirmed depended on disinterested aspiration to the truth.[21] At a banquet held in his honor in 1934, Pearson argued, "No man becomes great in science from the mere force of intellect, unguided and unaccompanied by what really amounts to moral force. Behind the intellectual capacity there is the devotion to truth, the deep sympathy with nature, and the determination to sacrifice all minor matters to one great end."[22]

While Pearson's moral/epistemological positions are astonishingly like those that I have ascribed to the entire epistemological tradition since Descartes, his work as a writer of real narrative makes him yet more useful in exploring the narrative of dying to know. For Pearson provides a narrative that helps make sense of the peculiar combination of ethics and epistemology, that strange commitment to self-restraint and intellectual ambition that I have thus far been exploring. *The Grammar of Science* is not so much Pearson's *Discourse on Method* as it is the consequence of a long narrative that preceded it. Like the conclusions of many such narratives, it effectively writes the end of the need for narrative. The point of knowledge, accessible only after long travail and travel, is the point of death (or a kind of death), after which there is no story. Pearson's novel throws an unusual light on his positivism, linking it directly to the romanticism I have been discussing. As novelist and scientist, he himself embodies the narrative of scientific epistemology.

The New Werther, published in 1880, when Pearson was a very young man, is not much of a novel, as he obviously knew, but its representativeness alone makes it interesting. It is a hyperserious quest novel of a familiar early nineteenth-century kind. In its preface, Pearson writes, "There is only one excuse for publishing the following letters, namely, they truly image the mind of him who has written them, and therefore necessarily to some extent the minds of the children of his generation, who are passing through a like struggle. They are beneath the notice of the critic, and appeal only to the sympathies of that young life which is cast creedless on the weary waters of nineteenth-century thought."[23] Pearson makes his narrative a stepping-stone to a "creed." While the novel echoes or anticipates Victorian fictions like *Sartor Resartus*, Froude's *Nemesis of Faith*, Newman's *Loss and Gain*, or Mrs. Humphrey Ward's *Robert Elsmere*, its author is determined to reject the weariness of the fruitless debates and worries of "the children of his generation." *The New Werther* looks like Pearson's renunciation of the ideal and the absolute.

It echoes in fervor for truth, justice, and meaning Goethe's deliberately adolescent and brilliantly sophisticated *Sorrows of Young Werther*. The story of a Pearsonlike young man named Arthur (Pearson gives some of his own journal entries to the protagonist), it is full of emotional overspill as Arthur tells the secrets of his epistemological soul to his beloved Ethel, with what romantic results one might imagine. Several scholars have discussed the relation of the novel to Pearson's work in statistics and eugenics.[24] While I want to draw on their arguments, I also want to emphasize the serendipitous fact of the novel's existence as it makes epistemology romance.

Like Goethe's novel, Pearson's is a series of letters from a highly sensitive, deeply romantic young man; but unlike Werther's, they are not to a male friend, Wilhelm, but to a lover, Ethel. The differences between the two novels are instructive, in particular because Werther is an artist, not a "thinker," though he speculates enough about large meanings, while Arthur seems to do nothing but think. Arthur begins his travels around Europe with no vocation but a longing for some ultimately clarifying knowledge. One has to believe that Pearson called his book the *new* Werther in part at least because he was imagining a different kind of aspiration. Werther's language moves to a world of feeling, a world where the irrational is fundamentally superior to the rational, where art, and nature, and romantic love matter more. These elements are also in Pearson's novel, but there is a peculiar abstractness to the imagination of the novel. Werther begins by getting away from home and away from books, while Arthur leaves home and his lover, Ethel, for his largely unnamed quest. But at the end of the first chapter one gets an inkling of why he is doing that: "I rush from science to philosophy, and from philosophy to our old friends the poets; and then, overwearied by too much idealism, I fancy I become practical in returning to science. Have you ever attempted to conceive all there is in the world worth knowing—that not one subject in the universe is unworthy of study?"[25] Arthur is, quite literally, dying to know. And he does.

Arthur's desire to put "the divine within our reach" is oddly a step in the positivist direction. It might at first seem anomalous that Arthur begins his journey feeling "those infinite yearnings which the reason itself cannot comprehend," and with "a soul straining to free itself from the empirical" (p. 3). He begins, as he says, "with feelings of man's divinity" (à la Feuerbach?). But while the rhetorical mode looks back to the intensities of German romanticism, the aspiration is remarkably prescient of the positivist position, which, though strongly empirical, is designed also to "free itself from the empirical," in the sense that positivism was always alert to underdetermination of empirical evidence.

"Ethel" is barely a shadow, and romantic love is rather a marker of the limits of knowing. The novel is not about romance but about how to find an equivalent for God after God's disappearance. When the foundation is shoveled away, how does one know anything? where does value come in? is there anything but the body? The story that supposedly shapes the novel echoes vaguely with the part of Carlyle's narrative that treats Teufelsdröckh's loss of "Blumine": he plummets into a negation of virtually everything, not so much because of the unrealized Blumine but because in a world from which God has been removed, all you need—as the Beatles put it many years ago—is love.

Another of Arthur's Pearsonian enterprises is to validate the "ideal." He seeks to reconcile materialism and idealism (one need not note the irony that the book itself seems to ignore, that he is doing this by denying his own physical relationship with his lover), and claims that "the Ideal world is not a world invented by the painter and the poet, but that it exists in every Actual" (p. 11)—"Here or nowhere is America," Teufelsdröckh proclaims. What Arthur will not surrender is the centrality of "reason." The novel embodies the radical skepticism that had already led many intellectuals to a complete naturalism. It was crucial for Pearson, and for his Arthur, and for the scientific naturalists that, as Arthur says, "the inexorable laws under which science asserts that the universe must for ever roll on, are not empiric, but deducible from the pure reason" (ibid.). And, finally, Arthur argues that "the sway of the intellect shall . . . be extended from the logical to the empirical, yet that the intellectual, the manly, shall itself be so bound up with feeling, the womanly, that the two shall be united in one being and in one life" (ibid.). "Everything is in unison with the Whole, which lives, evolves, and advances eternally throughout infinity. Nature gives one law of change and of life to all her children, and death but fuses the soul of the individual into the greater soul of all that lives; the unities are lost, the Whole alone remains, infinite and yet the same" (p. 21).

Though perhaps rather difficult to imagine at first, this kind of rhetoric is in a direct line with the work and the arguments of *The Grammar of Science*. Though *The Grammar* moves from an opening moral urgency to a commitment to statistics, this is no falling away from the moral imperative. Rather, statistics can then be read as almost like Carlyle's Everlasting Yea, self-consciously withdrawing knowledge from the realm of personal desire, fusing the individual in the unity of the whole. At the end of the epistemological romance, the most sublime pitch of knowing and of seeing the truth no longer through a glass darkly, narrative stops and denies itself. The end of narrative must then be death, literally and figuratively, or it can be God, the unmoved mover. When there is no more story to be told, life ends. Ending becomes the terminus that ever after denies narrative. However absurd it may seem, statistics take the place for Pearson of God—or death.[26]

Theodore Porter has argued that Pearson sought "the essence of things" (p. 4), and his novel dramatizes this quest. He quotes Pearson as saying in a letter to his brother that he had "no religion, but only religiosity" (p. 4). Like many other great Victorians, Pearson would have to find a substitute. The search begins most explicitly in *The New Werther*, where the quest for the divine turns out to be a quest for knowledge. As Porter says, "Faust's yearning for nature was relieved in some way by Gretchen, whom he seduced. Pearson was less lucky. He was too upright for seduction. He could not experience nature in anything like a direct way. The dusty existence of the positivist was all that was left to him" (p. 5). *The New Werther*, then, describes Pearson's prediction but also a monitory direction that he himself would refuse to take. It becomes a kind of cautionary tale, suggesting the grim possibilities of a totally "realist" approach to knowledge.

Arthur (a king in search of the ideal?) begins the narrative by renouncing Ethel (the noble?) in the requisite quest for knowledge, wisdom, truth. In the course of the narrative he discusses a wide variety of possible approaches to truth—philosophy, religion, science, in particular—only to find each of them wanting, incomplete, and potentially dangerous. At last, he decides to return to Ethel only to find that she has now married a friend Arthur made on his travels and with whom he had shared the secrets of his quest. Thus, betrayed by knowledge, friend, and lover, he commits suicide. The narrative is thin, and virtually nobody but Arthur is onstage, particularly not Ethel. In fact Gaspar, Arthur's dog, gets more attention and real affection than Ethel, whose answers are never recorded—although, of course, the narrative gives signs of moments when she disagrees with Arthur, when she has stopped writing to him, and that she misunderstands him. Structurally, that is, the story has the very quality of solipsism that threatens Pearson's eventual theory of knowledge.

Dying to know, Arthur renounces the claims of personality and the claims of the body; unable to know in the absolute and romantic terms that impel him, he can only die. In the first letter, Arthur announces the beginning of his epistemological pilgrimage. He leaves Ethel, not for another woman, but for some vague aspiration beyond the body. When Ethel asks him to write about his real self, Arthur virtually declares the leading motifs of late-century pessimism (such as Hardy's), and also of the epistemological difficulties of the dying-to-know narrative. He makes it clear that there is no escaping the "I": "Can I place myself outside the 'I,' and from thence give you a picture of what I am?" (p. 1) And in an obvious Carlylism, he talks of how each person is but "a drop in the stream of life," dashed and driven by circumstances "which not the most cunning of philosophers can find a law for" (intimations of Darwin and natural selection?). "The inner life ever flows between the banks of outer destiny," he says, and "Man is what his time, his nation, and his liver make him" (p. 2).

Arthur struggles with the traditional mind/body problem, and insists that to know his "true self" it is necessary to know the circumstances through which he moves. Arthur's narrative begins with this attempt to wrench free from the body, from the merely "empirical," but makes clear that it is impossible to dismiss the body. No sexual urges are to be allowed if he is to satisfy his spiritual need to know. So Arthur abandons his Ethel on this obscure quest. This is "dying to know" to the letter, and it is thus not surprising that Pearson has to kill Arthur at the end of the story.

Pearson signed the novel pseudonymously as Loki, a Norse god, the dictionary notes, "who created discord, especially among his fellow gods."[27] Given his interest in Germanic folklore and the explosion of interest in the subject throughout the culture, it is very likely that Pearson knew a great deal about Loki, and it is worth mentioning some of the things that might have attracted him to that pseudonym.[28] The notion of Loki as mischief maker would have to be primary and probably would have been available to whatever few lay readers the book found. But it is interesting also to note that Loki was rejected by the other gods because he tended to sell out to the giants. Loki was an outsider who as collaborator with the giants aided in their eventual overthrow of the gods. Moreover, he is one of the few figures in Norse lore capable of changing himself into other things, and he changes himself into a mare to seduce Odin. Oddly, then, he couples with Odin as the female partner and gives birth to, becomes the mother of, the goddess of hell. It is a strange and inconsistent history, but it plays into the work of the novel in many ways. Pearson transforms himself into Arthur, and Arthur, rejecting God, is ultimately expelled from divine presence and knowledge. He falls, at the end of the novel, with this melodramatic imprecation: "Spirit of Nature, or, if Christ had truth, eternal God, eternal hell, receive my soul! I summon you, I do not follow at your beck. Hell! hell! Oh, it was foul! base! Raphael, Ethel, I forgive you—you were but human. Ethel—beloved—I die" (p. 116).

Pearson plays seriously with the mischievous possibilities of Loki, his powers of self-transformation and his ambiguous relation to the gods. Inside the solipsistic space of the epistolary form Pearson adopts, it remains possible to read the book as a way of imagining Pearson's own intellectual biography. The descent into hell suggested at the end is a serious one, and the concern with the failure of the gods is equally so. In Arthur's death Pearson dramatizes the price of the epistemology of self-annihilation.

The novel begins and ends with a sense of the enormous moral and spiritual investment required of an adequate epistemology. For all its romantic extravagance, it retells the epistemological story as a Pilgrim's Progress ending in death, but without the redemptive divine illumination. When Arthur at the start speaks of how "not one subject in the universe is unworthy of study," he begins the Cartesian scientific romance; and when he tells Ethel that she "must picture my outer life to penetrate the inner" (p. 2), he makes a perhaps non-Cartesian but very Victorian point, that in order to know, one must understand how philosophy is completed by narrative.

Arthur quickly clears the ground as Bacon and Descartes did, distinguishing immediately between past superstition and the new possibilities of the divine. "When I write to you of man's religious nature," Arthur tells Ethel, "do not conjure up worn-out faiths and lifeless creeds" (p. 4). Engaging issues that made the substance of George Eliot's explorations in *Daniel Deronda,* Pearson registers awareness like hers of the primacy of feeling, although Arthur wants to assert the primacy of reason. The issues all resurface here, in certain respects, of course—and embarrassingly, in the question of the richness of the narrative—rather roughly simplified. But while George Eliot tried to imagine some way in which feeling and knowledge could mutually authenticate each other, in this epistemological romance, the tension between feeling and reason is explicit and irresolvable. The same issues are in play. Arthur understands that value derives from feeling, but feeling (as George Eliot, too, insisted) cannot determine belief. Arthur does try to blend feeling and intellect (he even tries to imagine his relation to Ethel as enacting that union [p. 11]). But untamed feeling must be kept under control.

This moves the epistemological romance to its second inevitable turn, the act of renunciation. Objectivity requires it. At the outset, Arthur exclaims, "Happy is the man who not only can say, 'I know nothing,' but can content himself with that nothing" (p. 6). His first move being the renunciation of the flesh in leaving his lover, he adopts renunciation in the name of the philosophers that Pearson discussed in his essays on the subject. Ethel tells Arthur that she does not wish to hear of his "outer life, but that "I want yourself, and yourself only" (p. 1). Here is a mind/body problem of the Cartesian kind, if ever there was one. Like Goethe's Werther (and like Carlyle), Arthur feels the phenomenal slavery, evoking his liver as a determining factor in shaping his identity. He struggles to free himself "from the empirical" and from Ethel's "swelling bosom" (p. 3), and he compares his renunciation to Wilhelm Meister's, giving precedence to his own, for, he tells Ethel, "we . . . have cast aside the bonds of a hypocritical society and the cant of custom" (pp. 5–6). The path to God and truth is through renunciation.

Because in much of Cartesian epistemology's history, God was the guarantor of truth and the way out of solipsism, Arthur's story constantly engages the religious question. "Be it true or not that there is a God," he says, "my nature requires at least that I shall believe there is one, and in this I have not advanced beyond the feeling of the great mass of my fellow men" (p. 61). While the Pearson of the early essays is militantly unbelieving, he has a characteristic Victorian intellectual's preoccupation with God, in the tradition of Feuerbach's religion of humanity, or that of George Eliot, who stopped believing in God but could not stop believing in divine ideals.

In two chapters on science, Arthur begins by seeing science as a new religion: "Is not Science the common field of all nations, as Religion used to be?" (p. 94). Science, Arthur believes (for a moment at least), unifies consciousness and matter, moving toward a series of "laws" that connect all things. Half playfully, he begins to set up a statistical model for Ethel, arguing that "many dimensional space" is a misnomer, for it is really "a pure relativity," a "trinity of coordinates" (p. 96). And then he insists on more than three, including coordinates to represent "my various feelings," and then those of other people. "There exists a multitude of points," he concludes, "on which any two beings . . . hold the same views," and every thing ends "by being recognized as an influence on everything else" (97).

These rather startlingly relativist ideas are also mystical, and they anticipate both Pearson's fully developed arguments about knowledge and the radical critique of science that, ironically, links *The Grammar of Science* with its enemies. In the following chapter Arthur assumes a voice similar to those of the very critics of science, such as Balfour, against whom Pearson was later to do combat. Listening to a lecturer on evolution who seemed ready to reveal "the secret of the universe," Arthur is deeply disillusioned—"they cannot tell us of the birth of life, of feeling, and of love" (p. 100). Here, at some length, is his fullest indictment:

> Modern Science wishes to dictate to and rule the world, because the world at large knows little about the *pro and con* of scientific truth. She floods the land with her primers full of unproved assertions, leaving the many in the belief that there exist proofs lying beyond their power in deep works of research which are for them sealed books. Religion once tyrannized over the world. Science has killed Religion; but instead of setting up a republic of thought, has instituted a worse tyranny in its place, the oligarchy of scientific specialists, who expect mankind at large to accept on the ground of authority whatever they choose to proclaim as truth! (p. 102)

In *The Foundations of Belief* (1895) Balfour strenuously attacks scientific dogmatism in this way (and Pearson would counterattack). But Pearson surely means what Arthur says, and in *The Grammar of Science* he accuses science of making unprovable metaphysical claims. His whole epistemology is devoted to establishing the limits of knowledge and deflating those claims. Yet there is a difference in the way Arthur articulates the question. Unlike the mature Pearson, Arthur is still looking for the large answers and angry that the scientists can't provide them. The mature Pearson knows that the big answers will never come. Arthur's disappointment may well have been Pearson's, but Pearson does what he does not give his protagonist to do—he finds meaning and consolation in the very limitations that ultimately drive Arthur to suicide.

Arthur has chosen his options mistakenly: either seek absolute truth beyond the senses, or relapse into the phenomenal world in the arms of his beloved. Pearson's Spinozan distrust of the body, which obfuscates and threatens pain, makes the second option a mistake. Somewhere between the realism of the seeker of ultimate truths and the resignation of the self into the consolations of love, Pearson needs to invent a life-giving possibility. The irony is that the life-giving possibility to which his novel drew him turns out to be another form of "self-elimination," a positivism that substitutes the discipline of statistical measurement for the embodied particulars of life.

The positions that Arthur ultimately finds inadequate in the novel are virtually all positions that in one way or another Pearson supported and developed. Pearson has to come to terms with the problem of solipsism and limitations, and his narrative of Arthur's death is in part a critique of the dying-to-know story. Arthur kills himself in the pursuit of purity of knowing. Yet the story reinforces another version of the dying-to-know narrative. If Arthur dies, epistemologically speaking, of an inadequate power of renunciation, Pearson emerges with a new epistemology, itself rigorously austere, insisting on the limits of consciousness, and on its powers. "Man," Arthur pontificates to Ethel (no wonder she left him!!), "is a divine atom, mankind a portion of the Whole; the Whole is God" (p. 9). Arthur's longing for absorption in another and larger thing marks Pearson's career, but Pearson finds the whole in consciousness, though a consciousness shared—in Feuerbachian communion—with the rest of his species.

Shortly after he finished his novel, Pearson wrote *The Trinity*, a massive passion play in blank verse, and here the quasi-religious need for a Feuerbachian religion becomes entirely explicit. Like many turn-of-the-century rewritings of the story of Christ (in England George Moore and D. H. Lawrence wrote such stories), Pearson's version humanizes God. Arthur's is the story of a clearing away of the debris of past superstitions and romantic aspiration. *The Trinity*, another effort at clearing away—in the line of the Baconian "idols" or Descartes's Great Doubt—attempts to follow Feuerbach in rescuing from superstition the human values at the center of the conception of God—and human values, it develops, are all there are.

In some ways, *The Trinity* imagines the world as Hardy imagined it in *Jude*. But Jude, like Arthur, is destroyed by the disparity between his aspiration and the meaninglessness of a merely physical world. Pearson's myth of Jesus lowers expectations. Where Hardy dramatizes the self-destructiveness and, indeed, impossibility of the self-discipline required by the ideal of full knowledge, *The Trinity* turns the divine into the human and celebrates the liberation of humanity from the superstitions of the old aspiration to the absolute. Pearson retells once again the Cartesian story, now updated by

Feuerbach, as the characteristic story of late nineteenth-century humanism—of the religion of humanity. In his brief preface to the play, he makes the point clearly: "Modern science and modern culture are freeing us from the old theological shackles. Let them take heed that in destroying a human divinity they do not forget a divine humanity."[29]

Through his fictional narrative and drama, Pearson makes clear the path from religion and ethics to scientific epistemology and science. The quest for absorption in some larger whole is worked out here. Early in *The Trinity* "The Demon of Doubt" speaks from a kind of prologue in hell to the presiding deity of "Negation." The entire prologue very neatly theologizes Darwin's theory of natural selection, emphasizing that life grows out of its negative, out of death, and that "through the sieve of evil testing passed / The pure, diviner elements of man" (p. 5). "The road to knowledge," says the Cartesian demon, "passes by the waste / Of doubt" (p. 8). But near the end of his speech, the demon seems to recapitulate Arthur's aspiration:

> While man gives credence to a form of god,
> Be it of stone, or that unchanging law
> Which runs like blood through Nature's giant frame,
> His quest for peace may p'rhaps be not in vain.
> If he allow no rule beyond his own,
> Beyond mankind's, or fate's necessity,
> Then will deep source of comfort fail
> His thirsty soul, and reason vainly seek
> A standard of unchangeable validity.
> The spirit of mankind must ever long,
> Amidst the shifting scenes of its small stage,
> To find that essence of tranquility,
> Which beyond time eternally remains
> Unchanged; to fix one spot, where the mind's eye,
> Fatigued by ceaseless motion, may repose
> Upon immutability's still peace. (p. 10)

But the aspiration to peace and tranquility here comes exactly from the surrender of the ideal of the absolute and acquiescence in the recognition of limitation. The religious form is the same, but the apparently transcendent "immutability" can only come "If he allow no rule beyond his own, / Beyond mankind's." To seek "A standard of unchangeable validity" beyond the human is inevitably to fail.

Pearson's philosophical problem will be to work out how, from the Darwinian flux and the inaccessibility of the thing in itself, to derive a form of knowledge that would provide such a standard. The world could never be unchangeable, as Pearson and Arthur had discovered; how, then, could knowledge? How gain access to that "essence of tranquility" if the attempt to fix nature, to achieve the stability of the eternal, unchanged thing, were chimerical and dangerous? Dangerous enough to drive Arthur to suicide?

The answer lies in the fact that humans are capable of projecting or imagining a stability to which the world may not correspond. Perhaps ironically, Pearson turns to science to find his "still point of a turning world." The play, ending with the death of the "son of the mortals," affirms a universe of love and knowledge. Jude, Arthur, and Christ all have to die before humans can learn how to join knowledge and life.

The dying-to-know story takes two different forms in Pearson's prescientific writing. There is the form of the novel, in which death follows on the recognition that no matter how disciplined and austere and honorable one's pursuit of knowledge, the human cannot achieve it with certainty. The end of that story is death, like Arthur's or like Jude's. The other form makes the transformation through the story of Jesus, now seen as a mortal, the divinity of humanity: in the pursuit of human ideals, the price might well be death, but the death is an embodiment of the ideal and points to knowledge after all, if a lesser, a merely probable knowledge.

Pearson's use of a "dying" narrative to open up alternative modes of knowing suggests once again how the act of renunciation can also be an act of affirmation of the self—which it is most distinctly in *The Trinity*. There Jesus approaches death and revelation at the same time, revelation that divinity is not out there in the cruel "elemental force" (p. 185), but in the self, in the power of will and love:

> *Self?* How at last we are thrown back on thee!
> Our parents, brothers, friends may help us o'er
> The little trials of life, but when our hour,
> The crisis of our fate arrives,
> All human kindness, e'er religion fails—
> The man is then dependent on *himself,*
> Must stand alone, trusting this one resource;
> For in the widest storms of life
> Self is the last sheet anchor of his hope. (pp. 182–83)

It is human consciousness that transforms the world, otherwise "dark, barren, harsh, and cruel" (p. 186), into a "paradise." It is a moment of revelation and resignation, and Jesus, like Everyman, like Bunyan's Pilgrim, approaches the end alone and humble—and, at least metaphorically, resurrected.

It is not only that self-effacement is required before the realities of nature, but that the very act of imagining and constructing reality entails a Christlike self-humiliation. Pearson attacks the old Cartesian story because he recognizes with late nineteenth-century despair that the ideal of clear and distinct knowledge is impossible and deadly. Yet as Feuerbachian humanism retained the forms of Christianity without a God, so Pearson's new narrative of scientific epistemology replaces the old realism that posited direct access to the truth of nature with a new idealism and constructivism that, in the end, retains the old forms. And the path to the new knowledge of limits turns out also to be the path of dying to know.

11

The Epistemology of Science and Art: Pearson and Pater

Pearson's early work reads like an apprentice's engagement with the narrative of scientific epistemology. On the one hand, it is a striking example of the way in which epistemological ideas do in fact live in narrative; on the other, it suggests how even so austere and anesthetic a philosophical mode as positivism is rooted in moral and spiritual urgencies. In this chapter I want to spell out more fully the details of Pearson's epistemological arguments as they emerge from the narrative difficulties discussed in the last chapter, while treating his philosophical arguments more directly to suggest other ways they are connected to broad cultural problems that extend beyond science and into our own moment. Pearson's epistemological work bears a remarkable similarity to the theories that lie behind late nineteenth-century aestheticism and engages issues in ways familiar to contemporary theorists of culture and consciousness.

But science and art are not, at the end of the nineteenth century, at war, as they seem to be so noisily at the turn of the twenty-first century. Close enough to their mutual sources in the quest for ideals that extend beyond the personal, they mutually informed each other, learned from each other, not least obviously in the work of Pearson himself. Pearson's thought is as surprising and original and culturally significant as the work of Walter Pater, now seen as a valued ancestor of many contemporary lines of thought. Looking at Pater and Pearson as unlikely twins working the same materials into structures only apparently antithetical might throw a different light on the way in which cultural criticism, art, and science have so antagonistically divided. The juxtaposition should suggest that many developments in modern cultural theory are far more closely allied to science and scientific epistemology than its exponents allow.

Pearson's work seems strikingly paradoxical. A passionate defender of science, he is an antirealist, a constructionist, an idealist. Austerely committed to "self-elimination" in the pursuit of knowledge, he does not, as we have seen, believe that knowledge of the outside world is possible. He wrote in the vein of the aggressive scientific naturalists, such as Huxley and Tyndall, and with a sophistication about philosophy that made the developing paradoxes of consciousness deriving from both Darwin and German post-Kantian philosophy a major concern of philosophers and psychologists. The paradox of consciousness itself was producing a new set of epistemological and aesthetic crises, and paradox was very much at the heart of them.[1] Darwinian science had transformed humans from those investigating to those investigated; and as animals, humans are obviously less trustworthy as commentators on that world. "Can the mind of man, which has, as I fully believe, been developed from a mind as low as that possessed by the lowest animal, be trusted when it draws such grand conclusions?" asks Darwin about his own "conclusions" about religion.[2] Self-consciousness about consciousness, increased awareness of other cultures through the activities of empire and of ethnologists and anthropologists, the very growth of authority of science itself as it extended its range from rocks, to stars, to people—all of these contributed to new efforts to imagine and understand the mind's powers. Moreover, all of these participated in the now long tradition that required, for objectivity—or, indeed, for the critique of objectivity—the stance of self-abnegation.

As Pearson engaged such issues in the interests of science, he spoke a language that might have been equally significant in the world of art. Self-elimination remained for him, as we have seen, a condition for knowing. Pater, precisely because he seems to belong entirely to the world of high and precious culture, and is ostensibly the solipsistic antithesis to the world of scientific epistemological authority, provides a particularly interesting point of comparison here. If it is agreed that cultural theory now owes much to the Paterian tradition, it is important to take into account the fact that Pater participated in the apparently antagonist tradition that also produced positivism. Cultural theory owes just as much to the positivist tradition that it has dismissed. If the ideal of self-abnegation seems an impossibility for science, so, too, must it be for Pater.

Some of the countercritiques of science that our new Werther, Arthur, anticipated and echoed became noisier as science extended its authority throughout the last half of the nineteenth century. These critiques emerged not only from the intellectually and politically conservative, like the duke of Argyll,[3] whose *Reign of Law* (1868) struck out against the chance-ruled world of Darwinian evolution, or, more effectively and popularly, A. J. Balfour, whose *Foundations of Belief* made something of a sensation when it appeared in 1895.[4] Nietzsche, at roughly the same time that Pater and Pearson were writing (though there seems to be no influence at work), detected the hidden narrative of "dying to know" in the theory and work of science: "This pair, science and the ascetic ideal," he says, "both rest on the same foundation— . . . on the same overestimation of truth (more exactly: on the same belief that truth is inestimable and cannot be criticized)."[5] And again: "modern science—let us face the fact!—is the best ally the ascetic ideal has at present, and precisely because it is the most unconscious, involuntary, hidden ally!"[6] With an iconoclastic energy stronger than that of either, he, like Pearson and Pater, attempted to expose the ethical and metaphysical underpinnings of science and scientific epistemology. Nietzsche is valuable to invoke here because, though he was no influence on either Pearson or Pater, and certainly not an empiricist, he was unblinkered in his view that asceticism was a form of the will to power; and unlike his English contemporaries he was content to know that science was both useful and a lie.

Pater and Pearson, from whom the alliance with the ascetic was certainly not "hidden," dodge the solipsism their empiricism would seem to entail, and dodge it precisely through asceticism, about which, unlike Nietzsche, they are undisturbed. They pursue without question, with, in fact, the peculiar enthusiasms of their disciplines, the goal of "self-elimination." For that elimination gave them that paradoxical power over experience that marks the ascetic, self-abnegating tradition from the start. Detachment, "dying," gives science its authority, the aesthetic its more than personal authority.

Modernist aestheticism and positivist science participate equally in the ascetic tradition, and equally so in empiricism. Pater and Pearson are separated philosophically not by any combat about realism—neither of them was a realist—nor by the thoroughness of their commitment to science and scientific method. In their acceptance of what they thought science and empiricism entailed, both of them seem at first reading, startlingly, to be virtual solipsists. Part of the incoherence of naturalism as Balfour analyzed it was that empiricism made impossible any assertions about reality and, in particular, any general assertions about it.[7]

Arthur Balfour's kind of complaint about "the comparative neglect of a philosophy of science,"[8] which allowed him to build his attack on naturalism, echoes with that other "Arthur's" complaints and partially explains Pearson's efforts in *The Grammar of Science* to rid science of its metaphysical incoherences. In an essay written in 1894, Pearson complains about Balfour's kind of strategy, attacking in that essay both Lord Salisbury and Gladstone, who professed to accept science but who

> emphasize our ignorance rather than our knowledge, and having learnt something of the critical spirit from its opponents, [are] able, not without effect, to point out the grave weaknesses in the present foundation of both physics and biology. It passes lightly from the true *Ignoramus!* of science to the *Ignorabimus!* of pseudo-science, and thence by an easy stage, the illogic of

which is scarcely noticed by the untrained mind, to the characteristic theological *Credendum est!*"[9]

And in "Reaction" (1895), his direct review of Balfour's book, Pearson first denies that science makes materialist claims and then says, "it may, perhaps, be argued that, although this materialism is hopelessly absurd, and although it is an error on Mr. Balfour's part to assert that it is the philosophical basis of Science, still he has done good service in showing that it is untenable."[10]

Pearson is happy to accept the claim that we *don't* know, but he will not have it that we cannot know. And he does not fall back on the solipsism that a theory of knowledge as "sensation" would seem to require (he engages the problem directly, as we shall see). In his response to Balfour's eagerness to return to a prescientific faith, he moved to positivism, as Pater, at least as skeptical as Pearson, moved to impressionism, and they traversed similar paths through long traditions of German post-Kantianism and of British empiricist thought.

The voices of science and of art speak from within consciousnesses they describe as closed off from the realities of the world around them. Their radical subjectivity is, however, no strong assertion of the primacy of the self, which in the empiricist tradition is always tenuous; it is, rather, a condition to be understood and a problem to be surmounted. Across the divide of their aesthetic and scientific concerns, they speak often in language strikingly similar and pointing to shared sources; they are separated primarily by the ways in which they seek to establish something like objectivity, something like stable and shareable knowledge. I say "something like" because Pater was rather less concerned with "objectivity" than Pearson, for whom the whole work of science was at stake. At the same time, Pater's voyage into aestheticism and into the discipline of ascesis entailed for him a plunge into the problems of empiricist epistemology. Moreover, the consequences of ascetic discipline, I want to argue, produce an aesthetic analogue of objectivity — a firm, even a "gemlike" reality that is not merely subjective, but allows the perceiver to stand outside the flux he is describing, if only in order to describe it.

Pearson and Pater, with all their grounding in German romantic philosophy, are inheritors as well of the great English empiricist tradition, from Locke through Berkeley and Hume, which concerned itself with the illusoriness of perception, the limitations of perspective, and, ultimately, the possibility of solipsism. Berkeley's philosophy avoided the threatened slide into solipsism rather in the way Descartes avoided it, by positing a God. But Berkeley's God turns perception itself into being. For many late nineteenth-century intellectuals, scientist and critic, that option was no longer available, and a secular version of Berkeley would produce a radical constructivism. Neither Pater nor Pearson accepted the possibility of a naïve realism that allowed for consciousness to have direct contact with the world "out there," and neither believed that the limited and embodied self could be eliminated as the dominant factor in knowing. Anti-antianthropomorphists both, they build their knowledge on the self, whose experiences turn out to be almost all there is.

If Pearson thought of himself as an idealist, Pater began his most famous book by taking on the radically objectivist arguments of Matthew Arnold, who saw the first object of culture as knowing the object as in itself it really is. Notoriously, Pater began his *Renaissance* by announcing that the first step in his project to "know the object as it really is" is "to know one's own impression as it really is, to discriminate it, to realise it distinctly."[11] But for neither artist nor scientist, despite Pater's apparent passion for the fleeting and unstable, was solipsism, or its cousin, an unconditioned relativism, a possible resting place. Pearson's early writings have shown that he could be the true Huxley-style advocate for the intellectual empire of science only because he had found some point of stability, something fundamental beyond the person and the personality on which he could build knowledge and in which he could rest. As he insists in *The Grammar of Science,* he seeks "judgments independent of the idiosyncrasies of the individual mind."[12] And the retiring Pater, whose formulation of what is usually taken as the solipsistic condition is perhaps the most famous in the language, sought just as persistently, though obviously in very different ways, to transcend the person and the personal in order to be able to make any claims at all for his history of art, and for his sense of its value. I do not mean to claim that my argument about Pater's indebtedness both to science and to empiricism is particularly new.[13]

Born eighteen years apart, into the heart of Victorian culture, Pater (1839–94) and Pearson (1857–1936) would seem to have had only an Oxbridge education in common. Unlike the combative and polymathic Pearson, third Wrangler in the Cambridge Tripos, some of whose activities I have already traced, Pater spent his entire adult life cloistered in Oxford. He was, also unlike Pearson, a notoriously unpolemical and retiring figure who was taken (almost, as it were, by surprise) as virtual father of late-century aestheticism, of the ideal of tremulously acute sensibility, of that apparently least scientific, least virile, most artful of intellectual movements. When his "Conclusion" to *The Renaissance* became notorious, he made no fuss but withdrew it from the next edition. Undoubtedly under pressure at Oxford both for the publication of *The Renaissance* and for the coming to light of some perhaps sexually compromising letters, Pater slipped quietly back into his always somewhat uncomfortable relationship with the university, now getting away on weekends to a house he decided to take in London. He stayed out of the polemics that his book promised to ignite.

Unlikely as at first sight it might seem, Pater read and absorbed a great deal of contemporary writing about science—far more, one would have to guess, than Matthew Arnold, who had nevertheless ventured (and quite successfully) to comment on scientific education in one of his famous lectures. Pater was particularly attuned to what he called "the sciences of observation." Conversely, as we have seen, the scientifically gifted Pearson was a surprisingly literary and spiritual positivist. After he took his Cambridge degree, he read intensely in Germanic literature and philosophy, did much European traveling—like his protagonist, Arthur—and wrote not only a novel and a play, but impressive philosophical and philological essays. Both men were steeped in the thought of Kant and post-Kantian German philosophy, and both found Darwin and evolutionary theory fundamental to their ways of thinking. As Peter Alan Dale argues about Pater,[14] Pearson was deeply influenced by German philosophy, particularly by Kant, but the basis of his epistemological position was almost certainly empiricist and recognizably compatible with the English tradition.

Pater, like Pearson, was obsessed with the question of how one gets to know, and he, too, drew strongly on the traditions of scientific epistemology. If Pearson was committed to "self-elimination," Pater insisted on ascesis—an austere, rigorous restraint of the self that, from the basis of an inevitable subjectivity, issued in an impersonality that opened both to art and to truth. Ascesis, though almost always taken as a condition for the disinterest and objectivity necessary for adequate scientific observation and experiment, is also a way beyond the entrapment of self, both a social and an aesthetic extension of the particulars of individual sensation to the shareable conditions of true knowledge. Ascesis is not, of course, equivalent to objectivity, but it is a condition for it. In extending Pater's concept of ascesis to a more explicitly epistemological enterprise, I am pursuing the Lorraine Daston-Peter Galison argument that scientific epistemology was intensely moralized in the nineteenth century and required of the scientist a total and highly disciplined surrender to experience equivalent to the disciplined restraint urged by Pater. When Pater insists on the necessity "to know one's own impression as it really is, to discriminate it, to realise it distinctly," he is pointing to a truly ascetic activity that is often mistaken for self-indulgent. Experiencing sensation is one thing; literally knowing it, discriminating it, entails repression of self, almost a denial of the sensation being experienced. Pater is arguing for a way to be objective about subjectivity, to find a position outside experience from which to experience it and posses it. I want to suggest that

Pearson's thought follows a similar path, and that the ascetic move has enormous theoretical importance both for empiricism and for the apparently more obvious relativist theories of our own time.

For the moment, however, it is sufficient to emphasize that both writers insist that true art like true science depends on "disinterest," and that is the obvious equivalent of selflessness. If for Pearson self-elimination was a condition for adequate scientific observation and for good citizenship, for Pater, commitment to the impersonal was a condition for adequate perception of the beautiful. Pater, to take just one example, claims that "self-restraint, a skillful economy of means, ascêsis, that too has a beauty of its own."[15]

The importance of science to Pater's thought is apparent in his essay on Coleridge, as it appears in *Appreciations*. There he argues that

> [m]odern thought is distinguished from ancient by its cultivation of the "relative" spirit in place of the "absolute." . . . To the modern spirit nothing is, or can be rightly known, except relatively and under conditions. The philosophical conception of the relative has been developed in modern times through the influence of the sciences of observation. Those sciences reveal types of life evanescing into each other by inexpressible refinements of change. Things pass into their opposites by accumulation of indefinable quantities.[16]

The language of Darwin echoes here, particularly those moments in *The Origin*, in which he talks of how species and subspecies "blend into each other in an insensible series."[17] Pater has picked up Darwin's gradualism, his dependence on vast accumulations of time, his obliteration of sharp divisions between species, his sense of constant transformation, his pervasive insistence on "imperceptible gradations." But he also picks up the historicist implications of Darwin's argument, the necessity to consider one's location as observer in space and time, his emphasis on the context in which individuals live, his substitution of historical for "absolute" explanation, his focus on the importance of the individual. Where Arnold had tried to separate Darwin's language and ideas from the language of value, Pater absorbs it without fuss into that language, self-consciously translating quantity into quality. He leans on the sciences of observation for the ideas that shape his views of art.[18]

In his late book, *Plato and Platonism*, Pater associates Darwin with a Heraclitean tradition that forms a critical part of the background of Plato's thought. Pater's Plato, perhaps not surprisingly, embodies the two fundamental aspects of Pater's vision, the recognition of the fleeting and historical nature of all phenomena, and the sense that beyond the flux (or through the flux) there is what Pater associates with Parmenides—the single, the fixed, the permanent, the stable. Here, however, it is the Darwinian-Heraclitean aspect of his vision that dominates: "'species,' the identifying forms of animal and vegetable life . . . are fashioned by slow development." And development, he claims, is "invading one by one, as the secret of their explanation, all the products of mind, the very mind itself, the abstract reason . . . and language is changing on our very lips."[19]

Empiricist sensationalism underlies this aspect of the Platonic and Paterian vision, and here I want to look at two representative examples of the formulation of this sensationalist epistemology in Pater and Pearson. In *The Grammar of Science*, Pearson writes:

> Turn the problem round and ponder over it as we may, beyond the sense-impression, beyond the brain terminals of the sensory nerves we cannot get. Of what is beyond them, of "things-in-themselves," as the metaphysicians term them, we can know but one characteristic, and this we can only describe as a capacity for producing sense-impressions, for sending messages along the sensory nerves to the brain. This is the sole scientific statement which can be made with regard to what lies beyond sense impressions. But even in this statement we must be careful to analyse our meaning. The methods of classification and inference, which hold for sense-impressions and for the conception based upon them, cannot be projected outside our mind, away from the sphere in which we know them to hold, into a sphere which we have recognised as unknown and unknowable. The laws, if we can speak of laws, of this sphere must be as unknown as its contents, and therefore to talk of its contents as producing sense-impressions is an unwarranted inference, for we are asserting cause and effect—a law of phenomena or sense-impressions—to hold in a region beyond our experience. We know ourselves, and we know around us an impenetrable wall of sense-impressions.[20]

This impenetrable wall famously occurs in Pater, some fifteen years before Pearson wrote, and it turned up there along with the radical antimetaphysical skepticism that would seem to cut off all access to the not-self. Although the passage is familiar to virtually everyone, it is important for my argument to review it:

> And if we continue to dwell in thought on this world, not of objects in the solidity with which language invests them, but of impressions, unstable, flickering, inconsistent, which burn and are extinguished with our consciousness of them, it contracts still further: the whole scope of observation is dwarfed

into the narrow chamber of the individual mind. Experience, already reduced to a group of impressions, is ringed round for each one of us by that thick wall of personality through which no real voice has ever pierced on its way to us, or from us to that which we can only conjecture to be without. Every one of those impressions is the impression of the individual in his isolation, each mind keeping as a solitary prisoner its own dream of a world.[21]

Pater's wall, like Pearson's, has been built from the great tradition of empiricism, but that tradition requires a special alertness to the conditions of the perceiving self. Such alertness is equally important to Pearson, for the first step in coming to terms with the awkward presence of the self in the search for knowledge is to make and keep yourself alert to it. The labor of self-awareness is just that labor of ascesis. It is one of ironies of this tradition that efforts to contain the overweening self are best accomplished by intense attention to its presence—it is just the kind of attention that narrative forces on the perceiver and the observer. Such attention allows a more precise view of the nature of one's own experience and thus of whatever it is that constitutes knowledge. (And in this respect, the strategies of both writers parallel the self-consciousness demanded by contemporary science and social science, particularly anthropology, where a crisis of representation has developed partly through the awareness that the cultural other is not transparently available to an observer, who inevitably brings to bear on what is observed unexamined cultural assumptions.)[22] Knowledge, then, can only be trusted if the perceiving self is aware of its own conditions of perception.

Pearson brings the lesson home most forcefully and graphically in a section of *The Grammar of Science* called "Outside and Inside Myself." Using an image created by Ernst Mach (figure) that resonates wonderfully for contemporary self-reflexive and destabilizing theories of self, Pearson argues that the distinction between outside and inside ourselves is "one merely of everyday practical convenience." The sketch is by Ernst Mach of himself "lying on his sofa, and having closed his right eye, the picture represents what is presented to his left eye."[23] Pearson's language is careful: the phrase "is presented to" replaces an expected "sees." "Sees," though an active form, implies that there is something out there impinging on the eye with which what Mach represents is to be compared. But "reality" in Mach's epistemology becomes a shadow world; a table is not "out there" but an imagination constructed out of "sense-impressions of hardness," and other sensations. "The *real* table lies for [us] in the permanent association of a certain group of sense-impressions, and . . . the shadow table is what might be left were this group abstracted" (p. 83). Mach and Pearson want to emphasize the degree to which what is "seen" is precisely what is presented—

Ernst Mach, *Study in Perspective*. Drawing, nineteenth century.
Reproduced in Karl Pearson, *The Grammar of Science*, 2d ed.
(London: Adam & Charles Black, 1900), p. 64.

there is nothing to correspond to the presentation, since outside and inside are one.

Here is how Mach himself describes his drawing:

> In a frame formed by the ridge of my eyebrow, by my nose, and my moustache, appears a part of my body, so far as it is visible, and also the things and space about it. . . . If I observe an element, A, within my field of vision, and investigate its connection with another element, B, within the same field, I go out of the domain of physics into that of physiology or psychology, if B, to use the apposite expression that a friend of mine employed upon seeing this drawing, passes through my skin. (pp. 78–79)

The image shows the left side of "Mach's" mustache more or less looming over a waistcoat that extends down into a pair of legs crossed at the ankle in diminishing perspective.[24] Within the same angle of perspective, one can see half a door with knob on the left, and a bookcase beyond, running to a wall directly above the legs; two windows; a little round table; and then, curved inside the walls of the eye socket on the right, part of an easel on which is set a piece of paper with a hand visible, drawing, one would have to assume, this very picture. The frame of the image is the eye socket itself, clearly defined with the shape of the nose above the mustache and somehow fading off on the left so that the socket is not defined there.

The image is both persuasive and strange.[25] It doesn't seem to correspond to what we unselfconsciously "see" as we go about our daily living. But what in ordinary vision we think we see, this image and the text attempt to persuade us, is only a psychological construction.[26] Indeed, Mach argues that insofar as we begin to think about the relation between any two objects actually in the picture, we have moved from physics to psychology. To put it differently, vision is determined by physical laws, but our understanding of what we see is determined by psychological ones. (It is worth noticing that the whole argument is based on an assumption that science, in one of its many aspects, can account for everything we experience—here, everything we see.) So insofar as in my description of the picture I talked about angles of perspective, I was in fact not describing what I was seeing (rather, what was presented to me) but generating out of my own consciousness things I had already learned about perspective in drawings.

The image, I take it, represents the "physics" of perception, all that we can actually see at any moment, and it requires a rather Paterian ascesis to know that that is what we are seeing. Framed in this way, the picture implies that the sense of the world our vision gives us (note how Mach chooses to draw what he sees with one eye closed—precluding stereoscopic effects—as a firm reminder of the degree to which what we "see" is controlled by the limits of our bodies) does not correspond to a single, stable, unperceived reality. If the image misrepresents the feel of human perception, it cleverly demonstrates that the feel of perception is misleading, rather a superstition. Everything we think we see really derives, "physically," only from what "is represented."

Here, positivism becomes both constructivist and idealist, and Pearson's arguments (and self-description) also make clear that constructivism is itself a form of idealism. The very subjectivity of the position becomes a kind of impersonality because it requires a position virtually outside the self to notice the presence of the self. To see as Mach's drawing represents seeing is to be ascetic (an activity almost exactly like that which Pater prescribes in his famous "Conclusion" to *The Renaissance*). The passional self disappears as, for Pearson *and* Pater, reality is reduced to a set of sensations, which are withdrawn from the complications of human feeling. For Pater, there is always a sensitive perceiver of the nuances of sensation, some consciousness that "prefers" one fleeting moment to another, but that perceiver is not much different from the self-observing Mach watching himself represent himself.

In the famous conclusion, Pater is insistent on the excitement, the felt quality of the experience. "Every moment some form grows perfect in hand or face," he says; "some tone on the hills or the sea is choicer than the rest."²⁷ But the aesthetic sensibility is curiously impersonal. The solipsistic engagement with sensations is, for Pater, a way to make life count, and yet the whole long passage strangely and almost inhumanly equates deeply different kinds of sensations, "strange colours . . . curious odors . . . the face of one's friend," for example. The "hard gemlike flame" with which Pater wants us to burn is not the engaged, sentimental, entirely human feeling with which one would think we normally engage the world. What counts is both human and oddly inhuman: "not the fruit of experience, but experience itself, is the end," he famously claims, and what matters is that we "be present always at the focus where the greatest number of vital forces unite in their purest energy."²⁸

This is not at all affectless, but it entails an ascesis as rigorous as that suggested by Pearson and Mach, a detachment from the norms of human feeling and interest. The expressed "interest" here is quantity, the sheer accumulation of experience, or rather of sensations, but the assumption is that it will be a quantity of the best. In its inhuman, quantitative, detached, and at the same time self-indulgent qualities, it is no wonder then that this passage offended so many people. What is perhaps more strange is that the arguments of Pearson and Mach were not greeted with equal alarm, since the ascesis they exemplify is drained even of Pater's sort of "affect." The whole accumulation of experience is registered with scientific curiosity and representational care rather than with any overt aesthetic or moral energy. It is fair and necessary to add, however, that since the point of the argument ultimately is the advancement of science for the betterment of humanity, it was less likely to cause problems on a large scale. Nonetheless, where Pater's preoccupations with the intricacies of perception in a world in flux seemed rhetorically pleased with the very texture of the work, Pearson's arguments are only, in a rhetorical mode characteristic of positivism, intellectually triumphant in their cleverness.

Pearson and Mach are only capable of being scientific—as they understand that word—because, as rigorous empiricists, they are almost uncannily aware of the fact of embodiment, of the way every instant of knowing is contingent upon the peculiarities of the perceiver. Ordinary representation of the scene Mach produces, for example, would have no mustache, no nose, no eyebrow in it. In other words, the positivist position, as opposed to the routine way in which it is regarded by its unselfconscious critics in our current intellectual climate, requires an intense awareness of the fact of embodiment and of the way embodiment inflects perception and thought. In this respect Pearson as much as Pater anticipates some of the main thrusts of contemporary theory, of feminism, of postcolonial critique, and of the critique of Enlightenment ideals.

These moves, as Herbert has pointed out, are Feuerbachian: the world that we think to be outside us is in fact our own consciousness; and when we talk about the world as being outside us, we are projecting our consciousness, reifying ideas—without rational justification. Pearson's description of how the process works echoes with Feuerbach's describing man's projection of his own ideals into God:

> So soon as man begins to form conceptions from his sense-impressions, to combine, to isolate, and to generalize, then he begins to project his *own* reason into phenomena.... He begins to confuse the scientific law, the product of his own reason, with the mere concatenation of phenomena, the natural law in the sense of Hooker and the Stoics. As he projects his sense-impressions outside himself, and forgets that they are essentially conditioned by his own perceptive faculty, so he unconsciously severs himself from the products of his own reason, projects them into phenomena, only to refind them again and wonder what reason put them there.[29]

There is nothing we "perceive" that we do not impose upon the world. Everything we "know" is constructed. We have no access to the outside world beyond the kind of thing that Mach has drawn, and we are locked into the frames of our eye sockets. In this move, the self virtually disappears:

> The distinction between ourselves and the outside world is thus only an arbitrary, if a practically convenient, division between one type of sense-impression and another. The group of sense-impressions forming what I term *myself* is only a small subdivision of the vast world of sense-impressions. My arm is paralyzed, I still term it part of me; it mortifies, I am not quite so certain whether it is to be called part of me or not; the surgeon cuts it off, it now ceases to be a part of that group of sense-impressions which I term "myself."[30]

It is useful to note in passing the rhetorical coldness of this passage, another version of Descartes with his dissected cat, except here, even in a writer who has denied objectivity in the nineteenth-century sense, there is a rhetorical gesture toward the hardness required to achieve knowledge—the kind of thing noted in Galton's autobiography. The self virtually disappears and the normal sentiments of human relations are nudged out of the discourse again.

But in this passage Pearson arrives at a kind of Paterian crux, for, as

Jonathan Loesberg has formulated it, Pater's problem was in finding a way to "define a position from which to savor sensation, one that accepts friction but that is not caught up within it, a position that transcends sensation."[31] For both Pater and Pearson the fact of the pervasiveness of flux and the continuity of that flux with the perceiving self sets up the question of how, then, the sensation can be fixed or, to be scientific, known.

In perhaps his most famous formulation of the idea, a modern version of Plato's Cave passage, Pearson claims that

> we are like the clerk in the central telephone exchange who cannot get nearer to his customers than his end of the telephone wires. We are indeed worse than the clerk, for to carry out the analogy properly we must suppose him *never to have been outside the telephone exchange, never to have seen a customer or any one like a customer—in short, never, except through the telephone wire, to have come in contact with the outside universe.*[32]

The totality of this is stunning, and the italics make clear that Pearson intended that effect. His formulation has the feel of the indifference of "natural selection," the ascesis of the "scientist" investigator, as it tames the disorder of psychological "superstitions" and the multiplicity of "sensations."

The most striking aspect of this empiricist epistemology, graduating, as it were, to a radical positivism, is that it leads logically to the sorts of deconstructions of selfhood and individuality that mark today's antipositivist discourses. Overcoming the "false distinction" between inside and outside requires a discriminating sensitivity to language, and to its implicit metaphysics—just as deconstruction requires it. It entails a rigorous self-reflexiveness, attention to the ways in which knowledge is determined by the limitations of the body, and an analytic intensity that makes dubious the positing of unproblematic entities, like the self.

There are no "laws of nature," only ideas of regularity imposed by consciousness on nature. "The laws of science," says Pearson, "are products of the human mind rather than factors of the external world."[33] This is an antirealism of quite radical proportions. This purging away of reality and metaphysics leaves positivism hovering over an abyss of nonknowing. While one could cite William James impressively resisting the chilling of human life implicit in Pearson's kinds of naturalistic positioning,[34] one should note that the pragmatist tradition shares intimately with this positivist one the refusal to allow for science or language as "representation" and particularly as representation of some "reality" that lies outside of anyone's consciousness. For Pearson as for the pragmatists, we are all bound by the egocentric predicament.

It is striking to find such positions, so distinctly postmodern as they seem, developed by a theorist committed to the virtually absolute intellectual authority of science. Modern constructivist attacks on "objectivity" and on science itself are unquestionably cousins of the positivism they are aimed at annihilating. The antimetaphysical animus of Pearson makes him a nominalist, as modern antimetaphysical critics of culture tend to be: what is claimed to be out there as an urgent and irresistible fact is in here like an idea. Pearson would agree: the scientist strictly speaking is not a discoverer but an inventor. But if "the laws of nature" owe their existence, as Pearson argues, to human consciousness, there follows the inevitable question of which we have heard so much in our contemporary debates: what is to stop us from inventing any old law? Pearson, shockingly, begins to sound like Oscar Wilde. "There is more meaning," he says, "in the statement that man gives laws to Nature than in its converse that Nature gives laws to man."[35] So Wilde, himself a student of late-century science, affirms, "Life imitates art more than art imitates life." Here the strange conjunction between positivism and aestheticism begins to seem like something more than an aberrant conjunction of two distinctive thinkers, Pearson and Pater. The epistemology and the scientific, antimetaphysical impulses that drive their work are fairly widespread phenomena in the last stages of Victorianism and the development of modernist art and culture. They inform both science and art.

Insofar as they do, however, the problem would seem greater for Pearson than for Pater. Pearson, after all, has to defend the claims of science to intellectual authority; he has to find a way to gain access to a knowledge that transcends the limits of the personal and that to many scientists has always seemed not local but beyond the limits of history—virtually, that is, transcendent and universal. While it may be reasonable to suggest that Pater anticipates what we are now calling postmodernism, the "most startling fact" about which, David Harvey claims, is "its total acceptance of ephemerality, fragmentation, discontinuity and the chaotic,"[36] Pearson would not have been pleased to find his ideas confirming the happy chaos of postmodern epistemology. In fact, neither Pater nor Pearson was ready to celebrate the possibilities of indeterminacy that their epistemologies threatened. Behind the new subjectivist narratives they write, the image of the aspirant dying to know remains.

We have seen the path Pearson took to the epistemology of *The Grammar of Science*, and perhaps the most striking thing about it is that as we trace it back we can see that positivism emerges from a complex of personal and cultural forces that would seem, at the start, to have nothing to do with science. The process suggests an important community of feeling and interest between the world of science and the world of culture; it should also suggest once more both the historical variability of the relation between ideas and their apparent ideological implications and the crucial role that the ideal of ascesis has played in art and science (and in contemporary critical theory, where it lurks in disguise).

The Grammar of Science is the culmination of the journey undertaken by Arthur, the end of the quest narrative that in effect announces that no further quest narrative will be necessary. But Pearson always remained the man Arthur describes himself as being: "The impulses of my being, its untamed powers and aspirations, are too great to be bound or limited by what binds or limits those of the ordinary man. I must work, think, love, with a thousand times his energy. I *will* grasp the Unknown, the Infinite, though it struggle with and overthrow me times innumerable."[37] The Unattainable and Unknowable have an afterlife in his scientific writings, not as metaphysical possibilities but as unknowable realities to which scientific method gives a kind of shadow access.[38]

Those Victorians a-sea in a creedless world can only acquire a new creed (one must infer from Pearson and from his positivist and naturalist contemporaries) by surrendering impossible metaphysical and transcendental aspirations. Arthur's last letter confesses that "philosophy, science, and art can give no comfort, no creed in these later days." The metaphysical quest has failed and he is, as he says, "tired of striving after the Unattainable."[39] *The Grammar of Science* is that new creed, for it purges science of its hidden metaphysics. The crack in scientific theory, Pearson feared, opens a lane to metaphysics and God. In this work, he transforms limitations of knowledge into firm and unshakably untranscendent truths. He moves, that is, from potential empiricist solipsism to a strong scientific impersonality and objectivity. As Feuerbach had said, "Dread of limitation is dread of existence. All real existence, *i.e.*, all existence which is truly such, is qualitative, determinative existence."[40] In his review of Balfour, Pearson made particular hay of Balfour's claim that "there is no special characteristic which marks of the central truth of theology, 'There is a God,' from one of the fundamental presuppositions of science.... 'There is an independent material world.'"[41] But Pearson escapes that metaphysical trap: "Science in no way depends upon the truth or falsehood of the ... proposition" (p. 192). What science depends on "must be sought on the perceptional side of the veil of sense-impressions" (p. 193).

The antimetaphysical arguments of *The Grammar of Science* are built on the view that science never explains, only describes. The laws of nature are only "routines of perception," recurrent patterns of sensation that scientists generalize. The laws are themselves merely descriptions, and not of things as in themselves they really are, but of what they really seem to be. Pearson celebrates science because it does not (or need not) pretend to provide causal explanation or insight into the Unknown. Pearson and Arthur ultimately agree, with differing emotive inflections, that science will never, as Arthur complains, "satisfy this yearning after the Unknown."[42]

The pure positivism of the book, its assertion that science has no claim on the "Unknown," is the foundation of Pearson's creed. Romantic defeat leads Pearson to a positivism that is virtually postmodern in its acceptance (initially at least) of incompleteness, fragmentation, chaos. Not postmodern, though, because Pearson moves from the metaphysical quest, through the flux as Darwin had realized it in the *Origin*, toward a limited kind of objectivity that provides an alternative to absolute knowledge. What we know about the world can never be the world itself; we can only know what is *probably* true about it. Statistics become the means to transcend the limits of each experientially-based science, the limits of perspectivalism and personality, by edging toward probability, the best knowledge any science can attain.[43] Through statistics Pearson attempted to turn social study into a true science, and through it to unite all sciences. Ironically, as Peter Dale has argued, positivism, through statistics, becomes the replacement for a lost Christian totality.[44]

"Certainty," Pearson notes, "belongs only to the sphere of conceptions." In the world of perceptions there is only probability. Pater had noted that modern thought is distinguished by the "relative spirit." For Pearson, this notion becomes the foundation of a truer science, not built on the absolutes that Pater and he had abjured, but on what can really be known in a world of constant flux and perspectival shifting. Pearson's science moves from representation of reality to statistical correlation. In effect its problems always take the following form: "A certain order of perceptions has been experienced in the past, what is the probability that the perceptions will repeat themselves in the same order in the future?"[45] Pearson risked the chaos of the relativism of total subjectivity and epistemological relativism in order to establish a nonmetaphysical and thus less vulnerable knowledge.

Correlation is at the basis of statistics. How often does event C follow event B, or does characteristic A occur with characteristic Z? Or, to put it in a Pearsonian way, do these sequences of perceptions turn into "routines"? Statistics can become the basis of an antimetaphysical science because they allow the invention of laws that will describe in the most economical way the relationship of characteristics or events, or, again to be strictly Pearsonian, "perceptions." For Pearson, the best that a scientist can do is detect correlations between perceptions: If I perceive this, I am also likely to perceive that. Such language replaces the idea of cause, or rather comes to be subsumed under the word *cause,* which Pearson is willing to use in its revised sense. As Ian Hacking puts it, Pearson would have claimed that "he testified to the correlation between the taming of chance and the elimination of ordinary causality."[46] In his massive three-volume biography of Galton, Pearson (as quoted by Hacking) argues that "Henceforward the philosophical view of the universe was to be that of a correlated system of variates, approaching but by no means reaching perfect correlation, i.e. absolute causality."[47] Here is how Pearson begins his definition of "cause":

> We are now in a position to scientifically define *cause*. Whenever a sequence of perceptions D,E,F,G is invariably preceded by the perception C, or the perceptions C,D,E,F,G always occur in this order, that is, form a routine of experience, C is said to be a *cause* of D,E,F,G, which are then described as its effects. No phenomenon or stage in a sequence has only one cause, all antecedent stages are successive causes, and, as science has no reason to infer a first cause, the succession of causes can be carried back to the limit of existing knowledge, and beyond that *ad infinitum* in the field of conceivable knowledge. When we scientifically state causes we are really describing the successive stages of a routine of experience. Causation, says John Stuart Mill, is uniform antecedence, and this definition is perfectly in accord with the scientific concept.[48]

Cause without metaphysical baggage retains the force of the old-fashioned *cause,* as probability becomes the means to "provability."

Pearson's statistical approach also undermines the determinism that had been governing both scientific and philosophic thought. Detailed examination of any phenomenon always would reveal a mass of correlations, the variety being such that it would be impossible to attribute any one event to any one cause. That is to say, the world of correlative statistics is much more complicated and subtle than the world of causation; it is a world in which dualism and either/or dichotomies are—or ought to be—invariably recognized as reductive. This is another way in which Pearson's positivism anticipates modern anti-positivist thought. There is, as Hacking interprets C. S. Pierce's view, "spontaneity in the world, of which our sense of freedom is a minor element."[49]

Pearson's theories of causation, scientific law, and empirical perception are threatened by the problem of the particularity of subjectivities. If science is a description of the contents of the human mind, and if all minds are different, science loses its power to generalize and to formulate laws. The "causal" sequences, the "routines of perception," might be different for different consciousnesses, and then the project of science, to overcome the limits of the particular and to achieve an authoritative generality, would be severely damaged. Without laboring at the problem, in fact treating it almost in passing, he resolves it by invoking "normality." Scientific law, he says, "is valid for *all* normal human beings."[50] While he recognizes that human beings as a species change, and that within species there are aberrations and alternative modes of development, he holds to the concept of an unproblematic normality. Pearson thus manages to sustain a position that threatens to explode into an unmanageable relativism and that persistently resists the conventions of knowledge dominant in realist discourse. He even hedges about universality: "The universality, the absolute character, which we attribute to scientific law is really relative to the human mind" (p. 120). And he concedes that he talks about humans as a "class which in the normal civilized condition have perceptive and reflective faculties nearly akin.... The 'universality' of natural law, the 'absolute validity' of the scientific method, depends on the resemblance between the perceptive and reflective faculties of one human mind and those of a second" (p. 121).

But as with the word *normal* so with the word *civilized:* Pearson dodges problems. For those who are appalled by the worst of Pearson's views on race and international competition, it might be easy to trace their intellectual justifications to this dependence on similarity. Some reliable position beyond the self is irresistible (and it is always also, of course, dangerous). Sustaining an utterly antimetaphysical explanation of the means to knowledge, he encounters immediately the problems that face any antifoundational epistemologies: how to make knowledge shareable. The unquestioned and virtually unstated grounds for Pearson are the ideas of "normality" and "civilization":

> Human minds are, within limits, all receiving and sifting-machines of one type. They accept only particular classes of sense-impressions—being like automatic sweetmeat-boxes which if well constructed refuse to act for any coin but a penny—and having received their material they arrange and analyse it, provided they are in working order, in practically the same manner. If they do not arrange and analyse it in this manner, we say that the mind is disordered, the reason wanting, the person mad. The sense-impressions of a madman may be as much reality for him as our sense-impressions are for us, but his mind does not sift them in the normal human fashion, and for him, therefore, our laws of nature are without meaning. (p. 121)

Pearson is aware of the virtual circularity of his definitions. The "normal"—the necessary ground for attaining shareable knowledge—are identified as those who arrange their perceptions "in practically the same manner." But since the idea of the normal itself is nebulous and so obviously determined by point of view, it needs a ground itself. That ground is Darwinian theory. How can the definition of *normal* be anything but arbitrary if there is not some hard evidence for it, some recurring signs that allow the border between normal and abnormal to be drawn for everyone? The evidence of abnormality is not an opinion or a point of view, Pearson would argue, but an inability to adapt to life. Failures of adaptation can be measured, in death rates, for example, or rates of illness, or in income. For Pearson, as for Galton before him, Darwinian theory can be firmly tested scientifically, for heredity can be measured, quantified, turned into statistics.

Rejecting metaphysical grounds, Pearson is not attending to the problem or worrying about the fact that his definition ends up being circular—the normal are those who adapt. Adaptation can be measured by working with an idea of normality. Whatever one thinks of the way this argument takes shape, it suggests again how indispensable some "ground" is for even the most ferociously antimetaphysical theory. Transcending the personal is again a condition for a valid epistemology. And if Pearson's way of transcending it is not to everyone's taste, it should be clear that he gets to that point by way of a radically relativist and constructivist theory. While Balfour finds that a deep skepticism leads to a necessary act of faith and to a very conservative politics, Pearson's exuberant and radical relativism leads to constructivism and socialism and imperialism, both freethought and the empire of science.

The strong empirical antimetaphysics of Pearson's theory is not only a way for him to establish the authority of science. It is his alternative to the lost faith. And it is as much his argument for a social and moral program as it is a strict (and it is very strict) scientific epistemology. Renunciation (or ascesis) was for him the condition of a correct epistemology and of a new socialist community; it represented a personal ideal and virtually a religious and transcendental vocation. Renunciation, remember, is the means to freedom from "phenomenal slavery."[51] When Pearson left metaphysics and romantic fiction behind, he did not leave behind these aspirations for the impersonal, for a "tranquility" that might release him from the stormy outer world of phenomena. He escapes from the solipsism and idiosyncrasy of individual consciousness through renunciation of the romantic (and Enlightenment) ambition for full knowledge. Statistics allow him

> To find that essence of tranquility,
> Which beyond time eternally remains
> Unchanged; to fix one spot, where the mind's eye,
> Fatigued by ceaseless motion, may repose
> Upon immutability's still peace.

"Science," Pearson argued, "so far from having in the popular sense materialized the world, has idealized it; for the first time rendered it possible for us to regard the universe as something intelligible rather than material" (p. 23): "the laws of the physical universe," he says, "are logically related to each other, flow rationally the one from the other. This is really the greatest result of human experience, the greatest triumph of the human mind. The laws of the physical universe follow the logical processes of the human mind. The intellect—the human mind—is the keynote to the physical universe" (p. 41).

Pearson was not being disingenuous when he rebuffed with bitter ironies the charge that he was a "materialist," an accusation made by George Mivart and Arthur Balfour, among others. He was, in the quite literal sense he claimed, an idealist. His assertion marks another and characteristic irony in the narrative of scientific epistemology, for what Pearson argues is that with the appropriate actions of self-elimination there comes an ultimate satisfaction, a sense of power. For the true satisfaction of his recognition of the ultimate limit of the egocentric predicament, the self's incapacity to get beyond itself, is that the world turns out to be exactly what Hardy treated it as not being and George Eliot struggled to make it—completely compatible with and containable by human consciousness. Positivism becomes an ultimate idealism, confirmation that the "world" not only conforms to the contours of our consciousness but is indeed consciousness itself.

This was not finally Pater's way, but the paths and even the results are somewhat similar. At all odds, Pater resisted formulae. In language that only seems opposed to the work Pearson chose to do with his form of solipsism, he asserts, "the theory or idea or system which requires of us the sacrifice of any part of . . . experience, in consideration of some abstract theory we have not identified with ourselves . . . has no real claim on us."[52] For Pearson, too, however, the sacrifice of "experience" is a sacrifice of a piece of knowledge, of at least a statistic that might shift our sense of the real. But Pater was not overtly interested in generalizing the particulars of experience, and he seems a long distance from aspiration after the universal. Yet his rhetoric makes claims it does not quite speak. Pater simply assumes the validity of the perceptions that he officially renders as valuable. What are those "exquisite passions"? Would they be recognized as such by everyone? Who decides whether they were exquisite? Even if in its celebration of particularity and transience Pater's aestheticism has something deeply life-giving about it, it must remain entirely personal until—and here is an irony of Pater's representations—it becomes comprehensible and credible to others. Identifying the "hard gemlike flame," the power to detect and respond to the utter uniqueness of each passing sensation, Pater creates a formula (against his own passion to deny all formulas) that influenced several generations and remains alive today.

Pater, too, aspired to some kind of stability within the world in flux. His characteristic narratives are thick with nostalgia for stabilities lost, in which history and science have convinced him he can no longer believe. But he is a historian and critic rather than a scientist and thus can linger on the past instead of attempting to supersede it. He can relish the charms of that past, of the irrational superstitions and outworn faith without committing himself to them, only to incorporating them as some of those sensations the aesthetic critic is committed not to miss.

Pater and Pearson ended by seeing the world very differently and embracing two different forms of knowledge, but they were equally constrained toward what Pater was to call a new "faculty for truth." For Pearson, that faculty, registering carefully and with something of a Paterian ascesis the particularities of human experience, transforms into statistics itself, the only way to capture at all the multiple and transient nature of experience. Pearson turns the flux, and the power to observe it in its nuanced and multiple nature, into knowledge. But it is not as though Pater surrenders the possibility of knowledge in his engagement with the flux. As Carolyn Williams puts it, Pater's relativism and intense sensitivity to the Darwinian evanescing of experience, the indefinable shades of difference, the subtleties of movement and multiplicity in a world that never held still, "does not mean that nothing can be rightly known." That "'new faculty for truth' must be employed if it is to operate within a field of relations so fluid and almost inconceivably complex."[53]

Opening themselves up to multiplicity and skepticism, Pater and Pearson continue to require some point of stability that will allow them to capture in the forms of knowledge some aspects of a world they recognize that they can never fully know. Pater's famous arguments for the aesthetic condition depend on a stability derivable from the flux and from the limitations of consciousness. Ascesis is crucial here as renunciation was for Pearson. As in the famous passage from the "Conclusion" to *The Renaissance*, Pater asserts that somehow one can actually isolate, locate, particularize "moments" from the flux. While an ironic undercutting of Arnold's commitment to the thing in itself hovers over the passage, Pater quite seriously employs in the "Preface" a language that echoes with scientific implications and implies a stability beyond impressionism. The function of the critic, Pater argues, is to understand the "virtue" of the object by determining "what effect does it really produce on me.... As in the study of light, of morals, of number, one must realise such primary data for one's self." The critic explains the "special impression of beauty or pleasure" by "reducing it to its elements," as a chemist might.[54]

The resort to the language of science is neither accidental nor gratuitous. That language, too, eschews metaphysics and at the same time implies a mode of knowledge not merely relativist. The strategy by which the critic isolates the elements of impressions is, as Williams describes it, totally consonant with the ideal of ascesis and yet also with historicism and perspectivalism, for it is the result of "the effort to separate the present moment from the past, and the effort to separate the object of perception from its context in the present viewer."[55]

As with Pearson, Pater insists that this effort requires great discipline and at the same time belies the full complexity of the reality—that is, the sensation. In effect, Pearson's statistical correlative method, isolating a set of relevant questions, disentangling the phenomena from the whole range of complications of life itself, parallels Pater's aesthetic strategy. And while each strategy is vulnerable to epistemological and practical critique, what matters for my argument here is that in their very differences they are engaged in fundamentally similar enterprises to salvage the knowable from the flux of experience by way of a radical ascesis.

In this one late-Victorian case, at least, the overlapping intellectual commitments of hard-line scientist and famous aesthete belie the inevitability of our current divisions between scientists and students of culture. Both share a commitment to empiricist validation; both are in effect constructivist, Pearson explicitly so. Both anticipate later developments in the name of intellectual and political liberation. The alignments here simply will not match the alignments that govern our current divisions. But it is striking that the way into potential solipsism is through the ascetic; the way out of that solipsism is a yet more intensified discipline of self-restraint. On either side of the divide, in the interests of truth, the self-restraint and intellectual and emotional discipline that Arnold's "disinterest" (trusted by nobody these days) also required becomes necessary.

It turns out, then, that Pearson, the scientist, finds in science the alternative to the absolute with which metaphysics (and romantic philosophy and literature) had failed to supply him. Positivism becomes a shadow child of romanticism. And Pater, the amoral aesthete who formulated for modernism a vision of art and reality unstable, flickering, self-preoccupied, based his vision on the knowledge that science proffered. These diverse and complicated efforts at stabilizing the instabilities released by the empirical tradition rely on the ascetic discipline of intellect and feeling, which, in Nietzschean readings, is a last expression of the will to power and which, despite the radical distrust of self-abnegation that marks contemporary thought, inescapably marks the work of all sides of current cultural conflicts.

EPILOGUE

Objectivity and Altruism

> To a living being then there can be no absolute. . . .
> The remarkable thing, however, is that though we very well
> know absolute truth to be unattainable,
> we are inevitably constrained to act as if it were attainable.
> SAMUEL BUTLER

The narrative of scientific epistemology, as I have been rather arbitrarily defining it throughout this book, traces the struggles and the triumphs of the quest for absolute truth without divine assistance (although Descartes, to be sure, requires God as a backup). It has the shape of the quest story—modified by the bildungsroman—as it brings the careful and increasingly alert quester through great difficulties to the maturity of self-sacrifice and the ultimate attainment of the truth. The story holds both for the method itself, that is, for the Cartesian quest for an authoritative knowledge, and for the movement of Western thought from its promising opening among the ancients, through the Dark Ages, in which superstition and inherited authority determined what could be believed, into modernity, when trust was given not to authority but to experience, and Western science developed in a deferential and self-eliminating way before the authority of the natural world. Comte told this story as the movement from the theological, to the metaphysical, to the positive stages; but all who told it saw it as progressive, and all who told it, in one form or other, required for its success the abasement of the self. The West, in order to know, had to die to desire, had to die to its human interests. The result, ironically, was power and the Baconian satisfaction of human needs coming as the reward for self-abnegation. Truth, it is true, produced its discontents. But truth was, after all, *true*, accessible only to the deserving, the moral, those who were courageous and disciplined and dedicated enough to sacrifice desire and self in its pursuit.

The story of "objectivity" is longer and more complex than this simplified account that has had such purchase in our culture, particularly in the nineteenth century. The problem of objectivity was the problem we have seen worried by Pearson and Pater: the relation between things out there and the consciousness that perceived them. Lorraine Daston has traced the history of the word *objectivity*, locating its current confirmed usage only as late as 1856. But as she claims, "long before the 'objective' had come of age in 1856, the thing existed, a close cousin if not an identical twin of our current notion."[1] Daston maintains that the attitude called "indifference" or "impartiality," the disinterest of modern objectivity, was not, at the time of its birth in the seventeenth century, self-evidently superior to other modes of knowing. But the feeling for its superiority has become the dominant form of the story of adequate scientific knowledge in our times—and that is the story of "dying to know." Daston sees in the "passionless passion" of Descartes, his "impartiality to desire or revulsion," an "intimation of the enlarged sense of objectivity that took hold in the nineteenth century,"[2] and that has been the subject of this book.

The propagation of that story—along, of course, with socially visible technological advances—allowed science to displace religion as the ultimate authority for all knowledge. And in the nineteenth century, with the most overt and thorough moralizing of the quest for knowledge and of scientific activity, science extended its reach from the inorganic to the organic to the human world. Comte established the idea of a science of society and of "man," in which most of the writers I have examined believed—even the women, such as Harriet Martineau and Beatrice Webb. The narrative of scientific epistemology, already in the centuries before its full and overt moralization, carried ethical freight; and in the nineteenth century, despite overt attempts to separate the epistemological from the ethical, the epistemological was latent with ethical force.

Perhaps the most florid and presumptuous representation of the set of attitudes surrounding this narrative emerges at the end of T. H. Huxley's relatively early essay, "On the Advisableness of Improving Natural Knowledge" (1866). There Huxley argues, in a rhetoric rich with ethical energy, that the pursuer of natural knowledge rejects authority for experience: the scientific spirit is at one with the ethical spirit, the spirit of resistance and independence, and at the same time with the necessity of ultimate submission to nature as it speaks through experience. The argument is a declaration of intellectual independence, a reenactment of Descartes's initiating rejection of all inherited knowledge, and it is also an assertion of new dominance. So Huxley writes:

> If these ideas be destined, as I believe they are, to be more and more firmly established as the world grows older; if that spirit be fated, as I believe it is, to extend itself into all departments of human thought, and to become

co-extensive with the range of knowledge; if, as our race approaches its maturity, it discovers, as I believe it will, that there is but one kind of knowledge and but one method of acquiring it; then we, who are still children, may justly feel it our highest duty to recognise the advisableness of improving natural knowledge, and so to aid ourselves and our successors in our course towards the noble goal which lies before mankind.[3]

The passion here has also a touch of the passionlessness that marks the history of scientific objectivity and the story of dying to know. The "impartiality to desire and revulsion" that Daston notes already in Descartes became a major part of the story as the insistence on eliminating all but intellectual passion dominated the arguments of scientific epistemology. The sacrifice of the questing scientific self is paralleled by the quester's indifference to the violence and brutality of the world perceived.

The highest morality is the willingness to face the worst. And though Huxley always sustains the high moral tone, the story easily veers off toward the amoral. The world of natural knowledge Huxley describes so stirringly is one in which the quester for knowledge invariably discovers "that our earth is but an infinitesimal fragment" of the visible universe; "that man is but one of innumerable forms of life," and that he will be succeeded by others; that "there is a definite order of the universe," knowledge of which will "loosen the force of men's belief in spontaneity, or in changes other than such as arise out of that definite order itself."[4] Thus the narrative Huxley writes teaches independence and resistance at the same time as it requires the thinking mind to accept its own powerlessness before laws of nature it cannot control. The consequence of knowledge is the obliteration of the self as agent. There is further irony in the fact that as he demonstrates the powerlessness of the self within the system of nature, he celebrates the heroic resistance to convention that allows the self to accept the reality of the amoral order of the natural world. And while Huxley ultimately developed this vision into the argument of *Evolution and Ethics* in which he claims that the responsibility of the human is to resist the forces of nature as long as possible, that version of the story ceases to be "comic" and progressive. Huxley, like Hardy, found in the end that the application of "natural knowledge" to human affairs could not be acquiescent and self-eliminating. Hardy's story ends in a useless and painful death. The social Darwinist version of the story, which Huxley finally resisted, turned the amoral into the moral and therefore indulged a high-minded ruthless indifference to human suffering.

These are elements that make unequivocal endorsement of the dying-to-know story suspect. The pain of the narrative and the cost of its triumph in, for example, John Stuart Mill's self-portrait, or Dickens's complicated celebration of female self-abasement, or Hardy's bleak rendering of the cost and possibilities of knowing, are demonstrably unacceptable. The dying-to-know story is at the heart of the value system by which Western epistemology and Western ethics get their official sanction. At the same time, it carries the weight of the ethical and the righteous at the cost of countenancing the unethical, and too often it disguises, beneath the mask of humility, ruthless thrusts at power.

The question for those who distrust official stories is whether all gestures of self-elimination are mere disguises for deep interest. The hermeneutics of suspicion makes straight reading of any version of the Cartesian theory of scientific epistemology virtually impossible, and it is partly for that reason that I have focused not on the logical coherence of arguments for objectivity but on forms of its embodiment in narrative as the dying-to-know story. The narratives make evident the surreptitious ethical energy that drives the epistemological quest: to reach the valued goal of knowledge, objectivity is required, and objectivity requires resistance to desire. Resistance to desire can itself be questioned, just as the possibilities of disinterest have been questioned, and the result is a rather odd alignment of critics. For resistance to desire, in evolutionary terms resistance to the personal interests of survival and procreation, makes altruism as well as objectivity possible; and with the admixture of evolutionary thinking, the crisis of this discourse today is, precisely, "altruism." Can those self-interested organisms usually identified through the hermeneutics of suspicion, in fact act in the interest of others at the possible sacrifice of themselves? (This is formally the same question as whether it is possible to surrender one's desire to allow nature to speak unmarked by the individual consciousness that perceives it.) Thus objectivity is in one sense only epistemological altruism. To deny the possibility of objectivity, the capacity to be in that nowhere from which truth at last will be visible, is, logically speaking, to deny the possibility of altruism. The fictional stories I have explored in this book worry out that problem by embodying the abstract epistemological quest in ethically complicated lives.

These narratives rarely take the dying-to-know story straight. Without rejecting the need for self-abnegation, they each in different ways explore the possibilities of alternatives to "dying," of making the pursuit of knowledge compatible with human satisfaction and fulfillment—as, I tried to show in the second chapter, such diverse thinkers as Donna Haraway and Francis Bacon explicitly do. But they need to do this at the same time that they resist the powers of satisfaction to distort knowledge. This is Pearson's project, and Pater's turn from the apparent self-indulgence of the aesthetic to the absolute necessity of the ascetic makes another version of that attempt. The official story of dying to know is always compromised and complicated by the narrative that haunts it and the narratives that reenact it. For that story is always laden with ironies, paradoxes, self-contradictions, as Samuel Butler perceived and argued in his various essays on science and evolution. As Butler knew and all of these narratives show, the narrative of "dying" always reemerges, but it never quite suffices.

None of these writers, except perhaps Hardy, completely give up on the ideal of self-elimination. Each finds that some mode of detachment, some way to move beyond the personal and to recognize a responsibility to community and to knowledge, is essential both to the work of knowing and to the work of living meaningfully. The self-elimination Pearson preaches is an aspect not only of epistemology but of socialism. While it is possible to trace developing critiques of the epistemological model among these writers, a critique I have been wanting to emphasize, it is also important to note the ambivalence implied in these narratives as they play out the paradoxical and self-contradictory nature of the work of knowing.

Butler is a particularly useful figure with whom to conclude these long ruminations on objectivity because his deep distrust of the idea, his profound disbelief in its possibility, was combined with a sense that its alternative is equally untenable. On the one hand, he virtually writes the dying-to-know narrative. He argues, for instance, that "if we could get absolute knowledge upon any, even a single, subject we should ere long cease to exist, or at any rate cease to be conscious of existence."[5] And yet in the same paragraph, he insists on the inevitability of continuing to act as though the absolute (with the accompanying "death") were inevitable. With his characteristic post-Darwinian irony, Butler assures his reader that while, "happily, there is no fear that people will leave off aiming at this [absolute knowledge] . . . happily also, there is not the remotest chance that we shall get what we are striving after" (ibid.).

Butler's ironic stance in relation to the unironic scientific imperialism of Pearson's constructivism finds its echo in Amanda Anderson's recent work. There she examines the way the hermeneutics of suspicion have undermined belief in disinterested knowledge, but she also explores not perhaps the "inevitability" of pursuing the ideal of objective knowledge but its social and moral importance. She considers the mixed strategies of the Victorians, who have been all too easily condemned for a provincialism their work had often tried to overcome; and she finds in their work important possibilities for contemporary thinking about cosmopolitanism and "altruism," although she does not use that word. Anderson, as I have noted earlier, views the ideals of disinterest and detachment not as a fulfilled or fulfillable possibility but as an "aspiration." Ironic and questioning about the tradition of "detachment," which she sees marking important elements of Victorian discourse, Anderson yet finds possibilities for expanding both epistemological and social imagination beyond the limits of personal or national interest. That is, without surrendering to the idea that the "absolute" is attainable, Anderson sketches out a tradition that combines fundamental skepticism about the possibility of getting past the limits of interest with a recognition that an attempt to do just that is both "inevitable"—as Butler put it—and desirable, as Anderson's commitment to cosmopolitanism makes it. In a different register, she pursues the problem of all thinkers in this tradition: how to get beyond the local and the particular to the general.

Butler's salutary emphasis on the self-contradictoriness of every assertion of truth can help in understanding the paradoxes of the dying-to-know tradition itself, and of the double relation to it that virtually all the writers examined in this book adopt. Butler says, in an astonishingly anticipatory way that one can only call deconstructive, "Every proposition, nay every idea, carries within itself the seeds of its own undoing" (18:112). On the one hand, the discourse of "dying to know" produces an almost intolerable self-humiliation in which the desiring self is abased and denied and, in many cases, literally dies, so that the act of knowing becomes the act of ceasing to know anything, or to care about it; there is also another road we have seen traced, particularly in Galton's autobiography and in Hardy's dark revulsion from it: relieved of the responsibility to satisfy the self, which is reduced to a "point" in an epistemological diagram (like Descartes's "ego"), the narrative registers with cold and righteous certainty its tough-minded confrontation with the worst. In the name of truth, it can register a world governed ruthlessly, requiring for its own health the infliction of pain and the refusal of the humane. The epistemological stakes of this absolutist game are very high—and narrative very often makes that clear.

But, on the other hand, the ideal of self-humiliation is also "inevitable" or, as Anderson sees it, a valuable "aspiration." If the emphasis on the local and the particular honors the diversity of nature and societies, it also tends to leave value and meaning entirely in the individual and disallows the possibility of action and thought that gets beyond the individual self. Although the dying-to-know narrative has led often to real deaths, it has also led to the possibility of knowing at all in a way that is shareable and life giving. I would suggest that the hermeneutics of suspicion, burrowing its way into the unspoken and often unrecognized tunnels of interest that underlie all human activity and thought, has no necessary connection to the political subversions, liberations, and openness it is often employed to validate.

The dangers of premature generalization, the perils of too full a surrender to the "real," are clear enough. We all know how Matthew Arnold's insistence on "disinterest" and an unconditioned "curiosity" was closely connected for him with the authority of the "State." It is clear enough how the ideal of scientific objectivity has been invoked to endorse self-evidently racist projects. I have tried throughout this book to emphasize the way the dying-to-know story turns the "death" into a moralized assertion of power and authority that seems unchallengeable. It has confronted the idea of objectivity with the reality of the death it metaphorically requires. I have tried, in addition, to trace the way, in the writings of Karl Pearson, an originating absolutist and idealist ambition transforms with the recognition that the self will *not* die and that all knowledge comes filtered through consciousness, into a belief that all of "reality" is only an elaborate construction of human consciousness. For Pearson, the transformation is not to the radical skepticism and historicism that his constructivism might seem to imply to thinkers of our own time, but to a new manifestation of that initiating absolutist and idealist ambition. It is tempered, of course, by "probability" thinking and by the skepticism that marks all positivist discourse. But it is unambiguous about the power of science to get at what truth can be achieved. "The material of science is coextensive with the whole life, physical and mental, of the universe," he claims boldly, and "wherever there is the slightest possibility for the human mind to *know,* there is a legitimate problem of science."[6] Unable to rest with the potential instability of constructivism, he finds a "scientific" way to establish the usefulness if not the epistemological validity of human knowledge. This requires, once more, "self-elimination," the power to recognize the degree to which consciousness participates in the creation of reality, and the power to generalize the particulars of consciousness—through probability and statistics.

As Pearson develops it, constructivism reinstates the imperial authority of science. The problem is, as Butler understood, that while the use of the strategies of self to disrupt the narrative of self-elimination makes a superb critical tool, it cannot sustain itself without the counterweight of generality. Butler, undeterred by the theoretical impossibility of getting it right, usefully notes how the necessary self-contradiction of all language leads to a seesawing of preferences. I would call it a choice of preferences between the dying-to-know narrative and the solipsistic one. Perhaps less pretentiously, Butler writes, "We can only escape the Scylla of having a different name for everything, by falling into the Charybdis of calling everything by the same name."[7]

I chose the metaphor of "dying" precisely because it won't allow the blurring of margins that is the normal and appropriate condition of human thought and language. The alternative of solipsism is equally extreme, excluding everything except the constructions of consciousness. But for Butler and for much of European culture, the Darwinian imagination eroded the sharply defined boundaries not only between species, but between all things in nature. When Butler claims that all language involves self-contradictions, he means that language always classifies, and no classification can honor the blurred margins of the class. The hermeneutics of suspicion builds on a sense that all language does indeed imply a contradiction in terms of the sort that Butler discusses—where the scientist struggles to find relations, the critic insists on individual distinction: "the world always has and always will oscillate between these two opinions."[8]

It oscillates now.

Epistemological discourse is always troubled by the need to break out of the raw individualism of its potential solipsism, to extend beyond the singular, beyond both the singularity of the perceiver (whose uniqueness is always emphasized by realist narrative) and the singularity of the data: for if all knowledge were only of singular objects or events or even relationships, the world would be like that giddying and dizzying one of Jorge Borges's "Tlon, Uqbar, and Orbis Tertius." This is all very well in the act of reading, but it is a living impossibility, an impossibility for living. Theorists of science have long insisted that there is no knowledge in singular facts. Knowledge begins as it rises to the level of generalization (and perhaps "law"), and the early positivist ideal was a program of reduction to more and more general laws. Comte's argument might be taken to summarize this perspective (though in the philosophical history that has followed, much of this has come under debate): "What is now understood when we speak of an explanation of facts is simply the establishment of a connection between single phenomena and some general facts, the number of which continually diminishes with the progress of science."[9]

Though Comte may have been something of a mad and vainglorious prophet of his own church, his insistence on generalizing laws is no aberration. That need has remained part of the epistemological tradition from its inception. When Hans Reichenbach laid out his positivistic vision of the narrative of scientific epistemology, he wrote first that the fundamental mistake of British empiricism was its failure "to examine knowledge . . . with the detachment of the disinterested observer."[10] Reichenbach's solution to the continuing problem of the empirical (that is, for him, inductive knowledge) is Pearson's, the development of "probable knowledge."

This struggle somehow to combine the theoretical force of abstract, particularly mathematical theory with the living embodied qualities of empiricism was a further imperial effort of the sciences to claim unique access to real knowledge in a world that is otherwise rather like that blurred and self-contradictory one that Butler describes. It is another version of the question that has recurred in different forms throughout this book: how to achieve a view from nowhere? how to be detached and yet care about knowledge (the problems of Mill and Daniel Deronda, for example)? how to gain unmediated access to a knowledge that must always be mediated by individual consciousnesses? And it repeats, in a new and perhaps disturbingly cocky way, the efforts of John Stuart Mill and Karl Pearson. As our contemporary critical theory has strongly emphasized the impossibility of achieving that generality, and the deviousness of all attempts at it, there has been a general assumption that such critiques of objectivity are ideologically necessary. But the virtually absolute distrust of the absolute, the utter conviction of the impossibility of disinterest, cuts more than two ways ideologically. In a certain sense it denies itself, because these critiques so frequently emerge from an ideological interest in reempowering the disempowered—perhaps the most disinterested of political moves.

I return here to what I hope will be seen as a leading motif of this book: setting ideas within narrative, embodying them, seeing them played out in stories and in history, demonstrate that there are no inevitable ideological or political or social implications to the abstract theories. All the narratives with all their variations that this book has considered make it clear that ideas get reshaped and redirected under the pressures of embodiment and history. Oscar Kenshur has argued against what he calls "ideological essentialism," and demonstrates persuasively that the "ideological valences of ideas are not intrinsic to them but are context dependent."[11] The dying-to-know epistemology of objectivity, for example, has been used to disguise racism, to denigrate cultural explanation, to sanction authoritarian action. Yet objectivity has also been a liberating, democratizing force, not only as Bacon and Descartes understood it, but as scientists in the positivist tradition lived it. From one perspective, the scientific naturalists' strategy of empowerment of the institution of science was a self-interested and ultimately dangerous elevation of science to an oppressive cultural authority. Historically, however, watching, for example, the rise of T. H. Huxley out of poverty to his astonishing success as a kind of scientific imperialist, the challenge to ecclesiastical authority was a true advance in the direction of democracy or at least meritocracy. The battle, couched perhaps artificially as science versus religion, had much to do with the ways in which the dying-to-know narrative replicated the narratives of religion and ethics as well. Adopting the stance of the heroic figures who would confront the truth at any cost, scientists like Huxley, Tyndall, Clifford, and Pearson could claim that their position represented a higher—though secular—morality than that of the religion they sought to displace.

Scientific ideas simply do not have unequivocal ideological implications. As Kenshur argues, they are usually recruited to do political work, but their connections with particular ideologies are historically contingent. To make this point, it is instructive to look briefly toward the logical positivists of the 1930s, because positivism has not only been dismissed as an epistemology but is regarded as a handy tool for reactionary politics. Yet the "ahistorical and totalizing" theories of the positivists, as Stephen Kellert describes them, were "politically progressive."[12] "In the European context, the scientific philosophers were socially every bit as radical as the American Pragmatists," says Ronald Giere. "Many had ties to European socialist and communist parties."[13] And Michael Friedman claims that the logical positivist movement "was not only identified with Einsteinian physics and modern abstract mathematics, but also with socialism, internationalism, and 'red Vienna.'"[14] But many critics have taken the ahistoricism and decontextualization of positivist epistemology as the fundamental requirement for passage to the nowhere from which one can finally touch reality. Carolyn Merchant has argued that the logical positivists' scientific decontextualism "makes possible the domination and degradation of nature." The effort of logical positivism, as sensitive as Pearson to the indeterminacy of "fact," to universalize knowledge seems to help produce a position akin to that potentially violent toughmindedness I have already discussed.

Yet Kellert points out that if one takes a contextualized view of the positivists, the story is somewhat different. Just as Pearson's early political writing celebrated the possibility of freedom within the framework of a secular humanism, so the manifestoes of the Vienna Circle were strongly progressive. Reichenbach's insistence on a distinction between the context of discovery and the context of justification, which might be seen as a scientific/epistemological echo of T. S. Eliot's distinction between the man who suffers and the poet who creates, is not ideologically conservative. For Eliot, the distinction authorized his contempt for popular focus on the author rather than the literature. But for the positivists, the discrimination worked under the threat of anti-Semitic fascism. For Reichenbach, "the denial that the character of a person proposing a scientific hypothesis has anything to do with the scientific validity of the hypothesis proposed . . . applies, in particular, to being a Jew, from which it follows that his dismissal from his position in Berlin, as well as the persecution leading to Einstein's resignation, had been in principle unwarranted."[15] Nazi Germany rejected "Jewish science," and drove its Jewish intellectuals out, regardless of the substance of their intellectual work.

Ironically, while positivism is now largely associated with efforts at hegemony, Reichenbach insisted that "science, surely, is not limited to national or racial boundaries; we prefer to stand for this historical truth, in spite of all the pretensions of a certain modern nationalism."[16] Just as Anderson locates in the Victorians a cosmopolitanism that would resist the worst aspects of nationalism, so Kellert and Giere find in the positivists an effort to avoid the exclusions and limits of nationalism for the sake of a cosmopolitanism open to all races.

The positivist hostility to historicism did indeed have political as well as epistemological implications. Epistemologically, the refusal of historicist explanation was an attempt to resist the relativism they feared would blur explanation with the conditions of the local and particular in which all ideas are produced. Such refusal allowed for the degree of generalization necessary to establish at least tentative truths. But it is important to recall that positivism was a deeply skeptical theory of knowledge. As Reichenbach puts it, "Scientific philosophy . . . refuses to accept any knowledge of the physical world as absolutely certain."[17] Historicism, from this point of view, focuses on unique events and even makes scientific ideas unique, rather than universal. As Friedman points out, "far from being naive empiricists, the positivists in fact incorporated what we now call the theory ladenness of observation as central to their novel conception of science."[18] Their efforts, in this context, to find a "neutral metaperspective" (the return of "dying to know") were therefore undermined by their own arguments. Without some kind of generalizing force that takes knowledge beyond the particular, knowledge itself would seem to be impossible.

Perhaps the most famous attack on historicism was Karl Popper's *The Open Society and Its Enemies,* a book with an explicit political agenda. Popper's work, though often linked to that of the positivists, includes significant critiques of the positivist system while suggesting sympathy with its focus on science. But it is even more sensitive than the positivists themselves to the tentativeness of knowledge and also deeply hostile to systems. It celebrates, rather, in a tradition already represented in this book by Pearson's passion for "free thought," democratic participation in knowledge, and Popper claims that "only democracy provides an institutional framework that permits reform without violence, and so the use of reason in political matters."[19] Its predominant animus is against forms of totalitarianism, but in combating it he invokes that tough-minded tradition we have seen in Huxley and Galton, the tough-mindedness characteristic of the dying-to-know tradition.

Why, asks Popper, are irrationalist and prophetic theories that are the foundation of totalitarianism so popularly effective in their rejection of the idea that science and reason are applicable to social life? "I am inclined to think," he responds, that the reason is that "they give expression to a deep-felt dissatisfaction with a world which does not, and cannot live up to our moral ideals and to our dreams of perfection" (1:5). Like Pearson, Popper takes this understanding of limits back to science, not as an infallible means to full knowledge but as the best that can be achieved. Popper brings with his arguments against the absolute a reliance on the purely human, a path of a Feuerbachian and even constructivist reading of reality. It certainly is not a naive surrender to the power of external nature.

"History has no meaning," Popper contends (and this, of course, is consistent both with a reading of Darwin's theory and with the sense that particulars do not create knowledge) (2:278). Against the meaninglessness of history, Popper proposes the ideal of reason, and toward the end of the book his rhetoric heightens: "We can interpret the history of power politics from the point of view of our fight for the open society, for a rule of reason, for justice, freedom, equality, and for the control of international crime," he says. "Although history has no ends, we can impose these ends of ours upon it; and *although history has no meaning, we can give it a meaning.* . . . It is we who introduce purpose and meaning into nature and into history" (ibid). I find it ironic that a position so skeptical and so alert to the power of the mind to create reality (at least social reality) has been so easily taken as politically reactionary. Ironically, Popper read "historicism" as totalitarian; now in great measure Popper's attack on historicism is regarded as totalitarian.[20] These are complicated issues, but out of the confusion it becomes clear once more that the narrative of dying to know has many political valences, and that when located historically, many ideas now associated with one sort of politics will be found to have worked in very different ideological directions in their moments. Current distrust of objectivity, disinterest, and the dying-to-know narrative is not justified by their essential relations to ideology (there are none); our current faith in what theorists are pleased to call the "progressive" implications of constructivism is belied by its uses, in different forms, in Pearson and Popper.

Popper's argument that we can impose on history the meanings we choose is interestingly analogous to the arguments Huxley makes in *Evolution and Ethics*. There Huxley—all too frequently linked to social Darwinism—takes out after the interpretation of evolution that conflates the ethical with the natural. Mill had preceded him in "Nature," where he famously claims that "[i]n sober truth, nearly all things which men are hanged or imprisoned for doing to one another, are nature's every day performances."[21] Against this "cosmic process," as Huxley calls it—the very process that clear-eyed science would reveal to those courageous enough to look carefully—he posits the "ethical process." Like Popper, Huxley refuses to take "the struggle for existence, which has done such admirable work in cosmic nature," as "beneficent in the ethical sphere."[22]

The tough-minded look at the worst, characteristic of the ideal of scientific epistemology, is not the end of the story for Mill, Huxley, Pearson, or

Popper. All three posit the possibility of resistance to nature, or the imposition of human ideals on natural processes. The discovery central to the history of the dying-to-know narrative is that the world has no telos, and has no meaning. Meaning is borne in human consciousness, as is value. But if the cosmic process determines what we are, how is it possible to resist it, using the equipment that the cosmic process has given—millions of years' worth of finely developed instincts and biological apparatus?

Huxley recognizes the difficulty of the position he takes, which implies the possibility of real human generosity and cooperation and altruism and love. He is horrified, however, by what he calls the "fanatical individualism" (p. 82) of those who take the cosmic process as a model for human relations. Throughout the "Prolegomena" to "Evolution and Ethics" he takes the metaphor of the garden as representative of the ways in which human society is always in fact resisting the cosmic process. The garden is an unnatural structure, designed precisely to keep out the forces of nature. Huxley's argument proceeds:

> I do not know that any one has taken more pains than I have, during the last thirty years, to insist upon the doctrine, so much reviled in the early part of that period, that man, physical, intellectual, and moral, is as much a part of nature, as purely a product of the cosmic process, as the humblest weed.
>
> But if, following up this admission, it is urged that, such being the case, the cosmic process cannot be in antagonism with that horticultural process which is part of itself—I can only reply, that if the conclusion that the two are antagonistic is logically absurd, I am sorry for logic, because, as we have seen, the fact is so. (p. 70)

The strenuous work of confronting the worst, one's own mere animality, is supplemented here by the yet more strenuous work of facing the incompatibility between human logic and the structure of the world. The work of altruism is both part of nature and hostile to it. (I don't want to force the idea, but I would suggest that Huxley could believe in altruism *because* he believed in the objectivity that allowed him to see its absence in nature.)

This problem haunts contemporary debate. Evelyn Fox Keller looks warily at recent developments in molecular biology. She traces the way the agent, the intelligent actor in the work of science and the world, is figured as an obstacle to knowledge, and she quotes J. D. Bernal, who argues that the embodied self is the obstacle to knowledge and rationality, is in fact the devil. So, Keller says, "the goal awaiting us is to rout desire from its hiding place and bring it into line with our objective aims."[23] And thus Keller sees alive and well, as it were, the paradox we have seen underlying the narrative of scientific objectivity from the start:

> [W]e can still see the endurance of a semblance of the original paradox . . . though the paradox is still present, its days are palpably numbered. As molecular neuro-biology extends its frontier ever deeper in the brain, promising to converge even on the problem of consciousness . . . , the scientist himself comes to be drawn ever deeper into the machine he has created. At such a point, the circle of scientific knowledge threatens to close in on itself, and the anchoring that has till now been provided by a residually anterior knowing subject threatens to disappear altogether. (p. 325)

The crisis, Keller argues, is being answered by what can only seem to be the absolute fulfillment of the narrative of scientific epistemology, a discourse that through computers "offers the promise of a mind that can indeed exist without the body, as Descartes once believed, but a mind that makes a mockery of all of Descartes' hope" (ibid.). The dream of the robot is the new narrative of dying to know, and wonderfully reenacts the romantic aspiration we have seen inspiring Pearson to positivism. Keller quotes Daniel Hillis, longing for an "artificial mind" that can "sustain itself and grow of its own accord" [from Darwin to Lamarck and G. B. Shaw]; then, says Hillis, "for the first time human thought will lift free of bones and flesh, giving this child of mind an immortality denied to us" (p. 326).[24]

We do not need robotics to see how powerful the dream of desireless, disembodied knowledge remains. But we need to see how that expulsion of desire in the name of immortal knowledge has always entailed enormous losses. And the greatest of these, I would argue, is the loss of the possibility of altruism. Ironically, the absorption of the scientist into the genetic material of the brain leads, in its very self-immolation, to the immolation of the power to act outside the reach of the body (hence the dream of the liminal figure of the robot).

Yet the question of altruism, of acting in the interests of another is, as I have argued, formally the same question as that of disinterested objectivity, of surrendering oneself entirely to the world beyond the self. In contemporary debate, some of the most strenuous public defenders of realism and objectivity are also the most critical of the idea that altruism is possible. As is well known, Darwin argued that no organism could act in the interest of another. To discover such an action in nature would be, he always claimed, "fatal to my theory": "natural selection can act only through and for the good of each being."[25]

Sociobiology and evolutionary psychology have worried out this idea with the Victorian toughness of the dying-to-know story. In an extremely useful essay, "A Sociobiological Expansion of *Evolution and Ethics*," George C. Williams points out that sociobiologists now believe that "altruistic behavior is limited to special situations," which need to be explained, and can be explained "by one or more of three possible factors, none of any use as a romantic's 'exemplar for human conduct.'"[26] The difference between this view and that of the positivist-oriented Victorians is, primarily, the apparent pleasure (and in many writers, not Williams, arrogant pleasure) in exploding the culture's sentimental myths about the freedom of humans from the constraints of nature that Huxley described. "Altruism," it seems, is never what it seems. Here is how Williams accounts for it: (1) nepotism ("cost-benefit relations, kinship effects, and information requirements are equally applicable to the bringing home of food for young," Williams explains); (2) manipulation, which implies "exploitation by a manipulator" ("anyone who makes an anonymous donation of money or blood or other resources as a result of some public appeal is biologically just as much a victim of manipulation as the snapper in the jaws of the anglefish"); (3) reciprocity ("whatever is given up by the donor costs it less than some repayment expected from the recipient").[27]

Here is a powerful hermeneutics of suspicion, like that which governs critiques of scientific claims to objectivity or of the U.S. government's claims to be defending justice and the oppressed when it intervenes around the world. Oddly enough, in this confusingly self-contradictory condition, and as ideology floats from idea to idea, depending on who is thinking them, the sociobiologists and the cultural critics share a method based on distrust of altruism. From many writers supporting the programs of sociobiology and evolutionary psychology, one hears an almost celebratory contempt for views that insist on the power of culture to determine human behavior or on the ways in which scientific and other knowledges are constructed (although, as we have seen, Pearson and Mach were constructivists).

Samuel Butler would have been delighted with so confused a picture. On the one hand, those who are most committed to the narrative of scientific epistemology are constructing by far the most powerful hermeneutics of suspicion operating in society today. All behavior will be explicable by natural selection, from which altruistic possibilities are excluded—except, of course, where "altruism" serves, paradoxically, for survival. The scientist becomes absorbed in the materials with which he works. And on the other, those who most overtly resist the model of dying to know—though the fact is unacknowledged—share the same hermeneutics. More ironically, the full commitment to local knowledges, the refusal of epistemological gestures toward the communal and the universal, implicitly endorses the sort of "fanatical individualism" that Huxley decried.

I close with a return to this book's dominant metaphor. Keller has argued the serious dangers of scientific epistemology's ambition to banish desire and the living, breathing human. And at the very foundations of Western epistemology, Plato's *Phaedo* explains Socrates' contentment with his death primarily through the argument for an epistemology uncorrupted by embodiment. "Is not purification," asks Socrates, "to separate as far as possible the soul from the body, and to accustom it to collect itself together out of the body in every part, and to dwell alone by itself as far as it can, both at this present and in the future, being freed from the body as if from a prison?"[28] The West has celebrated Socrates' death ever since—he died in order to know. And it has tended to instate officially, in the very heart of an empirical science ostensibly hostile to Platonism, the narrative of his progress. But I have tried to dramatize that body and soul cannot be separated in narrative, and narrative shows that there is something desperately missing from an imagination of knowledge that requires death for its acquisition. There is something desperately inadequate about a skeptical cultural theory that refuses the possibility of selflessness and of objectivity. What matters, to return to one of my favorite Victorians, is what avails for life.

The irony is that the altruistic politics of postmodern theory and postcolonial openness depend on the possibility of something like objectivity, the transcendence of the limits of the perceiving and desiring self. In the end, there is no escape for it: we are all of us dying to know, but now dying to know how not to deny; how not to resign ourselves to the inhuman in the quest for the knowledge that might humanize; how to be objective and altruistic at the same time. "Death," says Wittgenstein, "is not an event of life. Death is not lived through." And, he says, immortality would not, in any case, resolve the riddle because it would simply present another enigma.[29] The ethical is a construction. But the ethical lies always in the story buried and latent in epistemology, making it do more than explain (the limit Wittgenstein and Pearson allow for propositions), enjoining, insisting on the value of knowing and on the value of dying to know. The moral, then, of *my* story is that while one cannot live with the epistemology that points to that impossible liminal "nowhere," that space of deadly knowing, one cannot—and should not—live without it.

NOTES

INTRODUCTION

1. In John Kucich's *Repression in Victorian Fiction* (Berkeley and Los Angeles: University of California Press, 1987), there is a valuable discussion of this "desire for death" in the context of a study of Victorian fiction. The pattern I describe in epistemological discourse nicely overlaps with the pattern of "repression" Kucich analyzes in the novel. I am grateful to Kucich's discussion of George Bataille, whose theory of the sacred, sacrifice, and transcendence offers another, mythic version of the paradox I am exploring here. See Kucich, pp. 19–22.

2. John Kucich, *The Power of Lies: Transgression in Victorian Fiction* (Ithaca, N.Y.: Cornell University Press, 1994), p. 5.

3. Peter Galison, "Judgment Against Objectivity," in *Picturing Science, Producing Art*, ed. Peter Galison and Caroline Jones (New York: Routledge, 1998), p. 329.

4. Herbert's book, *Victorian Relativity* (Chicago: University of Chicago Press, 2001), modifies importantly the view of the Victorians as absolutist in epistemology. He shows that the relativist streak among Victorian thinkers is pervasive, and provides an important genealogy for current thinking. He points out that Mansel and Hamilton both articulated a clearly relativist epistemology for human thought, marking the profound limits of the possibility of knowing. For these writers, however, Herbert shows that the humanly inescapable relativism became grounds for asserting a theological absolutism.

5. William Whewell, *History of Scientific Ideas. Being the First Part of the Philosophy of the Inductive Sciences* (London: John W. Parker & Son, 1858), 1:46.

6. See Peter Galison and Lorraine Daston, "The Image of Objectivity," *Representations*, no. 40 (fall 1992): 81–128. Galison has filled in some of the details of the nineteenth-century moralizing of objectivity in this tradition of self-abnegation in his "Judgment Against Objectivity" (in *Picturing Science, Producing Art*, pp. 327–59; for a full analysis of the different kinds of objectivity and the distinction of the objectivity I will be calling "dying to know"—Daston calls it "aperspectival"—see Lorraine Daston, "Objectivity and the Escape from Perspective," in *The Science Studies Reader*, ed. Mario Biagioli (New York: Routledge, 1999), pp. 110–23.

7. For a valuable treatment of the status of Baconian method in the nineteenth century and the unofficial resistance to it, see Jonathan Smith, *Fact and Feeling: Baconian Science and the Nineteenth-Century Literary Imagination* (Madison: University of Wisconsin Press, 1994).

8. John Stuart Mill, *System of Logic, Ratiocinative and Inductive, Being a Connected View of the Principles of Evidence and the Methods of Scientific Investigation* (London: John W. Parker, 1856), 2:290.

9. Ibid., 2:291.

10. John Tyndall, *Fragments of Science* (New York: Appleton, 1899), 1:291.

11. Ibid., 1:292.

12. Perhaps the best known and most forceful is Peter Novick, *That Noble Dream: The "Objectivity Question" and the American Historical Profession* (Cambridge: Cambridge University Press, 1988).

13. James Moore has argued the case strongly. In the biography of Darwin coauthored with Adrian Desmond, there is convincing evidence of the connection between Darwin's theory and economic motives. After locating Darwin's thought on the subject within the context of Malthus and imperialism and Darwin's experience of General Rosas's genocide in South America, the authors trace the connections of the development of Darwin's biological theories to such events, and then claim, "At last he had a mechanism that was compatible with the competitive, free-trading ideals of the ultra-Whigs. The transmutation at the base of his theory would still be loathed by many. But the Malthusian superstructure struck an emotionally satisfying chord" (Adrian Desmond and James Moore, *Darwin* [New York: Warner Books, 1991], p. 297).

14. The best known and strongest case for such a reading is D. A. Miller, *The Novel and the Police* (Berkeley and Los Angeles: University of California Press, 1988).

15. Identifying the "great intellectual professions," Ruskin asserts that it is the duty of each to be willing to die for their ultimate aims; the doctor, for example, should risk death to keep the nation in health. The demonstration of the integrity of a professional is his willingness to die to fulfill his responsibilities. The pastor, for example, should be willing to die "rather than teach Falsehood" (*Unto This Last* [Lincoln: University of Nebraska Press, 1967], p. 25).

16. George Eliot, "The Influence of Rationalism," in *Essays of George Eliot*, ed. Thomas Pinney (New York: Columbia University Press, 1963), p. 413.

17. George Eliot, *Adam Bede* (Penguin: Harmondsworth, Middlesex: 1980), chap 17, p. 221.

18. Linda Peterson, *Victorian Autobiography: The Tradition of Self-Interpretation* (New Haven, Conn.: Yale University Press, 1986).

19. Albert Einstein, *Autobiographical Notes*, trans. Paul Arthur Schilp (La Salle, Ill.: Open Court, 1979), p.31.

20. Ibid., p. 5.

21. Thomas Nagel, *The View from Nowhere* (New York: Oxford University Press, 1986), p. 3.

22. Thomas Nagel, *The Last Word* (Oxford: Oxford University Press, 1997), p. 3.

23. In a discussion of objectivity, Lorraine Code conveniently summarizes the position: "It is by now a feminist commonplace that the epistemologies of modernity, in their principled neutrality and detachment, generate an ideology of objectivity that dissociates itself from emotions and values, while granting no epistemological significance to its own cognitive location. Ideal objectivity counts as the hallmark of knowledge worthy of the name. This ideology of neutrality is produced in positions of privilege that enable promulgators to believe that they are everywhere and nowhere" ("Who Cares: The Poverty of Objectivism for a Moral Epistemology," in *Rethinking Objectivity*, ed. Allan Megill [Durham, N.C.: Duke University Press, 1994], p. 180).

24. Taking a recent and forceful example, I would point to Amanda Anderson's argument, particularly in the last chapter of *Tainted Souls and Painted Faces: The Rhetoric of Fallenness in Victorian Culture* (Ithaca, N.Y.: Cornell University Press, 1993): Postructuralism's primary emphases, Anderson claims, have been "on indeterminacy and a radical alterity that functions at the symbolic or the structural level. But this privileging of disruption at the systemic level derives from a residual objectivism with poststructuralism itself, one that profoundly disables the political and ethical projects of those cultural critics who employ poststructuralist paradigms in an untransformed way" (p. 200).

25. John Dupré, *The Disorder of Things: Metaphysical Foundations of the Disunity of Science* (Cambridge, Mass.: Harvard University Press, 1993), p. 120.

26. Ibid., p. 13. Debates at this level go on endlessly, and I do not claim here to be covering all the possible options in them. I would only point out, as against the position that Dupré takes, the arguments of Barbara Herrnstein Smith in her *Belief and Resistance: Dynamics of Contemporary Intellectual Controversy* (Cambridge, Mass.: Harvard University Press, 1997), who insists that cultural critics of science do not deny the "constraining power of nature" on the dualistic terms that Smith believes are central to the objectivist/realist epistemological position. Using Bruno Latour as her primary example, Smith argues that what constructivists argue for is, rather, "not social interaction or discourse alone, and not social interaction or discourse simply *added to* empirical evidence, as the latter is classically understood, but a complex interactive process that is simultaneously dynamic, productive, and self-stabilizing" (p. 130).

27. In *Belief and Resistance* (Cambridge, Mass.: Harvard University Press, 1997), Barbara Herrnstein Smith, perhaps the most persistent, careful, and relentless supporter of the notion of "contingency" and constructivism, concludes that "some of the most notable, recurrent, and, from a skeptical perspective, problematic features of classic formulations and arguments (for example, their essentializing reifications and self-affirming circularities) seem to reflect cognitive tendencies that are, in certain respects, valuable or indeed indispensable" (p. xvii). One might take this as a more abstract and perhaps grudging way of putting my own argument, that the ideals of disinterest are strong "tendencies" that get played out and reaffirmed by narratives, even when they are put most radically to question.

28. Nagel, *The Last Word*, p. 7.

29. Ibid., p. 6. Barbara Herrnstein Smith, throughout her *Contingencies of Value: Alternative Perspectives for Critical Theory* (Cambridge, Mass.: Harvard University Press, 1988), rebuts the "self-refuting" argument against relativism. See particularly p. 157: "since a nonobjectivist, by definition, does not conceptualize 'truth' and 'validity' the way an objectivist does, she cannot give to 'true' and 'valid' the meanings those terms have for him, which are also the meanings that are required either for her position to be self-refuting or for her to be committed to that form of egalitarianism which, to the objectivist, defines relativism and entails quietism."

30. This little agon implies that people are persuaded by reasons. In narrative, this doesn't happen very often, nor, I suspect, does it in life. Certainly, a lot more than rationality goes into ordinary disputes. The decision to accept certain kinds of arguments is likely to be based on fundamental (and likely opposed) notions of what constitutes epistemological authority, and those notions are likely linked to social, personal, extralogical predispositions. The "dying-to-know" model assumes a culture that privileges intellectual over manual labor.

31. Kenshur's arguments on this issue are laid out in the following works: "Demystifying the Demystifiers: Metaphysical Snares of Ideological Criticism," *Critical Inquiry* 14 (winter 1988): 335–53; "An Exchange on Ideological Criticism: (Avoidable) Snares and Avoidable Muddles," *Critical Inquiry* 15 (spring 1989): 658–88; and *Dilemmas of Enlightenment: Studies in the Rhetoric and Logic of Ideology* (Berkeley and Los Angeles: University of California Press, 1993), chap. 1, "Ideological Essentialism."

32. Satya Mohanty, *Literary Theory and the Claims of History: Postmodernism, Objectivity, Multicultural Politics* (Ithaca, N.Y.: Cornell University Press, 1997), p. 165.

CHAPTER ONE

1. Alasdair MacIntyre, "Epistemological Crisis, Dramatic Narrative, and the Philosophy of Science," in *Paradigms and Revolutions,* ed. Gary Gutting (South Bend, Ind.: University of Notre Dame Press, 1980), pp. 54–74.

2. Ibid., pp. 59–60.

3. See the exchange about *Telling the Truth about History*, by Joyce Appleby, Lynn Hunt, and Margaret Jacob, "Truth, Objectivity, and History: An Exchange," in *Journal of the History of Ideas* 56 (October 1995): 675–80. See also Bonnie Smith, who in "Whose Truth, Whose History," speaks of traditional science as being on the "heroic model" (p. 661).

4. Myra Jehlen, "History beside the Fact: What We Learn from a True and Exact History of Barbados," in *The Politics of Research*, ed. Ann Kaplan and George Levine (New Brunswick, N.J.: Rutgers University Press, 1997), p. 127.

5. Evelyn Fox Keller, *Reflections on Gender and Science* (New Haven: Yale University Press, 1985), pp. 10–11.

6. John Herschel, *A Preliminary Discourse on the Study of Natural Philosophy* (Chicago: University of Chicago Press, 1987), p. 80.

7. For a discussion of this presence and the Victorian struggle with it, see Jonathan Smith, *Fact and Feeling: Baconian Science and the Nineteenth-Century Literary Imagination* (Madison: University of Wisconsin Press, 1994).

8. Francis Bacon, *The Great Instauration* (London: George Bell and Sons, 1839), pp. 7–8.

9. Peter Galison and Lorraine Daston, "The Image of Objectivity," *Representations*, no. 40 (fall 1992): 81.

10. Charles B. Paul, *Science and Immortality: The Éloges of the Paris Academy of Sciences, 1691–1799* (Berkeley and Los Angeles: University of California Press, 1980), p. 99.

11. Bacon, *The Great Instauration*, p. 399.

12. Ibid., p. 20.

13. Samuel Smiles, *Self-Control* (New York: William Alison, n.d.), p. 188.

14. William Whewell, *History of the Inductive Sciences from the Earliest to the Present Time* (New York: Appleton and Co., 1865), 1:42.

15. William Whewell, *The Philosophy of the Inductive Sciences*, reprinted sections in *Selected Writing on the History of Science*, ed. Yehuda Elkana (Chicago: University of Chicago Press, 1984), p. 249.

16. Whewell, *History of the Inductive Sciences*, 1:418.

17. René Descartes, *Discourse on Method* (Harmondworth, Middlesex: Penguin Books, 1968), p. 27.

18. Edward Dowden, "The Scientific Movement in Literature," in *Studies in Literature: 1789–1877* (London: Kegan Paul French and Trubner, 1892), p. 85.

19. Thomas Hardy, *Two on a Tower* (New York: St. Martin's Press, 1982), p. 56.

20. E. A. Burtt, *The Metaphysical Foundations of Modern Science* (New York: Doubleday, 1932), p. 18.

21. T. H. Huxley, *Methods and Results* (London: MacMillan and Co., 1893), p. 40.

22. Burtt, *The Metaphysical Foundations of Modern Science*, pp. 17–18.

23. Bertrand Russell, *Mysticism and Logic* (New York: W. W. Norton, 1929), p. 42.

24. Hardy, *Two on a Tower*, p. 58.

25. William James, *Pragmatism and the Meaning of Truth* (Cambridge: Harvard University Press, 1981), p. 118.

26. While this is not the place to talk about the science wars, it is interesting to note that the strongest, most extreme defense of science comes inside a rhetoric of absolute confidence in superpersonal grounding of scientific knowledge, and a quite passionate attack on those who have been arguing, in different ways, and with much different attitudes about science itself, that scientific knowledge is "constructed." Harré's arguments anticipate some of these events and make a case for science that is built on both that absolute confidence and an aware-

ness of the power of certain critical arguments about science's capacity to give an objective accounting of the "real" world.

27. Philip Kitcher, *The Advancement of Science: Science without Legend; Objectivity without Illusions* (New Haven: Yale University Press, 1993), p. 9.

28. Rom Harré, *Varieties of Realism* (Oxford: Basil Blackwell, 1986), p. 1.

29. In *Defining Science: William Whewell, Natural Knowledge, and Public Debate in Early Victorian England* (Cambridge: Cambridge University Press, 1993), Richard Yeo discusses the tension in early Victorian thought between the ideal of the scientist in solitude, "achieving truth by means of . . . separation from the distractions of society" (p. 136), and a developing ideal of a community of science which by virtue of its very quality as community avoided the problems Bacon attributed to scientific work in isolation, not allowing for cumulative knowledge. The community, too, is a moral institution, imposing a work methodology of personal restraint which, on the model of the scientist in solitude, depended entirely on personal will. See chapter 5, pp. 116–44.

30. See David L. Hull, *Science As a Process* (Chicago: University of Chicago Press, 1988). For an important critique of Hull, see John Dupré, *The Disorder of Things* (Cambridge, Mass.: Harvard University Press, 1993).

31. Harré, *Varieties of Realism*, p. 5.

32. Helen Longino, "Can There Be a Feminist Science?" in *Feminism and Science*, ed. Nancy Tuana (Bloomington: Indiana University Press, 1989).

33. Donna Haraway, *Cyborgs and Women: The Reinvention of Nature* (New York: Routledge, 1991), p. 187.

34. Steven Shapin, *A Social History of Truth: Civility and Science in Seventeenth-Century England* (Chicago: University of Chicago Press, 1994), p. 417.

35. Haraway, *Cyborgs and Women*, p. 186.

36. See Diana Fuss, *Essentially Speaking: Feminism, Nature, and Difference* (New York: Routledge, 1989). Her concluding remarks are to the point here: "I have argued from the start that essentialism underwrites theories of constructionism and that constructionism operates as a more sophisticated form of essentialism. This is simply another way of saying that constructionism may be more normative, and essentialism more variable, than those of us who consider ourselves poststructuralists have been willing to acknowledge" (p. 119).

37. James, *Pragmatism and the Meaning of Truth*, p. 7.

38. Keller, *Reflections on Gender and Science*, p. 117.

CHAPTER TWO

1. Lorraine Daston argues forcefully that Descartes did not propose an "aperspectival objectivity" and that our current interpretations of him as an objectivist (in our contemporary sense) is in fact anachronistic ("Objectivity and the Escape from Perspective," in *The Science Studies Reader*, ed. Mario Bagioli [New York: Routledge, 1999], p. 118). In an essay that places Descartes's ideas in the context of late medieval scholasticism, Calvin Normore in effect supports Daston's case by studying Descartes's theory of meaning, showing that Descartes's "attempt to explain why thinking is informative and why it is informative about objects" is an original theory, working with scholastic concepts. The implication, of course, for the purposes of my argument, is that Descartes was not using our contemporary notions of thought and object, of objectivity, and that his theory belongs squarely within the context of late medieval debate. Normore concludes that for Descartes, "an idea is a presentation of the very object represented." But it is Descartes, the "founder" of modern epistemology, the one who has been

read into the Cartesian anxiety, discussed below, that I am reading here. He, too, has a powerful historical presence. See Calvin Normore, "Meaning and Objective Being: Descartes and His Sources," in *Essays on Descartes' "Meditations,"* ed. Amelie Rorty (Berkeley and Los Angeles: University of California Press, 1986), pp. 223–24, 240.

2. Susan Bordo, *The Flight to Objectivity: Essays on Cartesianism and Culture* (Albany: State University of New York Press, 1987), p. 17.

3. Richard Bernstein, *Beyond Objectivism and Relativism: Science, Hermeneutics, and Praxis* (Philadelphia: University of Pennsylvania Press, 1983), p. 8.

4. Bernard Williams, *Descartes: the Project of Pure Enquiry* (Brighton: Harvester Press, 1978): "the trouble with Descartes' system is not that it is circular; nor that there is an illegitimate relation between the proofs of God and the clear and distinct perceptions; nor that there is a special problem about the proofs of God when they are not intuited. . . . The trouble is that the proofs of God are invalid and do not convince *even when they are supposedly being intuited*" (p. 210).

5. Stephen Gaukroger, *Descartes: An Intellectual Biography* (Oxford: Oxford University Press, 1995), p. 12.

6. George Heffernan, ed., René Descartes, *Discourse on Method A Bilingual Edition* (Notre Dame: University of Notre Dame Press, 1994), p. 159.

7. René Descartes, *Discourse on Method and Meditations,* trans. F. E. Sutcliffe (London: Penguin Books, 1968), p. 28. Subsequent references are to this translation and appear in the text. Quotations from the French will be from *Oeuvres de Descartes: Discours de la methode & essais*, vol. 6 (Paris: Librairie Philosophique J. Vrin, 1996).

8. The ideas of Georges Bataille are relevant here. Of course, Bataille does not talk directly about epistemology, but his sense that there is no fullness of satisfaction until death, and that death is the fulfillment of self stands as another way to read the phenomenon of "dying." It provides another way to account for the odd structure of Descartes's argument. Bataille constantly deploys the paradox of fulfillment in death; so to take just one example he claims that "[i]t is by dying, without possible evasion, that I will perceive the rupture which constitutes my nature in which I have transcended 'what exists.'" *George Bataille, Essential Writings* (London: Sage Publications, 1998), p. 203.

9. Susan Bordo argues that we should view Cartesian objectivism "as a defensive response to . . . separation anxiety, an aggressive intellectual 'flight from the feminine' rather than (simply) the confident articulation of a positive new epistemological ideal" ("The Cartesian Masculinization of Thought" in *Sex and Scientific Inquiry*, ed. Donna Haraway [Chicago: University of Chicago Press, 1987], pp. 248-49). Evelyn Fox Keller also gives a psychological explanation for the growth and power of the idea of objectivity. Her argument emphasizes the connection in the development of Western science between the epistemology of objectivity and power: "Just as objectivity is to be understood an interprersonal acquisition, . . . domination, even of nonhuman others, is an interpersonal project. To see how the twin goals of science—knowledge and power—are translated into objectification and domination, we need to explore the psychodynamic roots that link these goals together" (*Reflections on Gender and Science* [New Haven, Conn.: Yale University Press, 1985], pp. 71–72).

10. René Descartes, *Principles of Philosophy,* trans. Valentine Rodger Miller and Reese P. Miller (Dordrecht, Holland: D. Reidel Publishing Co, 1983), p. xxvi.

11. See my brief discussion of Huxley's resistance to authority in the introduction. In his essay, "On Descartes 'Discourse Touching the Method of Using One's Reason Rightly and of Seeking Scientific Truth,'" Descartes is identified as one of the "half-dozen men, endowed with intellects of heroic force and clearness," the thinker who stands as "stem" "towards the philosophy and science of the modern world" (in *Methods and Results* [London: MacMillan and Co., 1893], pp. 166–67). Huxley cites with enthusiasm Descartes's argument that "there is a

path that leads to truth so surely, that any one who will follow it must needs reach the goal, whether his capacity be great or small. And there is one guiding rule by which a man may always find this path. . . . This golden rule is—give unqualified assent to no propositions but those the truth of which is so clear and distinct that they cannot be doubted. The enunciation of this great first commandment of science consecrated Doubt" (p. 169). When Huxley called science "organized common sense," he obviously had the opening of the *Discourse* in mind.

12. Richard Rorty, *Philosophy and Social Hope* (London: Penguin, 1999), p. xxvii.

13. Thomas Nagel, *The View from Nowhere* (New York: Oxford University Press, 1986), p. 7.

14. Dalia Judovitz, *Subjectivity and Representation in Descartes: The Origins of Modernity* (Cambridge: Cambridge University Press, 1988), pp. 93–97. Judovitz's book, in its keenly analytical study of the way Descartes's means of "representation" modifies the abstract argument that will ultimately deny representation, anticipates many of the arguments and the direction of this chapter. Her objectives, tracing the movement from modernism to postmodernism, and her tendency to see Descartes as opaque to some of the crucial implications of his literary devices, are, however, significantly different from mine.

15. René Descartes, *Meditations on First Philosophy, with Selections from Objections and Replies*, trans. John Cottingham (Cambridge: Cambridge University Press, 1986), p. 3. The *Meditations* were written in Latin, and the French translation, although apparently approved by Descartes, is not obviously as authoritative as the original French of the *Discourse*. Where on occasion it might help, I will cite the French from *Oeuvres philosophiques* (Paris: Èditions Garniers Frères, 1967), vol. 2, *Meditations*, ed. Ferdinand Aliquie.

16. Judovitz, *Subjectivity and Representation in Descartes*, p. 87.

17. Richard Rorty, *Philosophy and the Mirror of Nature* (Princeton, N.J.: Princeton University Press, 1979), p. 61.

18. Williams, *Descartes*, p. 20.

19. Ibid.

20. For an excellent discussion of the morally conservative directions of the English bildungsroman, see Franco Moretti, *The Way of the World: The "Bildungsroman" in European Culture* (Thetford, Norfolk: Verso Books, 1987), esp. chap. 4. For a discussion of the self-annihilating forms of Victorian scientific autobiographies, see Regenia Gagnier, *Subjectivities: A History of Self-Representation in Britain 1832–1920* (Oxford: Oxford University Press, 1990), esp. chap. 6.

21. Williams, *Descartes*, p. 20.

22. Judovitz, *Subjectivity and Representation in Descartes*, p. 107.

23. Of course, the two texts must ultimately be read as independent, although there is no question that Descartes was, in both texts, describing the same experience. Against the view that the two are the same, Ferdinand Aliquie, editor of the *Meditations*, warns: "Nous voudrions seulement demander au lecteur de ne pas admettre sans examen que la métaphysique des *Meditations* est identique à celle de la quatrieme partie du *Discours* ou de la première partie des *Principes*. Chacun de ces textes a son originalité" (p. 380).

24. Of three translations, two allow the translation "persuade." I quote the Sutcliffe translation above (p. 150). John Veitch's older translation (*Discourse on Method and the Meditations* [Buffalo: Prometheus Books, 1989]) also talks of the "faculty of imagination" which is "sufficient to persuade" (p. 111). The Cottingham translation, however, puts it this way: "The conclusion that material things exist is also suggested by the faculty of imagination" (p. 50). The French translation has it this way: "est capable de me persuader leur existence" (p. 480). Cottingham seems closer to the Latin.

25. Marjorie Grene, *Descartes* (Minneapolis: University of Minnesota Press, 1985), p. 81.

26. Claude Bernard, *An Introduction to the Study of Experimental Medicine* (New York: Dover Books, 1957): "to learn how man and animals live, we cannot avoid seeing great numbers of

them die, because the mechanisms of life can be unveiled only by knowledge of the mechanisms of death" (p. 99). There is a remarkable passage in Bernard's discussion of the work of surgeons and physiologists that powerfully suggests the importance and the difficulty of the kind of intellectual stance Descartes argues for, and at the same time makes clear how much the work of detachment is premised on moral values: "A physiologist is not a man of fashion, he is a man of science, absorbed by the scientific idea which he pursues: he no longer hears the cry of animals, he no longer sees the blood that flows, he sees only his idea and perceives only organisms concealing problems which he intends to solve. Similarly, no surgeon is stopped by the most moving cries and sobs, because he sees only his idea of the purpose of his operation. Similarly again, no anatomist feels himself in a horrible slaughter house; under the influence of a scientific idea, he delightedly follows a nervous filament through stinking livid flesh, which to any other man would be an object of disgust and horror" (p. 103).

CHAPTER THREE

1. See John Stuart Mill, *A System of Logic Ratiocinative and Inductive, Being a Connected Views of the Principles of Evidence and the Methods of Scientific Investigation* (London: John W. Parker and Son, 1856), 1:337.

2. See the introduction to William Whewell, *Selected Writings on the History of Science* (Chicago: University of Chicago Press, 1984), particularly pp. xxi–xxiii. The quotation is from p. xxi.

3. Ibid., p. xxii.

4. William Whewell, *Theory of Scientific Method*, ed. Robert E. Butts (Indianapolis: Hackett Publishing Co., 1989), p. 58.

5. The view that the individual fact does not constitute knowledge at all is pervasive in thought about science throughout the nineteenth century. An early and succinct statement of it can be found in John Herschel's *Preliminary Discourse on the Study of Natural Philosophy* (Chicago: University of Chicago Press, 1987), a book that Darwin lists as one of the two most important in the shaping of his inclinations toward science: "The only facts which ever become useful as grounds of physical enquiry are those which happen uniformly and invariably under the same circumstance. This is evident, for if they have not this character they cannot be included in laws" (p. 119). All particularities in science aspire to the condition of universals. Left to their own peculiarities, they are, like the individual human body, the individual human cluster of feelings and desires, worthless.

6. For a discussion of Arnold's need for the general, the communal, the transcendence of the individual, see Ruth ApRoberts, *Arnold and God* (Berkeley and Los Angeles: University of California Press, 1983), pp. 147–49.

7. John Ruskin, *Sesame and Lilies, The Works of John Ruskin*, ed. E. T. Cook and Alexander Wedderburn (London: George Allen, 1905), 18:75.

8. Lorraine Daston and Peter Galison point out that the idea of "objectivity" as it is now broadly understood originated not with science but with aesthetics, an area of human experience that was from the first vulnerable to the disintegrating force of particularities: "de gustibus non est disputandum." For eighteenth-century aesthetic theorists, from Addison through Burke and Hume, the problem was to find a means to show that the personal was not singular, that there are universal standards of beauty.

9. Matthew Arnold, "The Buried Self," in *Poetry and Criticism of Matthew Arnold*, ed. Dwight Culler (Boston: Riverside, 1961), p. 114.

10. Friedrich Schiller, *On the Aesthetic Education of Man in a Series of Letters,* trans. Elisabeth M. Wilkinson and L. A. Willoughby (Oxford: Clarendon Press, 1967), letter 2, p. 17. Schiller's

location of the free self in the state fully anticipates Arnold's arguments for the state in *Culture and Anarchy*. Surprising and disappointing in Arnold's argument, they become much more comprehensible and somewhat less arbitrary and authoritarian when seen in the context of Schiller's.

11. But in his "Preface to First Edition of *Poems*" Arnold talks of the disappearance from modern life of "disinterested objectivity," and of the way the study of classical art puts its practitioners "under the empire of facts," and helps fulfill the artist's responsibility to "educe and cultivate what is best and noblest in themselves" (Culler, ed., *Poetry and Criticism of Matthew Arnold*, p. 212). These themes, normally part of the narrative of scientific epistemology, are here associated with Schiller's argument that "[a]ll art is dedicated to Joy, and there is no higher and no more serious problem than how to make men happy" (Ibid., p. 204).

12. Schiller, *On the Aesthetic Education of Man in a Series of Letters*, letter 19, p. 129.

13. Thomas Carlyle, *Sartor Resartus*, ed. Charles Frederick Harold (New York: Odyssey Press, 1937), p. 72.

14. Thomas Carlyle, "Characteristics," in *Critical and Miscellaneous Essays* (London: Chapman and Hall, 1899), 3:3.

15. Ibid., 3:5.

16. Carlyle, *Sartor Resartus*, p. 197.

17. Carlyle, "Characteristics," 3:26.

18. Particularly during the Second World War and in its aftermath, Carlyle was read as belonging to a dangerously authoritarian and antirationalist group of thinkers and writers, whose views assimilated easily—so the arguments went—to German fascism. One of the more interesting of the books on this subject is Eric Bentley's *Century of Hero Worship*. The emphasis on the body in this literature suggests some serious difficulties in contemporary arguments about the importance of embodiment in any politically "progressive" theory.

19. It is not necessary here to rehearse the overtold story of Carlyle's wedding night and the morning after, with signs of bruises on Jane Welsh. Fred Kaplan deftly summarizes the details of Carlyle's quite touching response to his wedding night experience in his *Thomas Carlyle: A Biography* (Ithaca, N.Y.: Cornell University Press, 1983), pp. 118–22.

20. J. A. Froude, *Thomas Carlyle: A History of the First Forty Years of His Life* (London: Longmans, Green and Co., 1908), 1:78.

21. Kaplan, *Thomas Carlyle*, p. 6.

22. Thomas Carlyle, "Signs of the Times," in *Critical and Miscellaneous Essays* (London: Chapman and Hall, 1899), 2:82.

23. Carlyle, *Sartor Resartus*, p. 191.

24. Carlyle, "Characteristics," 3:20.

25. Carlyle, *Sartor Resartus*, p. 53.

26. For a useful discussion of Carlyle's uses of etymology and its relation to the general function of metaphor in his thought and work, see G. B. Tennyson, Sartor *Called* Resartus: *The Genesis, Structure, and Style of Carlyle's First Major Work* (Princeton, N.J.: Princeton University Press, 1965), pp. 262–67. Tennyson has an excellent discussion of Carlyle's treatment of the word *pecuniary* in "The World Without Clothes" chapter of *Sartor*.

27. Thomas Carlyle, *On Heroes, Hero-Worship and the Heroic in History* (London: Chapman and Hall, 1901), p. 196.

28. Carlyle, *Sartor Resartus*, trans. C. F. Harrold, p. 130.

29. Carlyle's difficulties with his bowels and his health in general seem to have increased in the early stages of his marriage. After the marriage, as quoted by Kaplan, he wrote to his brother "about many things, *ut cum fratre, ut cum medico*," but he apparently found it difficult to be candid about the problems of his "outward man" (*Thomas Carlyle*, p. 118).

30. Froude, *Thomas Carlyle*, 1:384–85.

31. Kaplan, *Thomas Carlyle*, p. 118.

32. Again, Fred Kaplan has written about this essay in "Phallus Worship (1848): An Unpublished Carlylean Response to the Revolution of 1848," *Carlyle Newsletter*, no 2 (March 1980): 19–23.

33. Carlyle, *Sartor Resartus*, p. 168.

34. Carlyle, "Chartism," in *Critical and Miscellaneous Essays*, 4:130. For a still interesting argument that Carlyle's "mature thought constitutes a valuable contribution to radical political theory," see Philip Rosenberg, *The Seventh Hero: Thomas Carlyle and the Theory of Radical Activism* (Cambridge, Mass.: Harvard University Press, 1974).

35. The question is asked at critical moments in *Middlemarch*, but runs through much of the nineteenth-century bildungsroman. See Alan Mintz, *George Eliot and the Novel of Vocation* (Cambridge, Mass.: Harvard University Press, 1978). Mintz valuably juxtaposes Teufelsdröckh's crisis and fate to that of Dorothea Brooke, and focuses on the way their crises amount to crises of vocation, with the woman, of course, managing only to perform "unhistoric acts."

36. Carlyle, "On History Again," in *Critical and Miscellaneous Essays*, 3:167.

37. John Rosenberg, *Carlyle and the Burden of History* (Cambridge, Mass.: Harvard University Press, 1985), p. 31.

38. Carlyle, "On History Again," in *Critical and Miscellaneous Essays*, 2:89.

39. Ibid., 2:89–90.

40. Thomas Carlyle, *Past and Present* (London: Chapman and Hall, 1899), p. 38.

41. George Eliot, *Romola* (Harmondsworth, Middlesex: Penguin Books, 1980), p. 43.

42. Tyndall has a long memoir on his relations with Carlyle, in which he emphasizes the compatibility of Carlyle's thought with science. See John Tyndall, *New Fragments* (New York: Appleton and Co., 1892), pp. 347–91.

43 Frank Turner, "Victorian Scientific Naturalism and Thomas Carlyle," *Victorian Studies* 18 (1975): 330.

44. Ibid., p. 333.

45. Carlyle, *Sartor Resartus*, p. 186.

46. Ibid.

CHAPTER FOUR

1. See my essay, "Matthew Arnold's Science of Religion: The Uses of Imprecision," *Victorian Poetry* 26 (1988): 143–62, for a detailed discussion of Arnold's acceptance of the scientific ideal.

2. Among the most interesting are James Olney, *Metaphors of Self: The Meaning of Autobiography* (Princeton, N.J.: Princeton University Press, 1972); Linda Peterson, *Victorian Autobiography: The Tradition of Self-Interpretation* (New Haven, Conn.: Yale University Press, 1986); and Jonathan Loesberg, *Mill, Newman, and the Reading of Victorian Prose* (New Brunswick, N.J.: Rutgers University Press, 1986).

3. Peterson, *Victorian Autobiography*, p. 159.

4. Charles Darwin, *The Autobiography of Charles Darwin, 1809–1882*, ed. Nora Barlow (New York: W. W. Norton, 1958), p. 93.

5. Loesberg, *Mill, Newman, and the Reading of Victorian Prose*, p. 92.

6. Bernard Lightman has written extensively on this aspect of Huxley's work. See his *Origins of Agnosticism: Victorian UnBelief and the Limits of Knowledge* (Baltimore: Johns Hopkins University Press, 1987), pp. 131–34; see also his essay, "Pope Huxley and the Church Agnostic," *Historical Papers* (1983): 150–63.

7. T. H. Huxley, "On the Advisableness of Improving Natural Knowledge," in *Methods and Results* (London: MacMillan and Co., 1893), p. 41.

8. Susan Faye Cannon, *Science in Culture* (New York: Science and History Publications, 1978), p. 2.

9. Beatrice Webb, *My Apprenticeship* (Harmondsworth: Penguin Books, 1938), 1:153.

10. I am particularly struck by the kinds of sociocultural readings of fiction that have been made by D. A. Miller in *The Novel and the Police* (Berkeley and Los Angeles: University of California Press, 1988) and Franco Moretti in *The Way of the World: The "Bildungsroman" in European Culture* (Thetford, Norfolk: Verso Books, 1987). Both of these critics, in different ways and from different perspectives, focus on the relation between nineteenth-century fiction and the values of an emerging bourgeoisie. The novels can be taken as more or less complex exercises in the creation of bourgeois identity or, as Moretti takes the English bildungsroman, "fairy tales" that achieve their happy conclusions distributing earned moral rewards to the protagonists. I take my readings to be not inconsistent with these views, but rather an alternative and less sociological way to understand the patterns of the narratives. The major evaluative difference, I think, is that I see these narratives as, at least at their best, not complicit victims of an ideology but largely if not entirely sympathetic explorers of the possibilities and values of that ideology, which were built into the ethical and religious traditions that long predate the nineteenth century.

11. John Henry Newman, *Apologia pro sua vita*, ed. Martin J. Svaglic (Oxford: Clarendon Press, 1967), p. 18.

12. "our race's progress and perfectibility is a dream, because Revelation contradicts it, whatever may be plausibly argued in its behalf by scientific inquirers." John Henry Newman, *The Idea of a University* (New York: Holt, Rinehart and Winston, 1962), p. 54.

13. John Henry Newman, *Apologia pro vita sua*, ed. David J. DeLaura (New York: W. W. Norton and Co., 1968), p. 187.

14. Francis Darwin and A. Seward, *More Letters of Charles Darwin* (London: John Murray, 1903), 1:370–71.

15. Darwin, *The Autobiography of Charles Darwin*, p. 140.

16. John Stuart Mill, *Autobiography and Other Writings*, ed. Jack Stillinger (Boston: Houghton Mifflin Co, 1969), p. 20.

17. Darwin and Seward, *More Letters of Charles Darwin*, 2:41.

18. It is fair to note that Darwin did write, "I am inclined to agree with Francis Galton in believing that education and environment produce only a small effect on the mind of any one, and that most of our qualities are innate" (*The Autobiography of Charles Darwin*, p. 43). But as the quotation from the letters above shows, this view ran counter to Darwin's almost instinctively held original position.

19. Charles Darwin, *The Descent of Man and Selection in Relation to Sex* (Princeton, N.J.: Princeton University Press, 1981), 2:328–29.

20. T. H. Huxley, "The Educational Value of the Natural History Sciences," in *Science and Education* (New York: Appleton and Co, 1894), p. 45. "The vast results obtained by Science are won by no mystical faculties, by no mental processes other than those which are practised by every one of us, in the humblest and meanest affairs of life" (p. 45).

21. Samuel Smiles, *Self Help: With Illustrations of Character and Conduct* (London: John Murray, 1860), p. 71.

22. The problems of realism in Victorian fiction often had to do precisely with the question of heroism. George Eliot, to take the most obvious example of a writer directly confronting the issue, abjured notions of conventional heroism—operatic, stage heroism—but created a series of figures, from Adam Bede to Daniel Deronda, who, within whatever scope,

act out a kind of "realistic" heroism. For an interesting exploration of the possible contradictions, see U. C. Knoepflmacher's *George Eliot's Early Novels* (Berkeley and Los Angeles: University of California Press, 1968). See in particular pp. 24–37, "The Divided Aesthetic of 'Realism.'"

23. Stephen Jay Gould, *Time's Arrow, Time's Cycle: Myth and Metaphor in the Discovery of Geological Time* (Cambridge, Mass.: Harvard University Press, 1970), p. 120.

24. In a paper delivered at a conference on literature and science at Indiana University in February 1991, John Woodcock surveyed several modern scientific autobiographies, and what he found there confirmed strongly my argument that there is a close connection between a commitment to science and scientific method and a tendency if not really to diminish the self (the vanity of some famous scientists is rather hard to miss), at least to represent the self as insignificant. Woodcock identifies several characteristics of these works: their titles evade the idea that the books are about themselves (e.g., *Memoirs of a Thinking Radish, The Periodic Table, Haphazard Reality, Enigmas of Chance*). Their rhetoric implies that they are written by modest men whose selves are invisible to them, that life is governed by chance events, that individuality is illusory and agency beside the point. They do not present themselves as centers of their own narratives. Each scientist writes about his own life as if it were not personally interesting, as if it were conducted without his own agency, and fell out primarily by chance. However different the autobiographies became, they all tended to begin by effacing the self writing.

25. Charles Darwin, *Autobiography*, ed. Nora Barlow (New York: W. W. Norton, 1958), p. 21.

26. Ibid.

27. Mill, *Autobiography and Other Writings*, p. 2.

28. Anthony Trollope, *An Autobiography* (London: Oxford University Press, 1961), p. 1.

29. Ibid.

30. Mill, *Autobiography and Other Writings*, p. 132.

31. Loesberg *(Mill, Newman, and the Reading of Victorian Prose)* has impressively considered the book in its relation to Mill's theory of consciousness. This reading, I believe, works much more effectively than ones that focus on the only really interesting, or perhaps, dramatic, parts—the education and mental crisis. Olney's very sensitive reading of Mill distinguishes "the fleshless picture" Mill gives us from what might have been the case in life. But, he says, the narrative "bears the clear stamp of the school logic so carefully impressed on Mill's mind by his father some fifty years before he came to write it down. It is all very like a syllogism on private human experience—but no less interesting or significant for that" (*Metaphors of Self: The Meaning of Autobiography*, p. 259).

32. Trollope, *An Autobiography*, p. 1.

33. See Donald E. Fleming, "Charles Darwin, the Anaesthetic Man," *Victorian Studies* 4 (1959): 219–36, for a still useful discussion of Mill's crisis in relation to Darwin's.

34. Mill, *Autobiography and Other Writings*, p. 266.

35. This passage was canceled in the original manuscript, but the connection of the third stage with Harriet Taylor is in the text that survives. See the footnote on p. 137 of the *Autobiography*.

36. One of the great ironies of the story of the relation between Mill and Harriet Taylor is that the saint of rationalism had dangerously romantic tendencies and Taylor seems to have been rigorously rationalistic. E. A. Hayek approvingly quotes Knut Hagberg's observations on the ways Taylor influenced Mill's thought: "It is obvious that it was this woman who made him into a Radical rationalist. She has given the impress of her personality to all his greater works; to all her opinions Mill has given the form of philosophic maxims. But even in his most arid

reflexions on woman's similarity with man and on the nature of Logic, Mill is in reality a romantic" (Mill/Taylor Collection, XXVIII/143, 144, British Library of Political and Economic Science, London, quoted in E. A. Hayek, *John Stuart Mill and Harriet Taylor: Their Friendship and Subsequent Marriage* [London: Routledge and Kegan Paul, 1951], p. 18).

37. Mill, *Autobiography and Other Writings*, p. 3.
38. Ibid., p. 113.
39. William Whewell, *History of the Inductive Sciences from the Earliest to the Present Time* (New York: Appleton and Co., 1865), 1:63.
40. Ibid.
41. Trollope, *An Autobiography*, p. 29.
42. Perhaps the most stunningly grudging autobiography among the Victorians is T. H. Huxley's. His autobiography is barely a pamphlet, and in a prefatory letter he announces a "profound objection to write about myself" (Charles Darwin and T. H. Huxley, *Autobiographies* [Oxford: Oxford University Press, 1983], p. 100). He recounts austerely some bare facts leading to a final concession that it would be "mock modesty" to claim that he had not been successful in his life; but, he says, he was "driven" to success, and did not achieve it through his "own free will" (p. 109).
43. Ibid., p. 109.
44. Olney, *Metaphors of Self*, p. 184.
45. Evelyn Fox Keller, *Reflections on Gender and Science* (New Haven, Conn.: Yale University Press, 1985), p. 18.
46. Michael Polanyi, *Personal Knowledge: Towards a Post-Critical Philosophy* (New York: Harper and Row, 1964), p. 5.
47. Thomas Nagel, *The View from Nowhere* (New York: Oxford University Press, 1986), pp. 5, 9.
48. Darwin, *Autobiography*, p. 119.
49. David Kohn and Sandra Herbert, eds,. *Charles Darwin's Notebooks, 1836–1844* (Ithaca, N.Y.: Cornell University Press, 1987), p. 8.
50. Darwin, *Autobiography*, p. 140.
51. Mill, *Autobiography and Other Writings*, p. 3.
52. Trollope, *An Autobiography*, p. 1.
53. See Walter Kendrick, *The Novel-Machine: The Theory and Fiction of Anthony Trollope* (Baltimore: Johns Hopkins University Press, 1980).
54. Darwin, *Autobiography*, p. 67.
55. Ibid., p. 28.

CHAPTER FIVE

1. Richard Feynman's *Surely You're Joking, Mr. Feynman: Adventures of a Curious Character* (New York: W. W. Norton, 1985) is obviously, right from the title, not the sort of retiring, self-denigrating autobiography I have been describing as scientific. (I would suggest, however, that the pun of the title, which seems to be referring to Feynman himself as a "curious character," directs attention to the curiousness of the "adventures" more than to Feynman.) Nevertheless, while Feynman's flamboyance and playfulness are part of the anecdotal sequence of the book, it is noteworthy that the book is first of all a collection of stories gathered from Feynman by Ralph Leighton, who actually did the writing and whose interest in Feynman governs the book more than Feynman's interest in Feynman. While the anecdotes certainly provide a sense

of Feynman's intellectual daring and disregard of conventional decorum, they are almost scrupulously quiet about his private life. If they imply a Feynman of enormous stature, they are nevertheless far more interested in physics and the world (language, culture, technology, philosophy) than in Feynman.

2. See Alfred Russel Wallace, *My Life* (London: Chapman and Hall, 1905), 2 vols.

3. James Marchant, *Alfred Russel Wallace: Letters and Reminiscences* (New York: Harper & Bros, 1916), p. 7.

4. D. W. Forrest, *Francis Galton: The Life and Work of a Victorian Genius* (London: Paul Elek, 1974), p. ix.

5. Wallace, *My Life*, 2:17.

6. Ruth Cowan Schwartz, introduction to *English Men of Science: Their Nature and Nurture*, by Francis Galton (London: Frank Cass and Co., 1970), p. ix.

7. George Eliot, *Essays of George Eliot*, ed. Thomas Pinney (New York: Columbia University Press, 1963), p. 413.

8. John Stuart Mill, *Autobiography and Other Writings*, ed. Jack Stillinger (Boston: Houghton Mifflin Co, 1969), p. 101.

9. Ibid., p. 102.

10. Wallace, *My Life*, 1:89.

11. In a turn reminiscent of Darwin's argument that the shape of his nose determined his career, Wallace writes, "Had my father been a moderately rich man and had supplied me with a good wardrobe and ample pocket money; had my brother obtained a partnership in a populous town or city, or had established himself in his profession, I might never have turned to nature as the solace and enjoyment of solitary hours, my whole life would have been differently shaped, and though I should, no doubt, have given some attentions to science, it seems very unlikely that I should have undertaken what at the time seemed rather a wild scheme" (Ibid., 1:197).

12. In "The Progress of Science," Huxley discusses three unproveable postulates from which science starts: "the objective existence of the physical world," "the universality of the laws of causation," "the rules or so-called 'laws of Nature,' by which the relation of phenomena is truly defined, is true for all time" (in *Methods and Results* [London: MacMillan, 1893], pp. 60–61).

13. John Tyndall, "The Belfast Address," in *Fragments of Science* (New York: Appleton and Co., 1899), 1:181.

14. Note his famous comment in his lecture "The Physical Basis of Life": "all vital action may . . . be said to be the result of the molecular forces of the protoplasm which displays it. And if so, it must be true, in the same sense and to the same extent, that the thoughts to which I am now giving utterance, and your thoughts regarding them, are the expression of molecular change in that matter of life which is the source of our other vital phaenomena" (in *Method and Results*, p. 154).

15. John Tyndall, "Scientific Materialism," in *Fragments of Science*, 2:88, 86.

16. Huxley, "On the Hypothesis That Animals Are Automata, and Its History," in *Methods and Results*, p. 144.

17. Karl Pearson, *Life and Letters of Francis Galton* (Cambridge: Cambridge University Press, 1914), 1:61.

18. T. H. Huxley, "On Our Knowledge of the Causes of the Phenomena of Organic Nature," in *Darwiniana: Essays* (London: MacMillan and Co., 1893), p. 361.

19. Francis Galton, *Memories of My Life* (London: Methuen and Co, 1908), p. 10.

20. Note this example, as characteristic, on the use of the Welsh language:

Another indication of the wide use of the Welsh language and of the general education of the people, is the fact that the British and Foreign Bible Society now sell annually about 18,000 Bibles, 22,0000 Testaments, and 10,0000 special portions (as the Psalms, the Gospels, etc.); while the total sale of the Welsh scriptures during the last century has been 3$^{1}/_{2}$ millions. Considering that the total population of Wales is only about 1$^{1}/_{2}$ millions, that two counties, Pembrokeshire and Radnorshire, do not speak Welsh, and that the great seaports and the mining districts contain large numbers of English and foreign workmen, we have ample proof that the Welsh are still a distinct nation with a peculiar language, literature, and history, and that the claim which they are making for home rule, along with the other great subdivisions of the British Islands, is thoroughly justified. (Wallace, *My Life*, 1 : 168 – 69)

21. Galton, *Memories of My Life*, p. 141.
22. See Jack Morrell and Arnold Thackray, eds., *Gentleman of Science: Early Years of the British Association for the Advancement of Science* (Oxford: Clarendon Press, 1981): "To maintain a serious, academic tone it was thought prudent to ban women from the working Sections of the Association. Mary Somerville's anticipated absence from the meeting was important to Buckland when he went on to question whether ladies ought to attend at all: 'Mrs. Somerville's opinion as confirmed by her action is clearly in the negative'" (p. 150).

CHAPTER SIX

1. For important discussions of women's relation to science see Ludmilla Jordanova, *Sexual Visions* (Madison: University of Wisconsin Press, 1989); Barbara Gates and Ann B. Shteir, *Natural Eloquence* (Madison: University of Wisconsin Press, 1997); and Londa Scheibinger, *The Mind Has No Sex: Women in the Origins of Modern Science* (Cambridge, Mass.: Harvard University Press, 1989). Scheibinger neatly, if ironically, summarizes European culture's continuing refusal to allow women into science in this way: "In the seventeenth century, the English natural philosopher Margaret Cavendish spoke for many when she wrote that women's brains are simply too 'cold' and 'soft' to sustain rigorous thought. The alleged defect in women's minds has changed over time: in the late eighteenth century, the female cranial cavity was supposed to be too small to hold powerful brains; in the late nineteenth century, the exercise of women's brains was said to shrivel their ovaries. In our own century, peculiarities in the right hemisphere supposedly make women unable to visualize spatial relations" (p. 2). In this chapter, I will not be attending to these widespread cultural objections but attempting to view the prejudice in the light of the dying-to-know pattern.
2. Francis Galton, *Memories of My Life* (London: Methuen and Co, 1908), p. 293.
3. T. H. Huxley, "On the Advisableness of Improving Natural Knowledge," in *Methods and Results* (London: MacMillan, 1893), p. 40.
4. John Herschel, *Preliminary Discourse on the Study of Natural Philosophy* (Chicago: University of Chicago Press, 1987), pp. 67 – 84.
5. Deborah Epstein Nord, *The Apprenticeship of Beatrice Webb* (Amherst: University of Massachusetts Press, 1985), p. 58.
6. Linda Peterson, *Traditions of Victorian Women's Autobiography* (Charlottesville: University of Virginia Press, 1999), p. 2.
7. Florence Nightingale, *Cassandra* (Old Westbury, N.Y.: Feminist Press, 1979), p. 33.
8. Of course, men could and did take advantage of the freedom of illness, and Darwin is the most famous of these creative invalids. Janet Browne discusses this in "I Could Have

Retched All Night," in *Science Incarnate: Historical Embodiment of Natural Knowledge*, ed. Christopher Lawrence and Steven Shapin (Chicago: University of Chicago Press, 1998), pp. 240–87.

9. Nightingale, *Cassandra*, p. 40.

10. Linda Peterson treats *Aurora Leigh* at length in the light of traditions of women poets' autobiographies. Her conclusions point not to the dramatized renunciation inside the narrative, but to the way in which Aurora violates a feminine code against "*auto*biography." "Yet," writes Peterson, Aurora "characterizes her act not as self-violation but as self-improvement." But Aurora's expression of self is intended to "extend beyond the self and encompasses the civic body. . . . Except perhaps for the final section of Harriet Martineau's *Autobiography*, with its prophetic 'Last View of the World,' there is no autobiographical conclusion quite so engaged with the public good or quite so insistent about the need for the woman writer to move out of the private and into the civic realm" (*Traditions of Victorian Women's Autobiography*, p. 145). This focus on the future, on the condition of the country and the race rather than on a personal fate, is another form of the self-effacement that marks the moral direction of scientific epistemology.

11. *Harriet Martineau's Autobiography*, ed. Maria Weston Chapman (Boston: John Osgood, 1877), 1:33.

12. For a valuable discussion showing the Comtean hermeneutic structure of Martineau's *Autobiography*, see Linda Peterson, *Traditions of Victorian Women's Autobiography*.

13. Chapman, ed., *Harriet Martineau's Autobiography*, 1:59.

14. Beatrice Webb, *My Apprenticeship* (Harmondsworth, Middlesex: Penguin Books, 1938), p. 78.

15. As I have suggested in chapter 1, feminists often take the universalizing of Western epistemology as an effacement of the underlying masculine perspective. Evelyn Fox Keller has traced this way of thinking back to Plato and his homosocial community; Jane Duran attempts to develop a feminist epistemology with closer connections to the reality of the body by using recent "connectionist" thought that in effect makes philosophy a function of the structures of the brain. See her *Toward a Feminist Epistemology* (Savage, Md.: Rowman and Littlefield, 1991).

16. Chapman, ed., *Harriet Martineau's Autobiography*, 2:29.

17. Mary Somerville, *Mechanism of the Heavens* (London: John Murray, 1831), p. vi.

18. Mary Somerville, *The Connection of the Physical Sciences* (New York: Harper Brothers, 1853), p. 1.

19. Ibid., p. 390.

20. For an important discussion of the Unitarian influences on Martineau's life and thought, see R. K. Webb, *Harriet Martineau: A Radical Victorian* (London: Heineman, 1960), esp. pp. 77–89.

21. Chapman, ed., *Harriet Martineau's Autobiography*, 2:28.

22. Webb, *My Apprenticeship*, p. 111.

23. Chapman, ed., *Harriet Martineau's Autobiography*, 2:23.

24. Reginia Gagnier, *Subjectivities: A History of Self-Representation in England, 1832–1920* (New York: Oxford University Press, 1991), p. 271.

25. Webb, *My Apprenticeship*, p. 469.

26. Deborah Nord argues that the coherence and clarity of *My Apprenticeship* belies what Webb actually felt. "On the most profound level of experience and belief . . . the writing of this personal work reinforced Webb's sense of having failed to solve the problems of self and the world. The autobiographical text concludes in absolute affirmation and ringing optimism; the act of writing it ended in ultimate confusion and uncertainty" (*The Apprenticeship of Beatrice Webb*, p. 237). The "difference between what she wrote in the autobiography and what she

wrote in the diary" (ibid.) suggests that the narrative's acquiescence in the dying-to-know pattern was inadequate to the problems either of self or of society. The narrative affirmation through science does not work personally for its writer.

27. Estelle Jelinek, *Women's Autobiography: Essays in Criticism* (Bloomington: Indiana University Press, 1980), pp. 17–20.

28. *Personal Recollections from Early Life to Old Age of Mary Somerville with selections from her correspondence by her daughter, Martha Somerville* (Boston: Roberts Brothers, 1874), p. 54.

29. In Kitty Muggeridge and Ruth Adam, *Beatrice Webb: A Life* (London: Secker and Warburg, 1967), Muggeridge begins with a dryly satiric portrait of Beatrice and Sidney Webb through the eyes of her young niece, but from the perspective of the highly cultivated and fashionable. The portrait ends not unsympathetically. She notes how her aunt's visits were "astringent rather than pleasurable and no warm intimacy developed from them, only a kind of cool affection, as though tenderness were against her principles" (p. 15). The eccentric aunt with her frugal meals and her vulgar seeming husband (who sweat a lot) carried her passionate asceticism into her personal life, but clearly inspired in Kitty Muggeridge both fascination and affection. As Margaret Cole notes, Webb "did regard excess interest in, or excess time occupied in consuming food or drink as sinful waste" (Margaret Cole, *Beatrice Webb* [New York: Harcourt, Brace, 1946], p. 39).

30. Webb, *My Apprenticeship*, p. 15.

31. Peterson, *Traditions of Victorian Women's Autobiography*, p. 138.

32. Deirdre David, *Intellectual Women* (Ithaca. N.Y.: Cornell University Press, 1987), p. 30.

33. Somerville, *Personal Recollections from Early Life to Old Age*, p. 347.

34. Ibid., p. 141.

35. Alice Jenkins, in her essay, "Writing the Self and Writing Science: Mary Somerville As Autobiographer," quotes a passage that Somerville's daughter omits from her mother's reminiscences: "I have perseverance and intelligence but no genius, that spark from heaven is not granted to the sex, we are of the earth, earthy, whether higher powers may be allotted to us in another existence God knows, original genius in science at least is hopeless in this" (in *Rethinking Victorian Culture*, ed. Juliet John and Alice Jenkins [New York: St. Martin's Press, 2000], p. 163). In her self-conception, that is, Somerville accepts the central implication of the dying-to-know narrative—detachment from the physical, contingent world is a condition of knowledge. My thanks to Suzy Anger for calling my attention to this essay, which appeared after this chapter was completed.

36. Mary Somerville, while she remained faithful to her domestic responsibilities all the time she was doing important scientific work, strongly supported equality for women. In the *Recollections,* she complains about the fact that "British laws are adverse for women," and her daughter proudly prints a letter from Mill, with whom Mary Somerville had corresponded, about the petition to Parliament for the extension of suffrage to women, noting her thanks to him for writing *On the Subjection of Women* (*Personal Recollections from Early Life to Old Age*, pp. 344–45).

37. Keller, *Toward a Feminist Epistemology*, p. 12.

38. For major discussions of the limits of deconstructive critique and the necessity for a new epistemology that would disrupt conventional ways of thinking and knowing yet both recognizes a real (or many reals) and denies universality, see Haraway's two essays: "A Manifesto for Cyborgs: Science, Technology, and Socialist Feminism in the 1980's," *Socialist Review* 15 (March 1985): 65–142; and the essay discussed in chapter 1 above, "Situated Knowledges: the Science Question in Feminism and the Privilege of Partial Perspective."

39. Somerville, *Personal Recollections from Early Life to Old Age*, p. 129.

CHAPTER SEVEN

1. In *The Secular Pilgrims of Victorian Fiction* (Cambridge: Cambridge University Press, 1982), Barry Qualls reads the Victorian novel into the context of Bunyan, among others in religious and biblical traditions. Bunyan's pilgrim, like the protagonists of the great Victorian novels, seeks to answer the question: "What shall I do?" (see p. 6). Finding out what to do depends on the protagonists' discovery of what the case is—like the pilgrim, they are all dying to know.

2. As Ronald Thomas argues, the "sensation novel" led inevitably to the detective novel, both preoccupied with "establishing the identity of a shadowy personality . . . with explaining a secret from the past." The detective novel, in fact, "with its validation of a professional hero," uses the tools of science in this project of discovery (Thomas, "Wilkie Collins and the Sensation Novel," in *The Columbia History of the Novel*, ed. John Richetti [New York: Columbia University Press, 1994], p. 501).

3. See Audrey Jaffe, *Vanishing Point: Dickens, Narrative, and the Subject of Omniscience* (Berkeley and Los Angeles: University of California Press, 1991). The omniscient narrator actually occupies that nowhere. Jaffe points out how important it is for the inquirer to remain invisible, to see without being seen. The reader, of course, is invisible inside the narrative.

4. See John Kucich, *The Power of Lies: Transgression in Victorian Fiction* (Ithaca, N.Y.: Cornell University Press, 1994), p. 9; John Morley, *On Compromise* (London: Chapman and Hall, 1874), p. 3.

5. Morley, *On Compromise*, p. 34.

6. For a discussion of this realistic convention, see my *Realistic Imagination* (Chicago: University of Chicago Press, 1981), chap. 1.

7. John B. Farrell argues that Dickens's novels cannot be pinned down ideologically, but that they are actively engaged in shaping and reshaping. Farrell tries to show that, as Charles Taylor puts it, Foucault's arguments are flawed because he "cannot envisage liberating transformations *within* a regime." In Dickens's novels, Farrell says, the details of "micro-practices," social and personal interactions, dramatize a transforming "structure of feeling" (Raymond Williams's phrase) that cannot be identified as a hardened ideological regime: "As Dickens grew more penetrating and somber about the dominance and discipline of social institutions, even those that operated visibly in the Victorian state, his novelistic texts became more resourceful both in thickening the textures of his representations, and in locating the ingenerate grain of sociality in the practices of his characters and the performance of his art" ("The Partner's Tale: Dickens and *Our Mutual Friend*," *ELH* 66 [1999]: 773, 793).

8. *Dickens: The Critical Heritage*, ed. Philip Collins (London: Routledge, 1971), p.576.

9. Edgar Johnson, *Charles Dickens: His Tragedy and Triumph* (New York: Simon and Schuster, 1952), 2:132.

10. For a discussion of Dickens's attitudes toward science, see my *Darwin and the Novelists: Patterns of Science in Victorian Fiction* (Cambridge, Mass.: Harvard University Press, 1988), pp. 124–26.

11. John Kucich, *Repression in Victorian Fiction* (Berkeley and Los Angeles: University of California Press, 1987), p. 273.

12. Peter Ackroyd, *Dickens* (London: Sinclair-Stevenson, 1990), p. 955.

13. Charles Dickens, *Our Mutual Friend* (Harmondsworth, Middlesex: Penguin, 1971). Subsequent references are to this edition and appear in the text.

14. Patrick McCarthy, "Designs in Disorder: The Language of Death in *Our Mutual Friend*," *Dickens Studies Annual*, 17 (1989): 142.

15. John Ruskin, *Fiction, Fair and Foul, The Complete Works of John Ruskin*, ed. E. T. Cook and Alexander Wedderburn (London: George Allen, 1908), 34:271–72.

16. See *Dickens' Working Notes for His Novels*, ed. Harry Stone (Chicago: University of Chicago Press, 1987), p. 332.

17. Farrell argues against what he calls "a dead end to which Dickens is likely to be consigned even by astute critics [who] tend to suggest that the best Dickens can do is identify social problems and then haul off his heroes and heroines to some protected enclave where they may live a fulfilled but purely atomistic or eccentric life" ("The Partner's Tale," p. 762). I would argue that the strategy of disentangling the self from the deluding connections that society requires is part of the epistemological strategy of the book; in this respect, I disagree with Farrell's reading. But it is certainly true that in *Our Mutual Friend*, in more than any other Dickens novel, the narrative struggles to get beyond the society/individual dualism by means of its theme of resurrection. In his discussion of Jenny Wren's call, "Come up and be dead," Farrell claims that it is really an invitation to life. But surely it is an odd "life." It is like the epistemological strategy of self-obliteration in that one achieves fullness of vision as one is dissociated from ordinary community.

18. I don't as a layman want to invoke the "Heisenberg uncertainty principle" as would be expected here. I will only suggest that the conditions of producing an experiment in which the act of observation does not have some influence on what is being observed is extraordinarily difficult, and turning the idea of it into narrative, as *Our Mutual Friend* does, already suggests that difficulty.

19. John Herschel, *A Preliminary Discourse on the Study of Natural Philosophy* [1830] (Chicago: University of Chicago Press, 1987), pp. 79–80.

20. Claude Bernard, *An Introduction to the Study of Experimental Medicine* (New York: Dover Publications, 1957), p. 19.

21. Catherine Gallagher, "The Bio-Economics of *Our Mutual Friend*," in *Fragments for a History of the Human Body*, vol. 3, ed. Michel Feher (New York: Zone, 1989), p. 351.

22. Gallagher, "The Bio-Economics of *Our Mutual Friend*," p. 355.

23. Ibid.

24. Peter Brooks, *The Melodramatic Imagination: Balzac, Henry James, Melodrama, and the Mode of Excess* (New Haven, Conn.: Yale University Press, 1976), p. 4.

25. Elizabeth Deeds Ermarth, *Realism and Consensus in the English Novel*, 2d ed. (Edinburgh: University of Edinburgh Press, 1998), p. 210.

26. John Kucich, *Repression in Victorian Fiction* (Berkeley and Los Angeles: University of California Press, 1987)p. 222.

27. Gallagher, "The Bio-Economics of *Our Mutual Friend*."

CHAPTER EIGHT

1. See Herbert's *Victorian Relativity* (Chicago: University of Chicago Press, 2001), for a brilliant unearthing and exposition of the relativist ideas of Victorian intellectuals, from William Hamilton and John Stuart Mill to Herbert Spencer and Karl Pearson.

2. George Eliot, *The Mill on the Floss* (Harmondsworth, Middlesex: Penguin Books, 1979), bk. 4, chap. 1, p. 363.

3. Richard Rorty's pragmatist rejection of this epistemological dilemma leads him to a total commitment to the "contingent." "We should," he says, "drop the idea of truth as out there waiting to be discovered." This does not entail, he claims, the view that there is no truth. He rejects Cartesian dualism because he rejects all foundations and the correspondence theory of truth. Truth, he argues, gets shaped by human uses of human language. But "'The nature of truth' is an unprofitable topic." Rorty's important articulation of pragmatic contingency is most clearly and forcefully made in *Contingency, Irony, and Solidarity* (Cambridge:

Cambridge University Press. 1989). See particularly p. 8, from which this quotation has been taken.

4. F. R. Leavis, *The Great Tradition: A Study of the English Novel* (New York: Doubleday Anchor, 1954). As an appendix, Leavis reprints James's "Daniel Deronda: A Conversation."

5. See Amanda Anderson, *The Cultivation of Detachment: Ethics, Aesthetics, and the Victorian Response to Modernity* (Princeton, N.J.: Princeton University Press, 2001) for a strong representation of the genealogy of detachment and its value in the construction of a creative cosmopolitanism.

6. Dominick LaCapra, "The University in Ruins," *Critical Inquiry* 24 (autumn 1998): 48.

7. Christopher Herbert, *Culture and Anomie: Ethnographic Imagination in the Nineteenth Century* (Chicago: University of Chicago Press, 1989), pp. 154–55.

8. Ibid., p. 174.

9. See David Miller's influential work, *The Novel and the Police* (Berkeley and Los Angeles: University of California Press, 1988).

10. John Rosenberg, *Carlyle and the Burden of History* (Cambridge, Mass.: Harvard University Press, 1985), p. 16.

11. Thomas Carlyle, *Past and Present* (London: Chapman and Hall, 1899), p. 50.

12. For an interesting exploration of the phenomenon and its significance for the development of quantification and its relation to cultural history, see Simon Schaffer, "Astronomers Mark Time: Discipline and the Personal Equation," *Science in Context* 2, no. 1 (spring 1988): 115–46.

13. Theodore Porter, *Trust in Numbers: The Pursuit of Objectivity in Science and Public Life* (Princeton, N.J.: Princeton University Press, 1995).

14. George Eliot, *Middlemarch* (Harmondsworth, Middlesex: Penguin, 1965), p. 894.

15. Alexander Welsh, *George Eliot and Blackmail* (Cambridge, Mass.: Harvard University Press, 1985), p. 306. See all of chap. 14.

16. George Eliot, *Daniel Deronda* (Harmondsworth, Middlesex: Penguin Books, 1986), p. 35. Subsequent references are to this edition and appear in the text.

17. The complexity of her relation to the truth is well elaborated by Rosemarie Bodenheimer, whose analysis of George Eliot's work and letters exposes sympathetically the evasions and disguised interests that self-consciously mark the most solemnly moral writing and the behavior and pronouncements of her protagonists, particularly Deronda. See Bodenheimer's discussion of that book in *The Real Life of Mary Ann Evans, George Eliot: Her Life and Letters* (Ithaca, N.Y.: Cornell University Press, 1994), pp. 183–88, 257–66.

18. George Eliot, *Middlemarch*, p. 846.

19. Gertrude Lenzer, ed., *Auguste Comte and Positivism: The Essential Writings* (Chicago: University of Chicago Press, 1975), p. 213.

20. George Eliot, *Romola* (Oxford: Oxford University Press, 1965), p. 483.

21. Gordon Haight, ed., *The George Eliot Letters* (New Haven, Conn.: Yale University Press, 1955), 6:301–2. See the discussion of this letter and the general problem of the book's treatment of the Jewish theme in Jane Irwin, ed., *George Eliot's Daniel Deronda Notebooks* (Cambridge: Cambridge University Press, 1996), esp. p. xxix, where this passage is quoted.

22. Irwin points out the relation of Leavis's views on Deronda to the fact that when his wife Queenie married him her family "mourned her loss with the ritual *Shivah*" (*George Eliot's Daniel Deronda Notebooks*, p. xl). In a talk delivered at the Sadleir Conference at UCLA in February 2000, Claudia Johnson, in a remarkable analysis of F. R. Leavis's three essays on *Daniel Deronda*, makes a convincing case that in certain respects Leavis's attitude toward the Jewish sections was a result of his being so much like Deronda himself.

23. Haight, ed., *The George Eliot Letters*, 6:301.

24. Ibid., 6:196. The centrality of the Jewish question to the enterprise of *Daniel Deronda* is carefully laid out in Michael Ragussis, *Figures of Conversion: The "Jewish Question" and English National Identity* (Durham, N.C.: Duke University Press, 1995). Ragussis quotes another important letter by George Eliot: "A statesman who shall be nameless has said that I opened to him a vision of Italian life, then of Spanish, and now I have kindled in him a quite new understanding of the Jewish people. This is what I wanted to do—to widen the English vision a little" (p. 266, quoting *Letters*, 6:304).

25. In *Narrating Reality: Austen, Scott, Eliot* (Ithaca, N.Y.: Cornell University Press, 1999), Harry Shaw argues that the narrator of *Daniel Deronda* is partly historicized, remaining partly inside the "history" she is, from narratorial distance, understanding and analyzing.

26. Suzy Anger, "George Eliot's Hermeneutics of Sympathy," unpublished paper.

27. Franco Moretti, *The Way of the World: The* Bildungsroman *in European Culture* (London: Verso Books, 1987), p. 8.

28. John Stuart Mill, *Autobiography and Other Writings*, ed. Jack Stillinger (Boston: Houghton Mifflin Co, 1969), p. 84.

29. See Alan Mintz, *George Eliot and the Novel of Vocation* (Cambridge, Mass.: Harvard University Press, 1978).

30. "Life and Opinions of Milton," in *Essays of George Eliot*, ed. Thomas Pinney (New York: Columbia University Press, 1963), p. 156.

31. James Clifford and George Marcus, eds., *Writing Culture: The Poetics and Politics of Ethnography* (Berkeley and Los Angeles: University of California Press, 1986), p. 8.

32. Gillian Beer, *Open Fields: Science in Cultural Encounter* (Oxford: Clarendon Press, 1996), p. 77.

33. John McClure, *Late Imperial Romance* (London: Verso Press, 1994).

34. Welsh, *George Eliot and Blackmail*, p. 304.

35. Irwin, *George Eliot's* Daniel Deronda *Notebooks*, p. xlii.

36. Amanda Anderson, "George Eliot and the Jewish Question," *Yale Journal of Criticism* 10 (1997): 49.

37. Bernard Semmel, in his *George Eliot and the Politics of National Inheritance* (New York: Oxford University Press, 1994), reads *Daniel Deronda* as the expression of a more rigidly nationalist program than Anderson does, but he also shows that George Eliot was consistently hostile to imperial ventures and the denigration of other races.

38. See Gillian Beer, *Darwin's Plots: Evolutionary Narrative in Darwin, George Eliot, and Nineteenth-Century Fiction* (London: Routledge and Kegan Paul, 1983), pp. 181–209.

39. Bodenheimer, *The Real Life of Mary Ann Evans, George Eliot*, p. 265.

40. George Eliot, *Middlemarch*, p. 194.

41. Ibid., p. 668.

42. George Eliot, *More Leaves from George Eliot's Notebook*, ed. Thomas Pinney, *The Huntington Library Quarterly* 29 (August 1966): 364. Lewes argues at one point that "[i]deas are moving forces only in proportion to their emotional values, or, physiologically expressed, to the intensity and extension of the innervation they excite" (*Problems of Life and Mind*, 3d ser. [Boston: Houghton Mifflin, n.d.], p. 444). The epistemology implicit in the treatment of Mordecai's visionary knowledge is strongly connected to this ostensibly "physiological" theory.

43. George Eliot, "The Future of German Philosophy," in *The Essays of George Eliot*, ed. Thomas Pinney (New York: Columbia University Press, 1965) p. 153.

44. Ibid., p. 166.

45. Michael Ghiselin, *The Triumph of the Darwinian Method* (Chicago: University of Chicago Press, 1969), p. 4.

46. Jonathan Smith, *Fact and Feeling: Baconian Science and the Nineteenth-Century Literary Imagination* (Madison: University of Wisconsin Press, 1994), p. 19.

47. Steven Shapin, *A Social History of Truth: Civility and Science in Seventeenth-Century England* (Chicago: University of Chicago Press, 1994), p. 417.

48. James Clifford, *The Predicament of Culture: Twentieth-Century Ethnography, Literature, and Art* (Cambridge, Mass.: Harvard University Press, 1988), p. 23.

CHAPTER NINE

1. Thomas Hardy, *Tess of the D'Urbervilles* (Oxford: Oxford University Press, 1983), p. 226.

2. For a useful summary discussion of the relation in Hardy's work of the idea of Darwinian evolution and Herman Helmholtz and John Tyndall's exposition of the idea of the conservation of energy and of entropy, see Gillian Beer, "Hardy and Decadence," in *Celebrating Thomas Hardy: Insights and Appreciations,* ed. Charles P. C. Pettit (New York: St. Martin's Press, 1996), pp. 94–96. Beer argues that these ideas fully "enter the temper" of Hardy's "creativity," but that as always he refuses to allow his work to be appropriated to any system, including these systems.

3. At one point, engaging the familiar question of the relation of the ideal to the embodied particular, the narrator writes of Pierston's relation to Avice that "[o]nce the individual had been nothing more to him than the temporary abiding place of the typical or ideal; now his heart showed its bent to a growing fidelity to the specimen, with all her pathetic flaws of detail" (Thomas Hardy, *The Well-Beloved* [Oxford: Oxford University Press, 1986], p. 143). For an important analysis of the significance of the book for questions of art, see J. Hillis Miller's introduction to the New Wessex edition of the novel.

4. Dennis Taylor argues that "perhaps the most credible law of the universe invoked in the novel is simply that which decrees the painful opposition between desire and reality, between instincts and civilization" (p. xxviii); and he concludes, in a passage directly to the point of my argument, that Hardy rejected the letter of the biblical law "that suffering is a training of the soul, a purgatorial refinement, a testing of the spirit" (p. xxxi). Subsequent references to *Jude the Obscure* are to the edition edited by Taylor (London: Penguin Books, 1998) and appear in the text.

5. Jerome Hamilton Buckley, *Season of Youth: The Bildungsroman from Dickens to Golding* (Cambridge, Mass.: Harvard University Press, 1974), treats *Jude the Obscure* as "characteristic." Although he is right to locate the novel within the *Bildungs* tradition, he misses entirely the novel's deliberate undercutting of the patterns that normally lead to a moderately satisfying, comic resolution and accommodation to the conditions of society.

6. Ibid., p. 185.

7. Franco Moretti describes the new narrative circumstances: "in dismantling the continuity between generations . . . the new and destabilizing forces of capitalism impose a hitherto unknown *mobility*. But it is also a yearned for exploration, since the selfsame process gives rise to unexpected hopes, thereby generating an *interiority* not only fuller than before, but also—as Hegel clearly saw, even though he deplored it—perennially dissatisfied and restless" (*The Way of the World: The* Bildungsroman *in European Culture* [London: Verso Books, 1987], p. 4).

8. John Kucich, *The Power of Lies: Transgression in Victorian Fiction* (Ithaca, N.Y.: Cornell University Press, 1994), p. 201.

9. In 1888, Hardy transcribed this passage from an essay by E. A. Ross: "The final blow to the old notion of the *ego* is given by the doctrine of multiple individuality. Science tells of the conscious & the sub-conscious, of the higher nerve centers & the lower, of the double

cerebrum & the wayward ganglia" (quoted in *The Literary Notebooks of Thomas Hardy*, ed. Lennart A. Bjork [New York: New York University Press, 1985], 1:47–48).

10. Hardy's post-Darwinian narrative enacts in its own way the literalization of metaphor that Gillian Beer identifies as a characteristic of Darwin's argument. See Gillian Beer, *Darwin's Plots: Evolutionary Narrative in Darwin, George Eliot, and Nineteenth-Century Fiction* (London: Routledge and Kegan Paul, 1983). Darwin shows that the "family" resemblance among related organisms is a quite literal, material phenomenon; Hardy suggests that the Christian and scientific ideal of transcendence of the body is literally the death of the body (and, therefore, of the spirit as well).

11. Florence Emily Hardy, *The Later Years of Thomas Hardy* (New York: MacMillan and Co, 1930), pp. 13–14.

12. Elaine Scarry, "Work and Body in Hardy and Other Nineteenth-Century Novelists," *Representations* 3 (1983): 90–123.

13. Thomas Hardy, *Under the Greenwood Tree* (Garden City: Dolphin Books, 1965), pp. 32–33.

14. In such passages Hardy opens issues of class far less abstract and more overt than the issues I am considering here, but to which they are significantly related. Jude becomes an invisible man in Christminster society in ways similar to those of Ralph Ellison's protagonist. But the Christminster passages ironize the problem of the mind/body dualism by in effect reconfirming the "realities" of the Marygreen passages. That is, the spirituality is only sustained by distinctions of wealth and class that make it impossible for pure intelligence and spirit to thrive.

15. For discussion of Hardy's refusal to allow his protagonists physical consummation and of his distrust of the flesh, see my essay, "Shaping Hardy's Art: Vision, Class and Sex," in *The Columbia History of the British Novel*, ed. John Richetti (New York: Columbia University Press, 1994), pp. 533–59. In D. H. Lawrence's "Study of Thomas Hardy," Lawrence says about Jude, "there was danger at the outset that he should never become a man, but that he should remain incorporeal, smothered out under his idea of learning" (in *Phoenix: The Posthumous Papers of D. H. Lawrence* [New York: MacMillan and Co, 1936], p. 494).

16. See Alexander Welsh, "The Allegory of Truth in English Fiction," *Victorian Studies* 9 (1965): 21.

17. Patricia Galivan, "Science and Art in *Jude the Obscure*," in *The Novels of Thomas Hardy*, ed. Anne Smith (New York: Barnes and Noble), p. 134.

18. Although Nietzsche is not an "influence" on Hardy, it is relevant to Hardy's position to pursue Nietzsche's arguments about this issue, particularly in *The Genealogy of Morals*, ed. Walter Kaufman (New York: Vintage Books, 1976), essay 3, sections 24 and 25. Hardy did read Nietzsche later in his life, but he had already developed ideas consonant with some of Nietzsche's when he was writing his late novels. Nietzsche, for example, claims that all the problems of philosophy derive from the question, "how can something originate in its opposite, for example, rationality in irrationality, the sentient in the dead, logic in illogic, disinterested contemplation in covetous desire, living for others in egoism, truth in error?" And he claims that language and logic depend "on presuppositions with which nothing in the real world corresponds."

19. Kucich, *The Power of Lies*, p. 201. Kucich has written, "Hardy's affirmation of aesthetic consciousness as the single area in which honesty can survive—and then only by candidly denying, within art, the moral possibility of truth—anticipates a pattern of ethical insularity and self-reflexiveness shared by much modernist art" (ibid.).

20. For a valuable discussion of Hardy's view of science, especially in relation to *Jude the Obscure*, see Galivan, "Science and Art in *Jude the Obscure*," p. 172.

21. Bjork, ed., *The Literary Notebooks of Thomas Hardy*, 1:134.
22. Thomas Hardy, *The Life and Work of Thomas Hardy*, ed. Michael Millgate (Athens: University of Georgia Press, 1985), p. 239.

CHAPTER TEN

1. Auguste Comte, *Auguste Comte and Positivism: the Essential Writings*, ed. Gertrude Lenzer (Chicago: University of Chicago Press, 1975), p. 72.
2. John Stuart Mill, in his *Auguste Comte and Positivism* (Ann Arbor: University of Michigan Press, 1961), first published in book form in 1865, conveniently summarizes the fundamental epistemological position of Comte's Positivism:

> We have no knowledge of anything but Phaenomena; and our knowledge of phaenomena is relative, not absolute. We know not the essence, nor the real mode of production, of any fact, but only its relations to other facts in the way of succession or similitude. . . . The laws of phaenomena are all we know respecting them. Their essential nature, and their ultimate causes, either efficient or final, are unknown and inscrutable to us. (p. 6)

3. Christopher Herbert talks of the rhetoric of "objectivity" as bloodthirsty and sadistic. Another way to register the same phenomenon is that the style is part of that heroic narrative of scientific epistemology I have been discussing.
4. In his remarkable, clumsy, nobly intentioned *Problems of Life and Mind*, G. H. Lewes faced the fact that some of the outlawed metaphysical questions remained important even for empiricists. He ended by ruling out of court only those questions that he called "metempirical," that is, undecidable by empirical investigation.
5. Helen M. Walker, "Karl Pearson," in *The Encyclopedia of Social Science* (London: MacMillan, 1968), 11:502.
6. One of the most powerful expressions of the dying-to-know attitude in its aggressive form is that of Daniel Dennett in *Darwin's Dangerous Idea: Evolution and the Meaning of Life* (New York: Simon and Schuster, 1995). Dennett's brilliant account is deliberately and self-consciously polemical as well. His dismissal of traditional religion and alternatives to the Darwinian thesis is a fundamental aspect of his argument. Early on he affirms confidently:

> There is no future in sacred myth. Why not? Because of our curiosity. Because, as the song reminds us, *we want to know why*. We may have outgrown the song's answer, but we will never outgrow the question. Whatever we hold precious, we cannot protect it from our curiosity, because being who we are, one of the things we deem precious is the truth. Our love of truth is surely a central element in the meaning we find in our lives. In any case, the idea that we might preserve meaning by kidding ourselves is a more pessimistic, more nihilistic idea than I for one can stomach. If that were the best that could be done, I would conclude that nothing mattered after all.

The rhetoric is certainly very modern, but the formulation is a perfect example of the dying-to-know attitude underlying most post-Cartesian Western philosophy.

7. See George Levine, "Darwin and Pain: Why Science Made Shakespeare Nauseating," *Raritan* 15 (fall 1995): 97–114.
8. For a discussion of this quality of "natural selection," see Phillip Barrish, "Accumulat-

ing Variation: Darwin's *On the Origin of Species* and Contemporary Literary and Cultural Theory," *Victorian Studies* 34 (1991): 431–54.

9. Dennett, *Darwin's Dangerous Idea*, pp. 48–52.

10. Charles Darwin, *The Origin of Species* (Harmondsworth, Middlesex: Penguin Books, 1968), p. 129.

11. Charles Darwin, *The Descent of Man and Selection in Relation to Sex* (Princeton, N.J.: Princeton University Press, 1981), pp. 71–72.

12. Karl Pearson, *The Grammar of Science* (London: Walter Scott, 1892), p. 72. There were three editions of *The Grammar of Science* and a reissue, primarily of the first edition of 1892 in the Everyman Library. Subsequent references are from the first edition and appear in the text, although additions and some alterations in the later editions of 1900 and 1911 are often significant.

13. I am grateful to Professor Theodore Porter, who, in a talk at a science and culture workshop at UCLA in April 1997, first called my attention to the possible connection between these two works.

14. In his brilliant discussion of Pearson's "relativity," Christopher Herbert also makes the connection with Newman, whom he reads as an active developer of relativist thinking, and the title of whose major work on the subject, *The Grammar of Assent*, quite obviously influenced Pearson's title. The grounds for reading Newman as within an English skeptical/empirical tradition pointing back to Hume was formulated impressively in J. M. Cameron, "The Night Battle: Newman and Empiricism," *Victorian Studies* 4 (1960): 99–117.

15. Quoted in Daniel J. Kevles, *In the Name of Eugenics: Genetics and the Uses of Human Diversity* (New York: Columbia University Press, 1985), p. 29.

16. Karl Pearson, *The Ethic of Freethought* (London: Adam and Charles Black, 1901), p. 302.

17. Ibid.

18. See Bernard Semmel, *Imperialism and Social Reform: English Social-Imperial Thought, 1895–1914* (Cambridge, Mass.: Harvard University Press, 1960), pp. 30–52.

19. Pearson, *The Grammar of Science*, p. 433.

20. Pearson, *The Ethic of Freethought*, p. x.

21. In a review of the material in the second edition of *The Grammar of Science*, C. S. Pierce attacks Pearson's idea that science, like all other human activities, can be justified by its role in "strengthening social stability." To this, Pierce responds that this is not in fact what drives scientists to their work, that "the man of science has received a deep impression of the majesty of truth, as to which, sooner or later, every knee must bow" ("Pearson's Grammar of Science. Annotations on the First Three Chapters," *Popular Science Monthly* 18 [1901]: 297).

22. E. S. Pearson, *Karl Pearson: An Appreciation of Some Aspects of His Life and Work* (Cambridge: Cambridge University Press, 1938), p. 2.

23. Karl Pearson [Loki, pseud.], *The New Werther* (London: Kegan Paul, 1880), unnumbered prefatory page. Once again I wish to acknowledge here my debt to Theodore Porter, who was not only the first to alert me to the existence of this novel, but who troubled to photocopy his copy for my use.

24. In particular, Theodore Porter, in an as-yet unpublished work, has seen the connection between Pearson's positivism and statistics and his romanticism, and has emphasized qualities important to my argument: Pearson's anti-individualism and commitment to *Entsagen*. See also Bernard J. Norton, "Karl Pearson and Statistics: The Social Origins of Scientific Innovation," *Social Studies of Science* 8 (1978): 3–34.

25. Pearson [Loki, pseud.], *The New Werther*, p. 6.

26. One of Pearson's most important popular introductions to statistics is called "The

Chances of Death." The essay takes the subject "death" and literally discusses the chances of people for dying. Statistics lie behind the way in which insurance is priced, although this is not Pearson's task here. He tries to demonstrate that "chance" itself is not chancy, that in fact, if you take enough instances you can show that events like the throwing of the dice, the pulling of cards, or dying (which he also discusses in terms of the folk tradition of the dance of death) are actually quite ordered and predictable. Death "falls under the general laws of frequency" (Karl Pearson, "The Chances of Death," in *The Chances of Death and Other Studies in Education* [Cambridge: Edward Arnold, 1897], 1:21). And ostensibly random and frighteningly meaningless mortality (for individuals) is controlled and ordered by statistical procedures. Pearson does not minimize the human significance of death in his conclusion. He concedes throughout that the predictability applies to the group and not to the individual, so that everyone still feels subject to random and meaningless termination. The impersonality that leaves the individual still subject to chance disappears in the larger whole, in the species, whose meaning and value seem to be for Pearson a substitute for the consolations of religion.

27. *American Heritage Dictionary*, 1979 edition, s.v. "Loki."

28. I am grateful to Carol Clover of the University of California, Berkeley, for suggesting to me that Loki's name would likely have had broad cultural significance and for leading me to some of that mischievous god's adventures.

29. Karl Pearson, *The Trinity: A Nineteenth Century Passion-Play; The Son; or, Victory of Love* (Cambridge: E. Johnson, 1884), p. vi.

CHAPTER ELEVEN

1. The pervasiveness of paradox was an inheritance, in part, from Darwin's theory. A. Dwight Culler argues that this theory is "the dramatic reversal of otherdox thinking" by which an adaptation that looks to be intentional is shown to be the product of absence of intention. Culler traces a tradition of reversals of this sort back to Hume, in whom he finds a reversal more radical than Darwin's: an assertion "that it is not design which implies a designer but the appearance of design which gives rise in our minds to the conception of a designer." Such reverse thinking privileges irony and paradox as the forms of detachment that allow for an adequate perspective on knowledge. See Culler, "The Darwinian Revolution and Literary Form," in *The Art of Victorian Prose,* ed. George Levine and William A. Madden (New York: Oxford University Press, 1968), pp. 228, 230. See also Amanda Anderson's discussion of the privileging of irony as a mode of detachment in contemporary theory.

2. Charles Darwin, *The Autobiography of Charles Darwin, 1809–1882,* ed. Nora Barlow (New York: W. W. Norton, 1958), p. 93.

3. See George Gordon Douglass Campbell, duke of Argyll, *The Reign of Law* (London: A. Strahan, 1868).

4. A. J. Balfour, *The Foundations of Belief: Notes Introductory to the Study of Theology* (New York and London: Longmans, Green, 1895). Balfour's brilliant exploitation of the strategies of skepticism, which he had elaborated in a previous book, *In Defense of Philosophic Doubt,* left open the way to a reassertion of religious belief against what he took to be the metaphysical incoherence of scientific naturalism. For an excellent discussion, see Bernard Lightman, "'Fighting with Death,' Balfour, Scientific Naturalism, and Thomas Henry Huxley's Final Battle," in *Thomas Henry Huxley's Place in Science and Letters: Centenary Essays,* ed. Alan Barr (Athens: University of Georgia Press, 1997), pp. 323–50.

5. Friedrich Nietzsche, *On the Genealogy of Morals,* trans. and ed. Walter Kaufmann (New York: Vintage Books, 1989), p. 153.

6. Ibid., p. 155.

7. Attacking scientists for their failure to develop an adequate philosophy of science, Balfour ironically notes that "[i]t seems occasionally to have occurred to them that if their theory of knowledge were adequate, 'experimental reasoning,' as Hume called it, was in a very parlous state; and that, on the merits, nothing less deserved to be held with a positive conviction than what some of them are wont to describe as 'positive' knowledge. The self-satisfied dogmatism which is so convenient, and, indeed, so necessary a habit in the daily routine of life, has resumed its sway" (*The Foundations of Belief*, p. 97).

8. Ibid.

9. Karl Pearson, "Politics and Science," in *The Chances of Death and Other Studies in Evolution* (Cambridge: Cambridge University Press, 1897), 1:147.

10. Karl Pearson, "Reaction: A Criticism of Mr. Balfour's Attack on Rationalism," in ibid., 1:190.

11. Walter Pater, *The Renaissance: Studies in Art and Poetry* (Berkeley and Los Angeles: University of California Press, 1980), p. xix.

12. Karl Pearson, *The Grammar of Science* (London: Walter Scott, 1892), p. 7.

13. See Billie Andrew Inman, "The Intellectual Context of Walter Pater's Conclusion," reprinted in Harold Bloom, ed., *Walter Pater* (New York: Chelsea House, 1985), 131–50; Peter Alan Dale, *The Victorian Critic and the Idea of History* (Cambridge, Mass.: Harvard University Press, 1977), particularly pp. 173–85; and Jonathan Loesberg, *Aestheticism and Deconstruction: Pater, Derrida, DeMan* (Princeton, N.J.: Princeton University Press, 1991), particularly pp 18–23. Dale makes an interesting distinction between Pater's aesthetic formulation of the empiricist tradition and the positivist formulation (p. 181). Loesberg subtly analyzes the potential contradiction in Pater's empiricist commitment to the pervasiveness of flux and his "desire to hold onto life by abstracting sensation from that flux" (p. 19).

14. Dale, *The Victorian Critic and the Idea of History*, p. 179.

15. Walter Pater, *Appreciations* (London: Macmillan & Co., 1922), p. 17.

16. Ibid., p. 66.

17. Charles Darwin, *The Origin of Species* (Harmondsworth, Middlesex: Penguin Books, 1959), p. 107.

18. For a full discussion of the complications of Pater's historicism, of his use of Darwinian literature, of his commitment to the possibility of ascertaining truth despite an ostensible relativism, see Carolyn Williams, *Transfigured World: Walter Pater's Aesthetic Historicism* (Ithaca, N.Y.: Cornell University Press, 1989), pp. 49–57.

19. Walter Pater, *Plato and Platonism* (London: MacMillan and Co., 1907), pp. 19–21.

20. Pearson, *The Grammar of Science*, p. 68.

21. Pater, *The Renaissance*, p. 187.

22. The "crisis of representation" in ethnography is clearly related to the epistemological crisis implicit in Pater's and Pearson's versions of empiricism. The authority of any cultural representation has traditionally depended on transparence—the clean record of what actually happens in a culture, which depends on a fundamental realism. James Clifford and George Marcus have in a series of volumes explored the complexities of representation and power in anthropological writing, but a central part of the phenomenon they analyze is the inescapable recognition of the impossibility of transparence, the necessity to understand the conditions of the observer and his relation to the observed. See, among others, James Clifford and George E. Marcus, *Writing Culture: The Poetics and Politics of Ethnography* (Berkeley and Los Angeles: University of California Press, 1986); James Clifford, *The Predicament of Culture: Twentieth Century Ethnography, Literature, and Art* (Cambridge, Mass.: Harvard University Press, 1988); and George Marcus and Michael M. J. Fischer, *Anthropology As Cultural Critique: An Experimental Moment in the Human Sciences* (Chicago: University of Chicago Press, 1986).

23. Pearson, *The Grammar of Science*, p. 78.

24. There is an oddity about the illustration published in the three editions (2d and 3d ed., London: Adam and Charles Black, 1900, 1911) and the Everyman edition. The first edition's image is much more bare than the one published in the others. There are, in fact, significant differences that I cannot account for. The later image, reprinted here, seems closer to the one Mach actually describes. I have not been able to discover why there are different images because Mach, surely, drew only one. I describe the image that appears in the second and third editions.

25. An interesting problem, as Peter D'Agostino has suggested to me, is that the drawing itself is controlled by conventions of representation that Mach had obviously absorbed. The implication is that even the "sensations" that Mach is attempting to register purely, as they come to him apart from the conventions of perception that govern so much of our seeing, are "psychological" at least as much as they are physical.

26. An extraordinary essay by W. K. Clifford, whose *Common Sense of the Exact Sciences* Pearson edited and published after Clifford's early death, dramatically opens with an affirmation of the psychological basis of vision. Looking around the lecture room in which he is speaking (as Mach looks around his library), he describes it accurately and yet argues that "not one of these statements can by any possibility have been strictly true." He talks then of what impinges upon the two retinas of his eyes. What he does see is "two distinct surfaces, having no depth and no relief, and only a kind of distance which is quite different from that of the solid figures before." The question, he says, "is not about what is there, but about what I see" (W. K. Clifford, "Philosophy of the Pure Sciences," in *Lectures and Essays* [London: MacMillan and Co., 1901], 1:302–3).

27. Pater, *The Renaissance*, p. 188.

28. Ibid.

29. Pearson, *The Grammar of Science*, pp. 109–10.

30. Ibid., p. 80.

31. Loesberg, *Aestheticism and Deconstruction*, pp. 19–20.

32. Pearson, *The Grammar of Science*, p. 74.

33. Ibid., p. 44.

34. See a number of James's essays, including his reviews of work by G. H. Lewes and W. K. Clifford: William James, *Essays, Comment, and Reviews* (Cambridge, Mass.: Harvard University Press, 1987), pp. 303–7, 343–45. See also William James, *Collected Essays and Reviews* [1920] (reprint, Bristol, England: Thoemmes Press, 1996), pp. 137–46; and *Essays in Philosophy* (Cambridge, Mass.: Harvard University Press, 1978), pp. 23–31.

35. Pearson, *The Grammar of Science*, p. 104.

36. David Harvey, *The Condition of Postmodernity: An Enquiry into the Conditions of Cultural Change* (Oxford: Blackwell, 1990), p. 44.

37. Karl Pearson [Loki, pseud.], *The New Werther* (London: Kegan Paul, 1880), p. 53.

38. In one of his essays Pearson criticizes the later Kant's maneuvers to allow for the possibility of getting access to the thing in itself. See Karl Pearson, "Kuno Fischer's New Critique of Kant," *The Cambridge Review* (28 November 1883): 109–11: "Kant's attempt to build up a supersensuous system is like all previous attempts, mere dogma. . . . Is there not something almost immoral in a great intellect, which might confer endless blessings on mankind by a study of "conditioned causality," devoting itself to construction of a theory of the supersensuous. . . . Can we not once and for ever renounce all supersensuous cravings, in order to gain the fuller life and light of the sensuous world?" (p. 110).

39. Pearson [Loki, pseud.], *The New Werther*, p. 111.

40. Ludwig Feuerbach, *The Essence of Christianity*, trans. Mary Ann Evans (New York: Harper and Row, 1957), p. 15.

41. Pearson, "Reaction," p. 191.

42. Pearson [Loki, pseud.], *The New Werther*, p. 103.

43. Mary Poovey traces the development of the very concept of "fact" through a phase in the development of statistics as a science. The question was how to change raw facts into usable knowledge, into general laws. She quotes Adam Sedgwick: "By science, then, I understand the consideration of all objects, whether of pure or mixed nature, capable of being reduced to measurement and calculation . . . all phenomena capable of being brought under the semblance of law are legitimate objects of our inquiry" (quoted in Poovey, *A History of the Modern Fact: Problems of Knowledge in the Science of Wealth and Society* [Chicago: University of Chicago Press, 1998], p. 312). Pearson's task in the early twentieth century was to reduce everything, but particularly biological and mental phenomena, to measurement and calculation.

44. Dale, *The Victorian Critic and the Idea of History*, p. 12.

45. Pearson, *The Grammar of Science*, p. 168.

46. Ian Hacking, *The Taming of Chance* (Cambridge: Cambridge University Press, 1990), p. 188.

47. Karl Pearson, *Life, Letters and Labours of Francis Galton* (Cambridge: Cambridge University Press, 1914–30), 3A:1 ff, quoted in Hacking, *The Taming of Chance*, p. 188.

48. Pearson, *The Grammar of Science*, p. 154.

49. Hacking, *The Taming of Chance*, p. 203.

50. Pearson, *The Grammar of Science*, p. 105.

51. Karl Pearson, *The Ethic of Freethought* (London: Adam and Charles Black, 1901), p. 102.

52. Pater, *The Renaissance*, p. 189.

53. Williams, *Transfigured World*, p. 51.

54. Pater, *The Renaissance*, pp. xx–xxi.

55. Williams, *Transfigured World*, p. 53.

EPILOGUE

1. Lorraine Daston, "Baconian Facts, Academic Civility, and the Prehistory of Objectivity," in *Rethinking Objectivity*, ed. Allan Megill (Durham, N.C.: Duke University Press, 1994), p. 38.

2. Ibid., p. 58.

3. T. H. Huxley, "On the Advisableness of Improving Natural Knowledge," in *Methods and Results* (London: MacMillan, 1893), p. 41.

4. Ibid., p. 39.

5. Samuel Butler, *Collected Essays*, vol. 1 (New York: AMS Press, n.d.), reprint of *The Shrewsbury Edition of the Works of Samuel Butler*, ed. Henry Festing Jones and A. T. Bartholomew, 20 vols. (London: J. Cape, 1923), 18:126.

6. Karl Pearson, *The Grammar of Science* (London: Walter Scott, 1892), pp. 18, 23.

7. Butler, *Collected Essays*, 18:117.

8. Ibid., 18:118.

9. Auguste Comte, *Auguste Comte and Positivism: The Essential Writings*, ed. Gertrude Lenzer (Chicago: University of Chicago Press, 1975), p. 72.

10. Hans Reichenbach, *The Rise of Scientific Philosophy* (Berkeley and Los Angeles: University of California Press, 1951), p. 92.

11. Oscar Kenshur, *Dilemmas of Enlightenment: Studies in the Rhetoric and Logic of Ideology*

(Berkeley and Los Angeles: University of California Press, 1993), pp. 7–8. There is no space here to follow the intricate and brilliant arguments Kenshur makes against ideological essentialism. But he is particularly concerned with the ways in which appeals to objectivity and rationality are taken to imply "authority" and thereby threaten political oppression:

> [W]hen the critique of rationalism or other epistemological stances is linked to the critique of ideology, false consciousness tends . . . to be attributed to the strategies of domination. Thus the epistemological stance is deemed false, and those who have supported it are accused of serving those interests that wish to perpetuate oppression. And, by implication, the theory that unmasks objectivism, even if it is explicitly relativistic, is seen to be true and liberating and thus, paradoxically, takes on the role traditionally occupied by science. This is merely a reformulation of the paradox that I have identified earlier as undermining the absoluteness of absolute structural cooptation. The analyst is, in principle, incapable of achieving an Archimedian vantage point but implicitly claims one nonetheless. (p. 23)

12. Stephen H. Kellert is at work on a book on the subject, but I quote from a manuscript paper in which he lays out the general outlines of his argument: "Defending Decontextualization: Positivism, Ecology, and Rootless Cosmopolitanism." [Essay later published under the title "Never Coming Home: Positivism, Ecology, and Rootlessness," in *The Meaning of Being Human*, ed. Michelle Stoneburner and Billy Catchings (Indianapolis: University of Indianapolis Press, 1999), p. 192.]

13. Ronald N. Giere, *Science without Laws* (Chicago: University of Chicago Press, 1999), p. 233.

14. Michael Friedman, *Reconsidering Logical Positivism* (Cambridge: Cambridge University Press, 1999), p. xi. Friedman regards the positivists as true philosophical revolutionaries who are anything but the "foundationalist" justifiers of scientific specialization that they have too often been taken to be.

15. Giere, *Science without Laws*, p. 228.

16. Reichenbach, "Logical Empiricism in Germany and the Present State of Its Problems," *Journal of Philosophy* 33:60, quoted in Ronald N. Giere, *Origins of Logical Empiricism* (Minneapolis: University of Minnesota Press, 1996).

17. Reichenbach, *The Rise of Scientific Philosophy*, p. 304.

18. Friedman, *Reconsidering Logical Positivism*, p. 14.

19. Karl Popper, *The Open Society and Its Enemies* (Princeton, N.J.: Princeton University Press, 1962), 1:4.

20. In his recent biography of Popper, Malachi Haim Hacohen argues that "Popper and his interwar interlocutors recognized 'poststructuralist' dilemmas—they simply opted for different solutions" (*Karl Popper: The Formative Years, 1902–1945* [Cambridge: Cambridge University Press, 2000], p. 3).

21. John Stuart Mill, "Nature," reprinted in *Autobiography and Other Writings*, ed. Jack Stillinger (Boston: Riverside Editions, 1969), pp. 329–30.

22. T. H. Huxley, *Evolution and Ethics*, ed. James Paradise and George Williams (Princeton, N.J.: Princeton University Press, 1989), p. 141.

23. Evelyn Fox Keller, "The Paradox of Scientific Objectivity," in Megill, ed., *Rethinking Objectivity*, p. 322.

24. Quoting Daniel Hillis, in *The Artificial Intelligence*, ed. S. R. Graubard (Cambridge, Mass.: MIT Press, 1998), p. 28.

25. Charles Darwin, *On the Origin of Species: A Facsimile* (Cambridge, Mass.: Harvard University Press, 1964), p. 84.

26. George Williams, "A Sociobiological Expansion of *Evolution and Ethics*," in Huxley, *Evolution and Ethics*, p. 191.

27. Ibid., pp. 192-93.

28. Plato, *Great Dialogues of Plato*, trans. W. H. D. Rouse (New York: Mentor Books, 1956), p. 470. I am grateful to Suzy Anger for suggesting to me the importance of the "Phaedo" to my argument.

29. Ludwig Wittgenstein, *Tractatus Logico-Philosophicus* (London: Routledge and Kegan Paul, 1922), p. 185. I am grateful to Michael Wolff for turning me back to Wittgenstein.

INDEX

Adam, Ruth, 301n. 29
aesthetics and aestheticism: and objectivity, 292n. 8; and Pater, 247, 248; and Pearson, 244; and positivism, 258
Aliquie, Ferdinand, 291n. 23
altruism: and Darwin, 281; and disinterest, 272, 281; and hermeneutics of suspicion, 282; and Huxley, 280; impossibility of, 14; and objectivity, 271, 281, 283; possibility of, 271, 281
Anderson, Amanda, 14–15, 174, 175, 189, 272–73, 286n. 24, 310n. 1
Anger, Suzy, 183, 301n. 35, 315n. 28
Appleby, Joyce, 288n. 3
ApRoberts, Ruth, 292n. 6
Arnold, Matthew, 50, 175; on "best self," 68, 69, 78, 176; and community, 176; and disinterested objectivity, 293n. 11; and Pater, 248; and religion, 85; and scientific ideal, 294n. 1
ascesis or asceticism: and detachment, 255; and disinterest, 249, 266; and empiricism, 250; and Nietzsche, 246; and objectivity, 249; and Pater, 25, 249, 265–66; and Pearson, 263; and perspectivism, 252, 254; and relativism, 250; and self-awareness, 252; and solipsism, 246, 266; and subjectivity, 249
Austen, Jane, 13
authority: and autobiography, 145; and community, 34, 35, 36, 37, 42; and Descartes, 46, 49; and "dying to know," 287n. 30; in George Eliot, 191, 197, 198; and Francis Galton, 106, 114; and Huxley, 9, 29, 35, 126, 269–70; and intuition in Newman, 88; and Pearson, 263; and perspectivism, 270; and skepticism, 39, 42; and uniformitarianism, 91; and Wallace, 114; and women, 124, 145
autobiography: and authority, 145; and Descartes, 47, 56; as epistemology, 85–103; as narrative of discovery, 11; and perspectivism, 99; and self-denial, 87; and self-effacement, 93, 104, 124; self-deprecation in Darwin's, Mill's, and Trollope's, 85, 92, 101, 102, 103; and uniformitarianism, 95, 100

Bacon, Francis, 2, 6, 17, 20, 55; aggression disguised as self-effacement in, 22–23; and the Idols, 21; language of morality in, 22; language of pilgrimage in, 20–21, 24; method of, for acquiring knowledge, 55; rejection of tradition by, 29
Balfour, A. J., 245, 246, 259, 311n. 7
Barrish, Phillip, 308n. 8
Bataille, Georges, 290n. 8
Beer, Gillian, 188, 191, 306n. 2, 307n. 10
Bentley, Eric, 293n. 18
Berkeley, George, 247
Bernal, J. D., 280
Bernard, Claude, 52, 64, 65, 157, 291n. 26
Bernstein, Richard, 44
Bildung narratives, 11, 148, 268; and Descartes, 204; and Dickens, 203; disenchantment in, 204; and George Eliot, 184, 186–87; and Hardy, 203; and Mill, 186; and subjectivity, 204; and women, 167
Bjork, Lennart A., 306n. 9
Bodenheimer, Rosemarie, 191, 304n. 17
body, the: and Carlyle, 66, 70, 73–74, 76, 77, 79, 82, 84; and Descartes, 54, 57, 60, 61, 83; in Dickens, 151, 156, 164; in

317

body *(continued)*
  Hardy, 151, 202, 203, 204, 206, 208, 214; as object of study, 52; in Pearson, 236, 240, 257; in realist fiction, 151; and the senses, 51–52; and Tyndall, 112. *See also* dualism of mind/body; embodiment
Booth, Charles, 143
Bordo, Susan, 44, 290n. 9
Brontë, Charlotte, 130, 149
Brooks, Peter, 160
Browne, Janet, 299n. 8
Buckley, J. H., 203–4, 306n. 5
Burtt, E. A., 29, 30
Butler, Samuel, 268, 272

Cameron, J. M., 309n. 14
Campbell, George Gordon Douglass, 245
Cannon, Susan, 87
Carlyle, Thomas: and the body, 66, 70, 73–74, 76, 77, 78, 79, 82, 84; and cause-and-effect philosophy, 108; and Chartism, 78, 294n. 34; and consciousness, 72, 75, 83; on Descartes, 75; and dualism of mind/body, 73, 74, 82; and embodiment, 70, 79, 82; and empiricism, 82; and the ethical, 71, 73, 76; and history, 79, 80, 82; and Huxley, 81; marriage of, to Jane Welsh, 76, 293n. 19; and the material, 74, 76, 77, 80; and metaphor, 72, 75–76, 77, 78, 84; and morality, 84; and rationality, 75; rejection of tradition by, 73; *Sartor Resartus*, 20, 66–84; and "Selbst-Tödung" idea, 3, 173; and self-annihilation, 70, 80, 83; as sensualist, 73, 76; on sympathy, 175, 176; and Tyndall, 81
class: in Dickens, 153, 159, 166; and epistemology, 6; in Hardy, 204, 205, 206, 209, 307n. 14; and subjectivity, 175; and Wallace, 107, 298n. 11
Clifford, James, 188, 199, 311n. 22
Clifford, W. K., 312n. 26
Clover, Carol, 310n. 28
Code, Lorraine, 286n. 23
Cole, Margaret, 301n. 29
Coleridge, Samuel Taylor, 96–97
community: and Arnold, 176; and authority, 34, 35, 36; and detachment, 272; and George Eliot, 176; and human interchange, 41; and morality, 34; and objectivity, 36–37; and Pearson, 226, 230
Comte, August, 109, 136, 144, 179, 221–22, 268, 269, 275
consciousness: and Carlyle, 72, 75; and Darwin, 245; and Descartes, 55; in George Eliot, 184; and empiricism, 52; in Hardy, 200, 201, 210, 217; and the material, 54; and objectivity, 269; and positivism, 52, 264
constructivism (or constructionism): and essentialism, 40; and the ethical, 283; and idealism, 254; and objectivity, 12; and Pater, 266; and Pearson, 223, 232, 244, 263, 266; and perpectivism, 39, 256; and positivism, 254; and relativism, 39; and science wars, 288n. 26; and universalism, 40
cosmopolitanism: and detachment, 15; and disinterest, 272
Culler, A. Dwight, 293n. 11, 310n. 1

D'Agostino, Peter, 312n. 25
Dale, Peter Alan, 249, 311n. 13
Darwin, Charles, 15; and altruism, 281; and consciousness, 245; on dualism of mind/body, 99; epistemology critiqued by, 94; epistemology embodied by, 94; evolutionary theory of, 6, 65, 196, 249; on Francis Galton, 295n. 18; on John Herschel, 292n. 5; and humility, 26; and Pearson, 223, 224, 226; and religion, 86; and self-abnegation, 130, 245; and self-alienation, 94; and self-denial, 90; and uniformitarianism, 95, 100; and Wallace, 105. Works: *Autobiography*, 8, 27, 85, 86, 93, 94, 115, 125; *The Descent of Man*, 90, 108; *The Origin of Species*, 101, 108, 250, 281
Darwin, Erasmus, 113
Daston, Lorraine, 3, 21, 198, 249, 269, 270, 285n. 6, 288n. 9, 289n. 1, 292n. 8
David, Deirdre, 142
Dennett, Daniel, 104, 224, 308n. 6
Descartes, René, 2, 17, 44–65; and authority, 46; and autobiography, 47, 56; and the body, 54, 57, 60, 61, 83; cogito, 48, 51, 52, 58; and consciousness, 55; and detachment, 64, 65; on development of self-effacement through self-affirmation, 44;

and disinterestedness, 50–51; and embodiment, 18; and epistemology, 19; Great Doubt of, 9, 17, 21, 35, 45, 46, 53, 59, 62, 77, 127, 173; and hermeneutics of suspicion, 45; Huxley on, 290n. 11; on imagination, 57, 58, 60, 61, 62; on intellect, 60–61, 62; and the material, 50, 53, 60; and metaphor, 55, 56, 57, 58, 62, 63; method of, for acquiring knowledge, 53, 55, 58; moral impulse in, 63; and narrative, 19, 59; and objectivity, 51, 52; and possibility of knowledge, 46; rejection of tradition by, 17, 29, 46, 56; and secularism, 51, 53, 55; self-description of, 27; and the senses, 50, 58, 59, 60, 61, 62; and skepticism, 47, 49, 55; and solipsism, 172; and strategy of self-humiliation, 48, 62; and study of the human, 52; and subjectivity, 52; translations of, 291 nn. 15, 23, 24. Works: *The Discourse on Method*, 47–49, 203, 204; *Meditations*, 204

detachment: and ascesis, 255; and community, 272; and cosmopolitanism, 15; in Descartes, 64, 65; in Dickens, 153, 170; in George Eliot, 175; in Francis Galton, 116, 123; in Hardy, 210, 215, 218; and morality, 19–20

determinism: and George Eliot, 109; and free will, 110; and Francis Galton, 111, 114, 116, 119; and Huxley, 116; and Martineau, 109, 114, 131; and Mill, 109, 110; as moral problem, 108; and Pearson, 261; and self-denial, 114; and uniformitarianism, 109; and Wallace, 110, 111

Dickens, Charles: the body in, 151, 156, 164; class in, 153, 159, 166; death in, 152, 154, 155–56, 158, 159; detachment in, 153, 170; disinterest in, 167; embodiment in, 166, 169, 170; and Jewish question, 178, 180; objectivity in, 166, 170; and philosophy, 151; self-abnegation in, 168; self-annihilation in, 160, 166; self-denial in, 150, 169, 170; self-humiliation in, 169; self-surrender in, 154; transcendence in, 163, 164, 167–68, 169; women in, 155, 163, 166, 167–68, 169, 170. Works: *Bleak House*, 148, 152; *David Copperfield*, 6, 152; *Dombey and Son*, 152; *Little Dorrit*, 144; *The Old Curiosity Shop*, 152; *Oliver Twist*, 158; *Our Mutual Friend*, 148–70, 178; *Pickwick Papers*, 151

Didion, Joan, 188

disinterest and disinterestedness: and altruism, 272, 281; and ascesis, 249, 266; and cosmopolitanism, 272; and Descartes, 50–51; in Dickens, 167; in George Eliot, 187, 193, 199; as empowerment, 124–25; in Hardy, 203, 212; and hermeneutics of suspicion, 272; and Huxley, 113; and ideology, 279; impossibility of, 6, 13; and Martineau, 147; and Pater, 250; in Pearson, 232; rejection of, 10; and Somerville, 147; and uniformitarianism, 92; and Webb, 147. *See also* interest and interestedness

Dowden, Edward, 28

dualism of mind/body: in Carlyle, 73, 74, 82; in Darwin, 98–99; in Einstein, 8; in George Eliot, 7; in Hardy, 205–6, 207, 210, 213, 216; in Mill, 68, 98–99; and narrative, 26; in Pearson, 236, 238; and perspectivism, 206; in Trollope, 98–99; in Whewell, 25–26. *See also* body, the; embodiment

Dupré, John, 11, 289n. 30

Duran, Jane, 132, 300n. 15

"dying to know": and authority, 287n. 30; and Carlyle, 77; and Descartes, 4, 48, 50; and Dickens, 153; and Francis Galton, 117; and ideology, 279; as metaphor, 6, 8, 15, 23; and Mill, 96; narrative of, 9, 69; paradox of, 2, 4, 81, 85, 117; and politics, 279; as strategy of mortality, 50; as story of crisis of knowledge, 69

Einstein, Albert, 8, 9, 108

Eliot, George, 6–7; authority in, 191, 197, 198; the body in, 151; and community, 176; consciousness in, 184; and detachment, 175; and determinism, 109; and disinterest, 187, 193, 199; and empiricism, 182, 201; and feeling, 171, 181, 188, 192, 199; feminist critiques of, 6–7; and free indirect discourse, 182, 195; and Jewish question, 178, 180, 181, 183, 189–90; knowledge of other in, 176, 177, 180, 181, 182, 183, 199; and metaphor, 178–79; on Milton, 180; and objectivity, 174, 185,

Eliot, George *(continued)*
    192, 193, 195; and particularism, 194; perspectivism in, 7–8, 172, 182, 190, 191, 194; and positivism, 176; and rationality, 176, 186; and realism, 149, 176, 177, 178, 193, 200; and relativism, 172; on Riehl, 109; self-abnegation in, 171, 195; self-annihilation in, 177, 180; self-denial in, 150; self-obliteration in, 185; self-sacrifice in, 179; self-surrender in, 32; and skepticism, 173; and solipsism, 182, 198; and Stowe, 180, 181; sympathy in, 175, 181, 191; women in, 171–72. Works: *Adam Bede,* 7, 149; *Daniel Deronda,* 32, 150, 171–99; *Middlemarch,* 8, 28, 127, 177, 179, 192; *The Mill on the Floss,* 32, 109, 172, 176–77; *Romola,* 81, 179
Eliot, T. S., 9, 277
embodiment: and Carlyle, 70, 79, 82; and Descartes, 18, 45; in Dickens, 166, 169, 170; in Hardy, 208–9; and Mach, 255; and materiality, 26–27; and Pearson, 255; and perspectivism, 255; of philosophical ideas as metaphor, 8; and Schiller, 69; of truth, in novels, 12. *See also* body, the; dualism of mind/body
empiricism: and ascesis, 250; and Carlyle, 82; and consciousness, 52; and George Eliot, 182, 201; and Hardy, 201, 210, 216; and Mill, 67, 97; and Pater, 266; and Pearson, 222, 249, 259, 266; and positivism, 223; and the senses, 2; and solipsism, 182
Ermarth, Elizabeth, 161
essentialism: and constructivism, 40
ethics and the ethical: and Carlyle, 71, 73; as construct, 283; and epistemology, 19, 34, 35, 42, 271; and the human condition, 42; and realist fiction, 150. *See also* morality

Faraday, Michael, 24
Farrell, John B., 302n. 7, 303n. 17
feminist theory: and "heroic" epistemology, 18; critique of objectivity in, 10–11, 18, 39; of science, 36–37
Feuerbach, Ludwig, 86, 99, 256, 259
Feynman, Richard, 104, 297n. 1

Fischer, M. J., 311n. 22
Fleming, Donald E., 296n. 33
Forrest, D. W., 106
Freud, Sigmund, 15, 148
Friedman, Michael, 277
Froude, J. A., 73
Fuss, Diana, 40, 289n. 36

Gagnier, Reginia, 137, 291n. 20
Galileo, 8, 15
Galison, Peter, 3, 21, 87, 198, 249, 285n. 6, 288n. 9, 292n. 8
Galivan, Patricia, 215, 307n. 20
Gallagher, Catherine, 158, 163, 167
Galton, Francis, 90; and authority, 106, 114; Charles Darwin on, 295n. 18; and detachment, 117, 123; and determinism, 111, 114, 116, 119; and eugenics, 108, 111, 118; and family, 115; and free will, 116; and gender, 128; and generalization, 117; and natural selection, 117, 118; objectivist method embodied by, 123; and paradox of "dying to know," 117; and Pearson, 118, 261; and self-abnegation, 113; and self-effacement, 107, 121, 123; and statistics, 114, 115, 116, 117, 118; and subjectivity, 106–7, 115, 116, 119; and utopianism, 118. Works: *Hereditary Genius,* 113, 115; *Memories of My Life,* 106, 107, 113, 114, 115, 116–17, 120, 121, 122
Gates, Barbara, 299n. 1
Gaukroger, Stephen, 46, 47
Ghiselin, Michael, 196
Giere, Ronald, 277
Gissing, George, 153
Goethe, Johann Wolfgang von, 3, 233, 234
Gould, Stephen Jay, 92
Grene, Marjorie, 64

Haack, Susan, 11
Hacking, Ian, 261
Hacohen, Malachi Haim, 314n. 20
Hagberg, Knut, 296n. 36
Haight, Gordon, 304n. 17
Hamilton, William, 3
Haraway, Donna, 17, 38, 39, 40, 56, 145, 301n. 38
Harding, Sandra, 145

Hardy, Thomas: on art, 217, 218; the body in, 151, 202, 203, 204, 206, 208, 214; class in, 204, 205, 206, 209, 307n. 14; consciousness in, 200, 201, 210, 217; detachment in, 210, 215, 218; disenchantment in, 204; disinterest in, 203, 212; and dualism of mind/body, 205–6, 207, 210, 213, 216; embodiment in, 208–9; and empiricism, 201, 210, 216; and the ideal, 214, 215; irony in, 211, 213; and the material, 200, 201, 202, 205, 209, 217, 219; and morality, 216; and Nietzsche, 307n. 18; and rationality, 205; and realism, 202, 207; and religion, 202, 206; self-abnegation in, 30, 202–3; self-annihilation in, 29, 30, 205, 211; self-denial in, 150, 206; self-sacrifice in, 202, 212; and skepticism, 204, 207; and subjectivity, 205. Works: *Jude the Obscure*, 150, 200–219; *A Pair of Blue Eyes*, 207; *Tess of the D'Urbervilles*, 200; *Two on a Tower*, 28–29, 30, 114–15, 200; *Under the Greenwood Tree*, 207; *The Well-Beloved*, 201; *The Woodlanders*, 207
Harré, Rom, 32–36, 37, 50, 288n. 26
Harvey, David, 258
Hayek, E. A., 296n. 36
Heffernan, George, 47
Herbert, Christopher, 3, 14, 86, 172, 174, 175, 285n. 4, 308n. 3, 309n. 14
Herbert, Sandra, 101
hermeneutics of suspicion: and altruism, 282; and Descartes, 45; and disinterest, 272; and epistemology, 282; and language, 275; limits of, 14, 271
"heroic epistemology": in Bacon, 22; in Descartes, 27; feminist critique of, 18; in Hardy, 28–29; in Harré, 37; and realism, 90, 295n. 22; tradition of, 19; in Whewell, 25, 27
Herschel, Caroline, 124
Herschel, John, 19, 24, 79, 127, 157
Hillis, Daniel, 281
Hull, David, 34, 35, 289n. 30
human, the: and Darwin, 245; and Descartes, 52; study of, 52, 245
human condition: and the ethical, 42; and objectivity, 38, 39, 41
Hume, David, 247

Hunt, Lynn, 288n. 3
Huxley, T. H.: and altruism, 280; and authority, 9, 29, 35, 126, 221, 269–70; and Carlyle, 81; on Descartes, 290n. 11; and determinism, 116; and disinterest, 113; and free will, 112–13; and knowledge, 87, 216; and materialism, 112, 113; and method, 114; and naturalism, 111, 279; and science, 90, 298n. 12; and skepticism, 29, 49. Works: *Autobiographies*, 99, 297n. 42; *Evolution and Ethics*, 279, 280

ideology: and disinterest, 279; and interest, 3–4; and knowledge, 175; and objectivity, 279; of Zionism, 188–89
Inman, Billie Andrew, 311n. 13
interest and interestedness: and ideology, 3–4; and objectivity, 10; and skepticism, 273. *See also* disinterest and disinterestedness
Irwin, Jane, 180, 189

Jacob, Margaret, 288n. 3
Jaffe, Audrey, 161, 302n. 3
James, Henry, 173, 180
James, William, 31, 40–41, 42, 257, 312n. 34
Jehlen, Myra, 18
Jelinek, Estelle, 139
Jenkins, Alice, 301n. 35
Johnson, Claudia, 304n. 22
Johnson, Edgar, 151
Jordanova, Ludmilla, 299n. 1
Judovitz, Dalia, 52, 58, 59, 291n. 14

Kant, Immanuel, 222, 249
Kaplan, Fred, 73, 76–77, 293n. 19, 294n. 32
Keats, John, 3, 4. *See also* negative capability
Keller, Evelyn Fox, 18, 22, 41, 100, 132, 145, 280–81, 283, 290n. 9, 300n. 15
Kellert, Stephen, 277, 314n. 12
Kendrick, Walter, 102
Kenshur, Oscar, 14, 276, 287n. 31, 313n. 11
Kevles, Daniel J., 309n. 15
Kingsley, Charles, 88, 89
Kitcher, Philip, 33
Knoepflmacher, U. C., 295n. 22
Kohn, David, 101

Kucich, John, 2–3, 149, 150, 152, 166, 205, 216, 285n. 1

LaCapra, Dominick, 174
Latour, Bruno, 287n. 26
Lawrence, D. H., 73, 307n. 15
Leavis, F. R., 173, 180
Leighton, Ralph, 297n. 1
Lenzer, Gertrude, 304n. 19
Levine, George, 294n. 1, 302nn. 6, 10, 307n. 15, 308n. 7
Lewes, G. H., 151, 181, 182, 308n. 4
Lightman, Bernard, 294n. 6, 310n. 4
Loesberg, Jonathan, 86, 257, 296n. 31, 311n. 13
Longino, Helen, 36, 37, 145
Lyell, Charles, 91, 92, 93, 108

Mach, Ernst: differences in image by, 312n. 24; and embodiment, 255; and perspectivism, 252, 253, 254
MacIntyre, Alasdair, 17, 19, 42, 287n. 1, 288n. 2
Malthus, Thomas Robert, 6, 158
Mansel, Dean, 3
Marchant, James, 106
Marcus, George, 188, 311n. 22
Martineau, Harriet: and determinism, 109, 114, 131; and disinterest, 147; and dualism of mind/body, 144; and feminism, 143; and the material, 135, 136; and religion, 133–34, 135; and self-abnegation, 130; and self-denial as means to freedom, 132; and self-surrender, 145, 146; and positivism, 134, 144; *Autobiography*, 129, 130, 131, 132
material and materiality: and Carlyle, 74, 76, 77, 80; and Descartes, 50, 53, 54, 60; and embodiment, 26–27; and Hardy, 200, 201, 202, 205, 209, 217, 219; and Huxley, 112, 113; and Martineau, 135, 136; and modernity, 2; and morality, 108; and Pearson, 235; and Tyndall, 112
McCarthy, Patrick, 152
McClure, John, 188
Merchant, Carolyn, 277
Mill, John Stuart, 3–4, 6, 24, 136, 279; *Autobiography*, 85, 90, 93, 95, 96, 125, 185; and Coleridge, 96–97; on Comte, 308n. 2; and determinism, 109, 110, 114; and dualism of mind/body, 68, 98; and dualism of thought/feeling, 97, 98, 185–86; "dying to know" narrative critiqued by, 96; and empiricism, 67, 97; induction theory of, 67; and positivism, 109; and self-alienation, 94; and self-annihilation, 97; and self-effacement, 125; and Somerville, 301n. 36; and Harriet Taylor, 97–98, 296nn. 35, 36; and uniformitarianism, 95; and Wordsworth, 97
Miller, D. A., 286n. 14, 295n. 10
Miller, J. Hillis, 306n. 3
Mintz, Alan, 294n. 35, 305n. 29
Mohanty, Satya, 14
Montaigne, Michel de, 58
Moore, James, 286n. 13
morality: and Bacon, 22; and Carlyle, 84; and community, 34; and Descartes, 63; and detachment, 19–20; and determinism, 108; disguised in method, 55; embodiment of ideas and, 8; and Hardy, 216; and knowledge, 21; and the material, 108; and Pearson, 225, 226; and rationality, 54; and self-sacrifice, 20; and scientific stance, 5; and Tyndall, 5. *See also* ethics and the ethical
Moretti, Franco, 183–84, 204, 291n. 20, 295n. 10, 306n. 7
Morley, John, 149
Morrell, Jack, 299n. 22
Muggeridge, Kitty, 301n. 29

Nagel, Thomas, 50: and Cartesian tradition, 44; on circularity of antiobjectivist positions, 12, 14; and objectivity, 10, 11, 100–1; and perspectivism, 9, 10; on rationality, 13, 287n. 28
negative capability, 3, 4, 69; and self-interest, 4
Newman, John Henry, Cardinal, 47, 51; on intuition as authority, 88; and Kingsley, 88, 89; and probability, 226. Works: *Apologia pro vita sua*, 88–89; *Grammar of Ascent*, 226
Newton, Isaac, 25–26
Nietzsche, Friedrich, 1, 216, 245–46
Nightingale, Florence, 129
Nord, Deborah, 128, 130, 300n. 26

Normore, Calvin, 289n. 1
Norton, Bernard J., 309n. 24
Novick, Peter, 286n. 12

objectivity: and aesthetics, 292n. 8; Arnold on, 293n. 11; and altruism, 271, 281, 283; and ascesis, 249; Cartesian, 51; and community, 36–37; in contemporary social science and cultural theory, 5; and consciousness, 269; and constructivism, 12; in Dickens, 166, 170; in George Eliot, 174, 185, 192, 193, 195; and the feminine, 290n. 9; feminist critique of, 10–11, 18, 39; and gender, 124; and human condition, 38, 39; and ideology, 279; and interestedness, 10; and interpretation, 3; in nineteenth-century science, 3; and objectivism, 174; and otherness, 18, 42, 174; and Pater, 247; and Pearson, 225, 232, 259; and perspectivism, 289n. 1; and power, 290n. 9; and pragmatism, 10; and rationality, 9; and realist fiction, 149, 150; and relativism, 3, 11; and self-denial, 3; and skepticism, 283; and solipsism, 42; and subjectivity, 52
Olney, James, 86, 100, 296n. 31
otherness: and objectivity, 18, 42, 174; and perspectivism, 18, 173, 182
Owen, Robert, 110, 114

Pater, Walter, 3, 220; and aestheticism, 247, 248; and Arnold, 248; and ascesis, 25, 247, 249, 265–66; and Coleridge, 250; and constructivism, 266; and Darwin, 250; and disinterest, 250; and empiricism, 266; and impressionism, 247, 251–52, 266; and objectivity, 247; and perspectivism, 247, 250, 252, 254, 265, 266; and relativism, 248, 265, 266; and sciences of observation, 249, 250; and self-abnegation, 245; and the sensory, 249, 251, 254; and skepticism, 265; and solipsism, 247, 248; and subjectivity, 247, 252. Works: *Appreciations*, 250; *Plato and Platonism*, 250; *The Renaissance*, 248, 265–66
Paul, Charles B., 22
Pearson, Karl: and aestheticism, 244; and ascesis, 263; and authority, 263; and Balfour, 246–47, 259; the body in, 236, 240, 257; on causation, 261; on citizenship, 227, 229; and community, 226, 230; and constructivism, 223, 232, 244, 263, 266; and Charles Darwin, 223, 224, 226; on death, 309n. 26; and determinism, 261; and disinterest, 232; and dualism of inside/outside, 252, 256, 257; and dualism of mind/body, 236, 238; and embodiment, 255; and empiricism, 222, 249, 259, 266; on facts as sensations, 224; feeling in, 238; and feminism, 223; and Francis Galton, 118, 261; and humanism, 240–41; and idealism, 25, 235, 248, 264; on ignorance, 228; and Kant, 222; and the material, 235; and morality, 225, 226; on normality, 262–63; and objectivity, 225, 232, 259; and Pater, 220; and perspectivism, 225, 247, 252–53, 256, 260, 261; and positivism, 25, 113, 221, 229, 234, 247, 249, 260, 274; and probability, 226, 260, 274, 275; and rationality, 225; and relativism, 3, 239, 263; and religion, 231, 232, 238; on renunciation, 230–31, 238, 242, 263; and romanticism, 229; self-denial in, 229; self-effacement in, 227; self-elimination in, 227, 244, 264, 274; and the sensory, 251; and skepticism, 265, 274; and socialism, 223, 230, 263, 272; and solipsism, 240, 247, 259, 263, 264; and statistics, 260, 264; and subjectivity, 231, 252. Works: *The Grammar of Science*, 220–23, 226, 228, 229–30, 235, 248, 251, 258, 259, 260, 262; *The New Werther*, 233, 235, 236; *Trinity*, 240
perspectivism (or perspectivalism): and alterity, 18; and ascesis, 252, 254; and authority, 270; and constructivism, 39, 256; and dualism of mind/body, 206; in George Eliot, 7–8, 172, 182, 190, 191, 194; and embodiment, 255; and Francis Galton, 123; of inside/outside, 7, 9–10, 25; William James on, 31; and knowledge, 92; Mach on, 252, 253, 254; and Nagel, 9, 10; and objectivity, 289n. 1; and Pater, 247, 250, 252, 254, 265, 266; and Pearson, 225, 247, 252–53, 256, 260, 261; and self-descriptions in autobiography, 99; and statistics, 260–61

Peterson, Linda, 8, 86, 106, 129, 141, 145, 300nn. 10, 12
Pierce, C. S., 309n. 21
Pinker, Stephen, 104
Pinney, Thomas, 305n. 30
Plato, 1, 15, 283
Polanyi, Michael, 100
Poovey, Mary, 313n. 43
Popper, Karl, 278, 279
Porter, Theodore, 176, 226, 235–36, 309nn. 13, 24
positivism: and aestheticism, 258; and Comte, 109, 144; and consciousness, 52, 264; and constructivism, 254; and George Eliot, 176; and empiricism, 223; as idealism, 264; and Martineau, 134, 144; and Mill, 109; and Pearson, 25, 113, 221, 229, 234, 247, 249, 260, 274; as religion, 87; and romanticism, 267; and skepticism, 221, 278; and socialism, 277; and solipsism, 25
pragmatism: and objectivity, 10; and William James, 32
Pynchon, Thomas, 121–22

Qualls, Barry, 302n. 1

Ragussis, Michael, 305n. 24
rationality: and Carlyle, 75; and George Eliot, 176, 186; and Hardy, 205; and morality, 54; and narrative, 12–13; and objectivity, 9; and Pearson, 225
realism and realist fiction: critique of, 100; domestic heroism in, 90, 295n. 22; and George Eliot, 149, 176, 177, 178, 193, 200; and the ethical, 150; and Hardy, 202, 207; and objectivity, 149, 150; and self-representation, 150; and Trollope, 102; and uniformitarianism, 93; and women, 142
Reichenbach, Hans, 275, 277, 278
relativism: and ascesis, 250; and constructivism, 39; and George Eliot, 172; and objectivity, 3, 11; and Pater, 248, 265, 266; and Pearson, 3, 239, 263
religion: and Arnold, 85; and class, 6; and Darwin, 86; and Hardy, 202, 206; and Martineau, 133–34, 135; and Pearson, 231, 232, 238; science and displacement of, 4; and Somerville, 133, 135; and Webb, 134, 135
Rorty, Richard, 7, 50, 53, 303n. 3
Rosenberg, John, 79, 80, 175
Rosenberg, Philip, 294n. 34
Rousseau, Jean-Jacques, 99, 101, 102
Ruskin, John, 6, 69, 158, 286n. 15
Russell, Bertrand, 30, 202

Scarry, Elaine, 207
Scheibinger, Londa, 299n. 1
Schiller, Friedrich, 69, 292n. 10
Schwartz, Ruth Cowan, 108
secularism: and Descartes, 51, 53, 55
Sedgwick, Adam, 313n. 43
self-abasement: in Descartes, 56
self-abnegation: and Darwin, 130, 245; in Dickens, 168; in George Eliot, 171, 195; in Hardy, 30, 202–03; as ideal, 4; and Francis Galton, 113; and history of scientific epistemology, 5; and interpretation, 3; and Newman, 89; and Pater, 245; and self-interest, 4; in Somerville, Martineau, and Webb, 130; and Wallace, 106
self-alienation: as narrative strategy in Descartes, 56
self-annihilation: in Carlyle, 70, 80, 83; in Darwin, 103; in Dickens, 160, 166; in discursive form, 5; in George Eliot, 177, 180; in Hardy, 29, 30, 205, 211; in Mill, 97, 103; paradox of, 2, 4; in Trollope, 99, 103
self-denial: and autobiography, 87; and Darwin, 90; as determinism, 114; in Dickens, 150, 169, 170; and epistemology, 87; in Hardy, 150, 206; and objectivity, 3; in Pearson, 229; in religious form of resurrection, 5; in Somerville, Martineau, and Webb, 132
self-deprecation: in Darwin, 85, 92, 101, 102–3; in Mill, 85, 92, 101; in Trollope, 85, 92, 101, 102
self-effacement: and autobiography, 93, 104; in Bacon, 23; in Darwin, Mill, and Trollope, 103, 124–25; in Descartes, 44; as ideal, 15; in Francis Galton, 107, 116, 121, 123; in Pearson, 227; and women, 127–47

self-elimination: and Pearson, 227, 244, 264, 274

self-humiliation: in Carlyle, 80; in Descartes, 48; in Dickens, 169; in fiction, 148; as foundation of scientific truth claims, 17; secular reconstruction of, 34

self-interest: in George Eliot, 7; and negative capability, 4; and self-abnegation, 4

self-obliteration: in George Eliot, 185

self-renunciation: Tyndall on, 4

self-repression: tradition of, 9

self-sacrifice: in George Eliot, 179; in Hardy, 202, 212; and morality, 20; and women, 126, 139

self-surrender: absorption of, into procedures of mind, 55; in Dickens, 154; in realist fiction, 150; in Somerville, Martineau, and Webb, 145, 146

Semmel, Bernard, 305n. 37, 309n. 18

senses and the sensory: and Descartes, 50, 58, 59, 60, 61, 62; and empiricism, 2; and Pater, 249, 251, 254; and Pearson, 251; and solipsism, 255

Shapin, Steven, 38, 128, 197

Shaw, G. B., 153

Shaw, Harry, 305n. 25

Shteir, Ann B., 299n. 1

skepticism: and authority, 39; and Descartes, 47, 49, 50, 55; and George Eliot, 173; and Hardy, 204, 207; and Huxley, 29, 49; and interestedness, 273; limitations of, 42; and objectivity, 283; and Pater, 265; and Pearson, 265, 274; and positivism, 221, 278; and universalism, 173

Smiles, Samuel, 24, 90–91

Smith, Barbara Herrnstein, 287nn. 26, 27, 29

Smith, Bonnie, 288n. 3

Smith, Jonathan, 196, 285n. 7, 288n. 7

solipsism: and asceticism, 246, 266; and Descartes, 172; and George Eliot, 182, 198; and empiricism, 182; and objectivity, 42; and Pater, 247, 248; and Pearson, 240, 247, 259, 263, 264; and positivism, 25; and the sensory, 255

Somerville, Martha, 142–43

Somerville, Mary, 124; and disinterest, 147; and dualism of mind/body, 144; and Mill, 301n. 36; rejection of self-surrender, 145, 146; and self-abnegation, 130; on self-denial as means to freedom, 132; and religion, 133, 135. Works: *The Connection of the Physical Sciences*, 133; *Personal Recollections*, 129, 139, 142

Spencer, Herbert, 131, 134, 221

Stone, Harry, 303n. 16

Stowe, Harriet Beecher, 180, 181

subjectivity: and ascesis, 249; and class, 175; disembodiment of subject by, 52; and Francis Galton, 106–7, 115, 116, 119; in Hardy, 205; and objectivity, 52; and Pater, 247, 252; and Pearson, 231, 252; and Wallace, 120; and women, 142

Taylor, Charles, 302n. 7

Taylor, Dennis, 202

Tennyson, G. B., 293n. 26

Thackeray, William, 149

Thockray, Arnold, 299n. 22

Thomas, Ronald, 302n. 2

Thompson, William, 201

Trollope, Anthony: *Autobiography*, 85, 92, 93–94, 101, 102; on dualism of mind/body, 99; and realism, 102; and uniformitarianism, 95

Turner, Frank, 82

Tyndall, John: and the body, 112; and Carlyle, 81; and materialism, 112; and morality, 5; and naturalism, 111–12; on self-renunciation, 4

uniformitarianism: and authority, 91; and autobiography, 95, 100; and determinism, 109; and disinterest, 92; and realism, 93

universalism: and constructivism, 40; and knowledge, 68; and the particular, 81; and skepticism, 173

Walker, Helen M., 308n. 5

Wallace, Alfred Russel: and authority, 114; and class, 107, 298n. 11; and Darwin, 105; and determinism, 110; on free will, 110–11; and materialism, 107; and Robert Owen, 110; and self-abnegation, 106; and socialism, 111; and subjectivity, 120; *My Life*, 105, 106, 120, 298n. 20

Webb, Beatrice, 87, 114; and disinterest, 147; and dualism of mind/body, 144; *My*

Webb, Beatrice *(continued)*
  Apprenticeship, 129, 131, 137, 140–41 and religion, 134, 135; and self-abnegation, 130; on self-denial as means to freedom, 132; self-surrender rejected by, 145, 146; and socialism, 143, 144; and Herbert Spencer, 131, 134; and statistics, 137, 138
Webb, R. K., 133
Weber, Max, 188
Welsh, Alexander, 177, 188, 214
Whewell, William, 3, 24, 67; and dualism of mind/body, 68; and ideal of disinterest, 25; on Newton, 25–26, 27, 68, 98; on theory, 292n. 2
Wilde, Oscar, 216, 258
Williams, Bernard, 45, 49, 55–56, 290n. 4
Williams, Carolyn, 265, 266, 311n. 18
Williams, George C., 282
Wittgenstein, Ludwig, 283
women: and authority, 124; and *Bildung* narratives, 167; in Dickens, 155, 163, 166, 167–68, 169, 170; in George Eliot, 171–72; and realism, 142; and science, 299n. 1; and self-abnegation as mode of self-affirmation, 130; and self-effacement, 126–47; and self-sacrifice, 126, 139; and subjectivity, 142
Woodcock, John, 296n. 24
Wordsworth, William, 97

Yeo, Richard, 289n. 29